SYSTEMATIC ELECTRONIC TROUBLESHOOTING

SYSTEMATIC ELECTRONIC TROUBLESHOOTING: A Flowchart Approach

James Perozzo

Delmar Publishers Inc.®

NOTICE TO THE READER

Publisher does not warrant or guarantee any of the products described herein or perform any independent analysis in connection with any of the product information contained herein. Publisher does not assume, and expressly disclaims, any obligation to obtain and include information other than that provided to it by the manufacturer.

The reader is expressly warned to consider and adopt all safety precautions that might be indicated by the activities described herein and to avoid all potential hazards. By following the instructions contained herein, the reader willingly assumes all risks in connection with such instructions.

The publisher makes no representations or warranties of any kind, including but not limited to, the warranties of fitness for particular purpose or merchantability, nor are any such representations implied with respect to the material set forth herein, and the publisher takes no responsibility with respect to such material. The publisher shall not be liable for any special, consequential or exemplary damages resulting, in whole or in part, from the readers' use of, or reliance upon, this material.

Cover photos:
  Digital meter photo courtesy of Simpson Electric
  Tracker 2000 photo courtesy of Huntron.

Delmar Staff
Associate Editor: Cameron O. Anderson
Editing Manager: Barbara A. Christie
Project Editor: Christopher Chien
Design Coordinator: Susan C. Mathews
Production Coordinator: Larry Main

For more information address Delmar Publishers Inc.
3 Columbia Circle
Box 15-015
Albany, New York 12212-5015

Figures 2-11, 3-39, 3-46, 3-85, 4-79, 4-80, 4-81, 4-82, 4-83, 4-84, 4-87, 4-88, 4-89, 4-90, 5-10, 5-11, 5-20, 5-24, 5-35, 6-6, 6-7, 8-14, 9-2 and 9-15 from *Practical Electronics Troubleshooting*, by James Perozzo. Copyright 1985 by Delmar Publishers Inc.

All other interior photographs otherwise uncredited courtesy of Woodbury "Bud" Abbey.

COPYRIGHT © 1989 BY DELMAR PUBLISHERS INC.

All rights reserved. No part of this work covered by the copyright hereon may be reproduced or used in any form or by any means—graphic, electronic, or mechanical, including photocopying, recording, taping, or information storage and retrieval systems—without written permission of the publisher.

Printed in the United States of America
Published simultaneously in Canada
by Nelson Canada,
A Division of International Thomson Limited

10 9 8 7 6 5 4 3

Library of Congress Cataloging-in-Publication Data:

Perozzo, James.
  Systematic electronic troubleshooting: a flowchart approach/James Perozzo.
    p.   cm.
  Includes index.
  ISBN 0-8273-3288-2
  1. Electronic apparatus and appliances—Maintenance and repair.
  I. Title
TK7870.2.P48   1989
621.381—dc19

88-29775
CIP

CONTENTS

Preface vi
Phase 1 / Finding the Defective Equipment 1
Phase 2 / Isolating the Defective Card 25
Phase 3 / Locating the Defective Stage 45
Phase 4 / Identifying the Defective Component 129
Phase 5 / Replacing Defective Components 211
Phase 6 / Dead Circuit Troubleshooting 253
Phase 7 / Troubleshooting the Power Supply 261
Phase 7A / Troubleshooting the Transformer/Rectifier Power Supply 275
Phase 7B / Troubleshooting the Switching Power Supply 287
Phase 7C / Troubleshooting the Phasing Power Supply 293
Phase 7D / Troubleshooting Voltage Regulators 297
Phase 8 / Troubleshooting Software-Driven Circuits .. 303
Phase 9 / Tracking Hum and Unwanted Signals 321
Phase 10 / Repairing Profound Failures 347
Appendixes
Appendix A / The Frustration List 351
Appendix B / Electronic Schematic Symbols 353
Appendix C / Part Identification Prefixes 361
Appendix D / Answers to Odd-Numbered Review Questions 369
Index 375

PREFACE

This book is based on the assumption that the technician must locate and repair a problem that exists somewhere within an electronic system, to the board, and then to the component level. An orderly sequence, shown by flowcharts, is used to lead the technician to the problem in the most efficient manner. Only topics directly involved with the repair are presented in the flow of thought.

The typical electronic system used in this book consists of individual equipments. Each equipment is made up of either circuit cards or functional areas. These cards, or functional areas of the equipment, are in turn made up of stages. Finally, each stage has within it the individual components. Based upon this core, a repair is considered to consist of separate phases:

Phase 1 - Finding the Defective Equipment
  Narrowing the problem from system to equipment
Phase 2 - Isolating the Defective Card
  Narrowing the problem from equipment to card
Phase 3 - Locating the Defective Stage
  Narrowing the problem from card to stage
Phase 4 - Identifying the Defective Component
  Narrowing the problem from stage to component
Phase 5 - Replacing Defective Components
  Making the repair

Figure P-1 presents a diagram of this progression.

Any specific task may require modification of these phases. There are many cases where shortcuts can be taken, but this approach is the basis for all electronics troubleshooting.

Additional phases may be necessary if the problem requires a deviation from the classic troubleshooting approach. For instance, dead-circuit troubleshooting may be called for in some cases. These additional phases are provided in this text.

FLOWCHARTS

Flowcharts are used throughout this book to help the technician follow a logical path to the problem under consideration. These flowcharts begin with the assumption that the equipment at one time worked well but has failed in service.

A Quick Course in Flowcharts

Since the use of flowcharts may be new to some students, a sample chart can be used to understand the concepts behind their use.

Flowcharts are used extensively by computer programmers. In this text, we will be using only four of the flowchart's symbols: The terminator, the connector, the process, and the decision symbols.

Flowcharts begin with a *terminator,* horizontally **parallel lines connected on the ends with a semi-circle.** Look for these as you enter each flowchart to ensure that your entry point is correct.

Things to be done, or *processes,* are shown as **rectangles.** The flowcharts in this book have accompanying explanations in the text for those rectangles that may raise questions. These explanations are found by referring to the paragraph number near the rectangle.

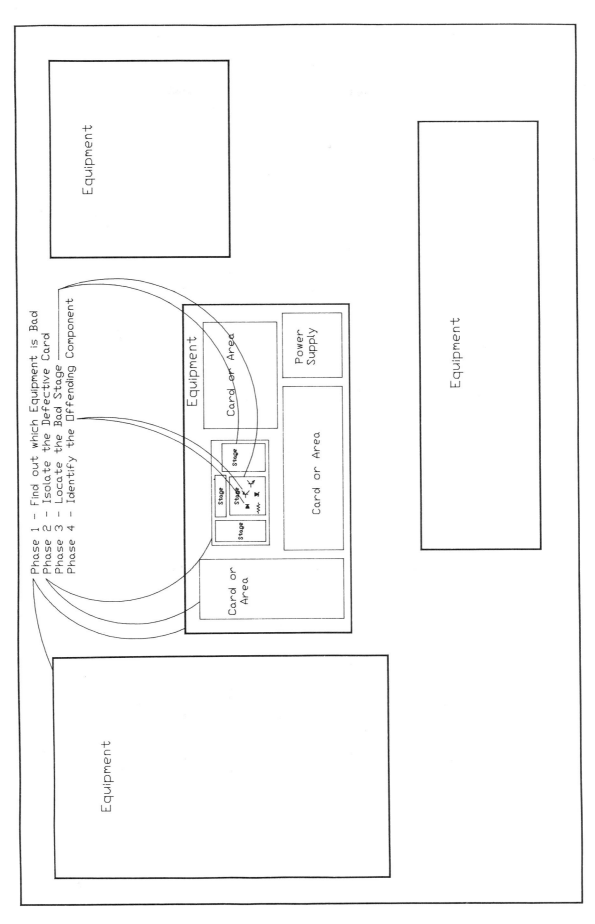

Figure P-1

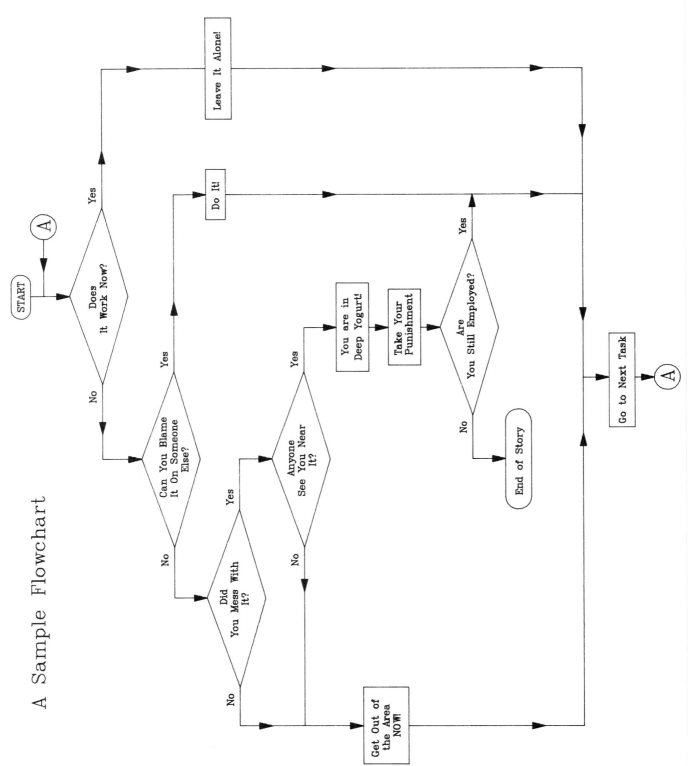

Figure P-2 A SAMPLE FLOWCHART

Decisions to be made and actions based upon them are shown as **diamonds.** There are only two possible ways to proceed when making a decision: "Yes" or "No." The Yes or No answer to a question determines the direction of exit from the diamond.

Connectors are alphabetically keyed circles. At the end of the sample flowchart shown in this foreword, for example you will note an "A" in a circle. This means to look for and continue with the corresponding lettered circle elsewhere on the chart. In our example, it is at the top of the chart, near the **START** terminator. These connectors are used to keep flowcharts neat and uncluttered by many crossing lines. The author assists the student in this text by providing an arrow at each terminator (an untraditional symbol) pointing to the direction to look for the matching continuation point.

A NOTE TO THE INSTRUCTOR

In the real world of electronic troubleshooting, today's electronic technician is expected to be competent in the diagnosis of a staggering variety of electronic failures. Sometimes, the repair must be accomplished without proper test instruments and under less than ideal working conditions. The technician may be required to work alone without technical assistance immediately available. An enormous diversity of test equipment also exists, requiring the technician to be familiar with analog and digital theory, discrete semiconductors, and integrated circuits. Add to these requirements that the technician must be able to read and interpret technical manuals and reference materials and to apply the information from these sources toward an effective repair, and it is easy to see that the modern electronic technician indeed must be a professional in every respect.

Learning how to repair electronic equipment has traditionally been accomplished by formally teaching the fundamentals of electronic theory and the laboratory use of test instruments. Students have been left mainly to their own devices to develop the logical train of thought that is necessary to repair real equipment. Learning this way often results in the development of inaccurate analysis of a problem and more guessing than knowing, and it can take many years before the technician can finally be called a "Supertech." While basic electronic theory and familiarity with test instruments are certainly necessary, the practical side of troubleshooting should also be taught.

What has been lacking until now is a truly logical approach to electronic troubleshooting, one sprinkled with liberal amounts of common sense and practical considerations. A structured approach to any given electronic problem reduces the total time necessary to effect a repair, particularly for the entry level technician. A logical progression of thought is necessary, from consideration of the symptoms to the final repair, to avoid countless possibilities for straying from the most direct path to the correction of the problem. Students need to learn the sensible way to begin and to progress through a repair job from start to finish. They need to be aware of the multitude of decisions that must be made to accomplish even the simplest of repairs. Although there are precious few hard-and-fast rules that will apply in every case, general guidelines can be given and followed in a majority of cases. This text explains not only when to make these decisions but explains why the decision must be made. These steps help the student visualize what is happening while troubleshooting. The flowcharts and text have been designed with the hope that they will enable a technician to repair at least 90 percent of the defective equipment that might be encountered.

There is no better method for showing the logical flow of thought than a flowchart. Only pertinent items are covered enroute to the result. The charts in the back of this book are the foundation of this course. *Please note that the text supports the flowcharts, not vice versa.* Consistent instructor guidance in referring to the charts during a repair sequence will instill an orderly approach to the great majority of problems. When a student gets "lost," the instructor should review the student's perceived position in the flowcharts and backtrack from there to a point of sharper focus, that point in the charts where the symptoms of the problem and instrument indications are *confirmed* before proceeding again toward the final repair. Note that, when using these charts, there is no need to begin the entire troubleshooting session from the beginning.

It is particularly important for the instructor to detect and point out to the student any perceived ambiguous feelings about a problem. The student and the instructor should beware of an insubstantial response such as "I think so" or "I assume it is." Successful troubleshooting is largely based upon positively verifying symptoms before acting upon that information.

Throughout the flowcharts, rectangles, printed in italics and often beginning with "Notes on..." are inserted. These subjects are temporary diversions from the progression of the job through the flowcharts. Notes are additional information that relate directly to the problem at hand, information which is often merely assumed in other texts.

No single instrument can troubleshoot in all instances. Only *generic*, modern test equipment is assumed: digital multimeters, high-frequency oscilloscopes, and logic analyzers are representative of the instruments covered. Limitations of older equipment are not addressed. Two exceptions to the generic approach are necessary: The current tracer manufactured by Hewlett-Packard and the Huntron tracker by Huntron Instruments. These top quality instruments are, to the best of my knowledge, made only by these two companies. Proprietary parts and circuits used within them preclude exact duplication by competing companies. Each instrument occupies a highly specialized niche in generic troubleshooting.

The acceptance by industry of the Huntron tracker as the standard solid state tester is evident by the use of the tracker in major companies, such as all five of the Armed Forces, the Bell Telephone System, and the Boeing Company. Every well-rounded technician should be familiar with the operation of this instrument because of the unique problems it can identify with ease.

Technicians also must be encouraged to read and understand the test equipment operation manual! Only by thoroughly understanding the test instrument can its full use be appreciated.

ACKNOWLEDGMENTS

I wish to thank those who helped make this text a reality. I received encouragement and help from my friends Bob Dudley, Ron Fleming of Starcrest Northwest, David Malland, and Bud Abbey, the photographer for most of the pictures in this text.

Of particular help in developing the material and the inevitable debugging process that must ensue after any such project were my co-teachers and dear friends, Art and Vickie Thompson and "CJ" and Donna Lemmon. The feedback from these people was instrumental in the accuracy of the flowcharts and the text.

I also wish to thank Mr. Mac McPherson of Radar Electric and Mr. Bill Hunt of Huntron Instruments for their encouragement. The following reviewers provided valuable feedback during the final development of the manuscript: Mr. Gary Boyington of Chemeteka Community College; Mr. Gary Webster of Cincinnati Technical College; Arlyn L. Smith of Alfred State College; Raymond Jordan of Alfred State College; and Richard Stedman of Mitchell Vocational Technical School.

ABOUT THE AUTHOR

James Perozzo has 30 years of experience in repairing electronic equipment of all types. After 20 years of service as a technician and officer in the Coast Guard, he broadened his experience by working in marine and aviation electronics, microwave voice carriers, and computer repair. A native of the Pacific Northwest, his qualifications include an FCC General Radiotelephone Certificate, Amateur Extra Class License, FAA Avionics Certification, and an Electronics Teaching Certificate. He currently teaches Industrial Electronics at Renton Vocational-Technical Institute in Renton, Washington. When not teaching, he is active in designing and building air cushion vehicles. He is also the author of *Practical Electronics Troubleshooting* (Delmar Publishers, Inc., 1985) and *Microcomputer Troubleshooting* (Delmar Publishers, Inc., 1986).

PHASE 1
FINDING THE DEFECTIVE EQUIPMENT

☐ **PHASE 1 OVERVIEW**

Phase 1 will attempt to make the technician aware of typical system problems and provide guidelines on how to handle them. Techniques presented in this phase identify the equipment that is causing a malfunction of an overall system.

The first part of any repair job is to analyze all symptoms of the problem to see if there is a simple cause. Many repair calls have been made because of a plug coming loose or an operator error. Intermittent problems should also be recognized when present.

1.0 COULD THIS BE A CASE OF OPERATOR ERROR? — As technology advances, the operation of equipment, particularly computers, has become more and more specialized and difficult. These days, even microwave ovens have plenty of "bells and whistles," such as the time of day or programmable recipes that are available at the touch of a button. Training personnel in the use of complex company equipment is becoming a substantial investment for industry. Generally speaking, the more controls there are—and this includes software controls—the easier it is for the operator to make a mistake that is often blamed on the equipment. The servicing technician must be constantly aware of this possibility.

1.0.1 Interview Operator to Eliminate This Possibility — A few minutes of consultation with the operator should give the technician an indication of that person's general level of expertise. One of the most effective ways of determining if the problem is caused by the operator is to have the operator demonstrate the exact nature of the reported problem. By watching carefully, the technician can determine whether or not an operator error is the real problem. This situation must be handled with a great deal of tact so as not to offend the operator.

1.0.2 Check All Operating Controls — Operating controls include knobs, switches, a keyboard, or mechanical inputs to the system. Manipulation of these controls helps localize the problem or proves the system is not defective at all—for the moment, at least. "Operating controls" refers only to those controls normally operated, not to the controls used for calibration or any internal adjustments. A control that is slipping or that has no effect is an immediate clue to a problem. It might also be incidental, something unrelated to note and fix after the main problem is solved.

Consider a complex, modular stereo installation as an example of a system. This system might include a tuner, a turntable, a tape deck or two, a power amplifier, an equalizer, and several sets of speakers. If the owner reported a problem with this type of system, it would be to the technician's advantage to get an overall feel for what-feeds-what in the system and then to manipulate the controls to understand which piece of equipment within the system is defective. If the tuner is defective, everything except the tuner would work fine, but attempts to receive radio stations with the tuner would not be successful.

1.0.3 Teach Operator the Proper Way

— If the operator has made an error, the technician may need to point it out. In order to do this, the technician must first be familiar with the system and its operation. While there is little need to memorize special operations, the technician must be able to use the system at a fundamental level and be reasonably familiar with the proper printed references to look up and perhaps point out the proper way to accomplish a given task. Once the equipment is operating normally, give the operator an opportunity to learn and to demonstrate the proper way of operating the system.

1.1 IS THE REPORTED PROBLEM STILL PRESENT?

— After determining that the problem is not caused by operator error, the next task is to get all the facts available about the problem. If possible, the technician should observe a demonstration of the problem, preferably by the person who made the initial call for service. Whether or not the symptoms of the problem appear will determine the technician's next course of action.

If the problem can be demonstrated it will be easier to fix. On the other hand, if the problem has apparently "fixed itself," there is more work to be done.

1.1.1 DO YOU RECOGNIZE THE PROBLEM?

— A familiarity with a specific kind of electronic equipment will eventually lead to a recognition of certain symptoms. For example, if a "Model XYZ" power supply usually fails by shorting one of the rectifiers, the technician experienced with this equipment will undoubtedly go directly to this as the most probable defect if the power supply is blowing fuses. After perhaps a single verifying check—in this case an ohmmeter check of the rectifiers—the problem can often be found without further analysis.

1.1.1.1 Try the "Usual" Repair for These Symptoms

— Replacing components based entirely upon familiar symptoms should be done with caution. It is hardly worthwhile to keep a list of all the possible failures of a particular piece of equipment and then blindly match symptoms and replace parts accordingly without verifying the suspected component failure. For instance, the case of the shorted rectifiers just mentioned could certainly have several variations. If a particular rectifier is usually the one to go bad and this is the one that you replaced, it is all too easy to assume that this is the only thing wrong. In some cases this may be true; in others it might not be. It may be caused by failure of another component. A shorted filter capacitor could be the cause of a rectifier shorting, or perhaps more than one rectifier may have shorted at the same time.

Every case of "recognized symptoms" should be verified before replacing components, and each case should be investigated for additional failures, either primary or secondary to the first failure discovered. See Phase 5 for more information on component replacement and what to watch for before reapplying power.

1.1.2 CAN YOU SEE ANY INDICATIONS OF POWER APPLIED?

There are usually obvious indications when primary power is applied to electronic equipment. Indicator lights may be on, the cathode-ray tubes of televisions and computer monitors have at least a flashing cursor, and the sound of blower motors might be heard.

1.1.2.1 Check the Simple Things: Switches, Controls, and Cables

Night cleaning crews, gremlins, or "nobody" sometimes goes about turning off obscure switches or slightly loosening plugs. A favorite target seems to be the intensity control of computer monitors. When one of these seldom-used controls gets moved into the apparent "off" position, it can make the equipment appear defective. While watching for indications of power, wiggle the power cord at both ends, where it goes into the equipment and at the wall socket. Wall sockets are prone to wear and after much use may make poor contact, even when the plug is all the way into the receptacle. A momentary flicker of the power indicator while moving the power cord is a sure indication that the power connection is in need of repair.

Be sure that power is available where it should be: at the wall receptacle. When in doubt, a different indicating load, such as a lamp, should be used to check for power. It can be very embarrassing to begin troubleshooting by removing the equipment covers, only to trace the lack of power right back to the wall socket. The technician should have checked *that* first. See also Paragraph 1.1.3.

1.1.2.2 NOTES ON TROUBLESHOOTING COAXIAL AND OTHER CABLES —

Short cables with multiple wires or coaxial cables are best tested by simply substituting a known good section of cable. If the problem is gone with a new cable, obviously the old one is either bad or the connection at either end was not firmly seated.

Testing multiple-wire cables for shorts — Testing multiple-conductor cables means making tests for the two possible kinds of failures: shorts between conductors and opens of the wires from end to end. Both tests involve tricks the technician can use to make the job easier.

How many tests must be made on a cable of six wires to be sure that a possible short between any of the conductors will definitely be found? A haphazard guess might be anywhere from 6 readings to 50! The actual answer is 15. Figure 1–1 shows the procedure.

Summarizing the procedure of the figure, the cable conductors should be laid out in a row to organize them. The first wire is tested to all others. Then the second wire is tested to all the wires *after* the second (not back to the first). The third wire is then tested to all after the third, and so forth until the next-to-the-last is tested only to the last wire. Using this system prevents taking the same reading more than once and prevents missing possible combinations that might be shorted.

Testing multiple-wire cables for opens — Two people may be required to make end-to-end checks of a cable for continuity. Some sort of communications may be necessary at great distances either by radio or telephone. There are two possible ways to provide the return path necessary for the ohmmeter: an external conductor (which can be earth-grounded), such as a metal railing or the metallic structure of a ship or aircraft, or one of the wires within the cable can be designated as the return. The ground path must, of course, have good continuity from one end to the other for this method to work. After agreeing on the return line to be used, one of the wires is shorted to the return line at the far end, and the near end is tested for continuity between the suspect wire and the ground. See Figure 1–2.

Testing wired circuits — There are still a great many circuits that are not built exclusively on printed circuit boards. Electrical panels for higher-powered equipment are an example. The technician may have to troubleshoot these circuits. Two kinds of problems are possible for equipment that, we must assume in this text, was working properly at one time. The wiring can become shorted because of pinching or rubbing against a sharp edge, or the wiring can develop an open circuit.

It is a simple matter to troubleshoot shorts in wiring where the cabling is not tied into bundles. The problem magnifies when the wiring problem develops where wiring goes into bound cables or into cable raceways for such long distances that it cannot be easily traced by physically following the length of the wire.

Some sort of wiring diagram will be necessary to efficiently troubleshoot wiring problems. Wiring problems are usually best done "hot," with power applied. The ohmmeter

4 □ Finding the Defective Equipment

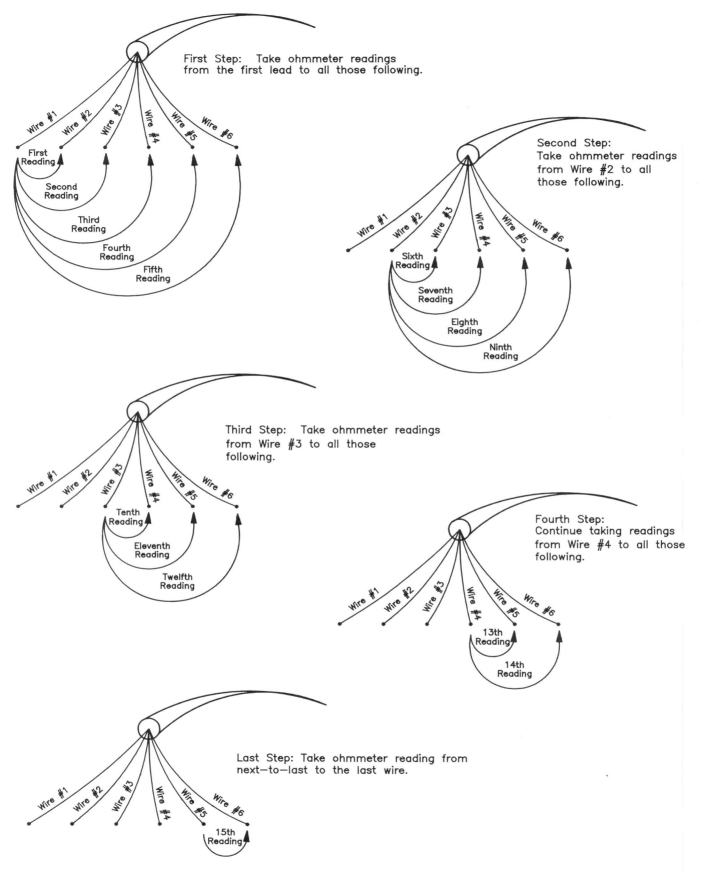

Figure 1-1 Testing a multiple-wire cable for shorts

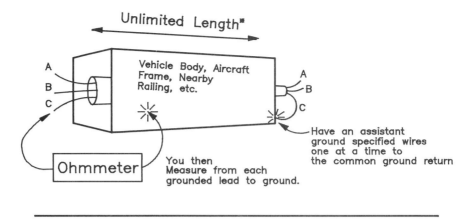

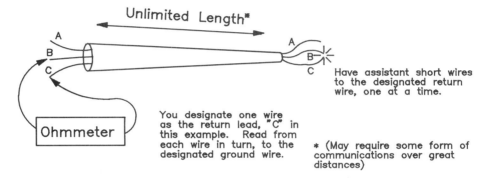

Figure 1-2 Two methods of providing returns for long cables during end-to-end testing for continuity

is not as effective as a voltmeter in finding wiring problems because of the great possibility of unseen paralleling paths when attempting to find an open circuit. One principle fact should be kept in mind when troubleshooting wiring problems: *If there is continuity and current is flowing, there should be only a few millivolts of voltage drop from one end of a wire to the other.* More than this indicates an overloaded wire (excessive current flowing through it), a bad connection, or an open wire between the two points being measured with the voltmeter.

One of the problems encountered when working on wiring problems is that the wiring diagram or schematic may not bear much resemblance to the physical wiring layout. The diagram may indicate, for instance, that four electrical points are tied together with wiring. The purpose of wiring is to keep these points at essentially the same potential. In doing so, the wiring can be physically laid out in many different ways.

An example here will involve three wires, connecting four electrical points. This means that, regardless of how the circuit is actually wired, we should account for six wire ends. If you can account for only five of them, one is missing—probably the one that is disconnected, causing the problem.

Another problem is that the Digital AC Voltmeter can be too sensitive for use in some AC power circuits. The mere presence of strong AC magnetic fields can easily result in misleading readings on a digital meter. Two completely different, isolated windings on the output side of a transformer can be shown to have perhaps 44 VAC between them. A VOM, even an inexpensive one, will give a more realistic reading of a volt or two and be more easily recognized as a "stray voltage." A resistance of about 10,000 ohms for each volt present in the circuit may be used in parallel with a digital voltmeter to reduce this error.

Wiring can break inside insulation, but this is very rare. Wiring is most likely to break where connected to terminals, wire lugs, soldering lugs, or at the point of joining any metal attachment at the end of the wire. It is at these points that the wiring will receive repeated sharp bends.

Testing coaxial cables — Coaxial cables can be tested for shorts between the inner and outer conductors and for end-to-end continuity with the ohmmeter. See Figure 1–3.

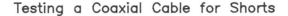

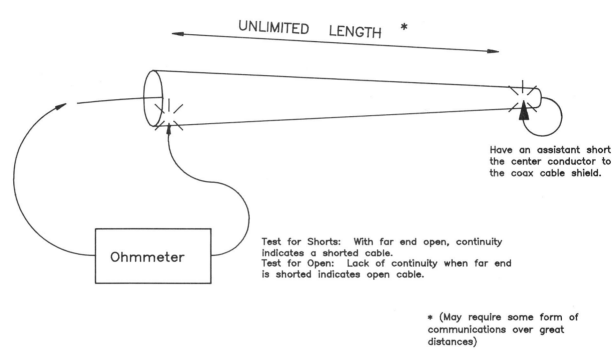

Figure 1-3 Simple testing of a coaxial cable for shorts and opens using the ohmmeter

Crushing a coaxial cable or tearing the outer shielding can make the cable ineffective at high frequencies. These faults will not show on an ohmmeter if the inner and outer conductors are not shorted together and if there is still DC continuity through a few remaining strands of the outer shield. Worse yet, if there is a long run of this cable through many clamps, such as used on ships, the location of the problem along the length of the cable will not be evident. A more sophisticated way is needed of identifying the kind of problem that a coaxial cable might have and locating that fault along the length of the cable. A tool is available that uses the technique of Time Domain Reflectometry or TDR.

TDR, simply stated, pulses a cable and watches for any reflections caused by discontinuities in the cable. Discontinuities might be caused by crushed cable, damaged shielding, or in a very sensitive instrument, even a normal cable connector. Figure 1–4 shows what might be expected on good and bad cables. Note that the distance of the reflected echo along the horizontal is proportional to the length of cable between the instrument and the fault detected, thereby giving the location of the problem along with the nature of the failure.

Figure 1–5 shows a TDR instrument that is used to generate the TDR waveform and to receive the reflections caused by coaxial cable problems.

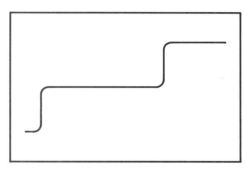

Cable shorted at the far end

Cable open at the far end

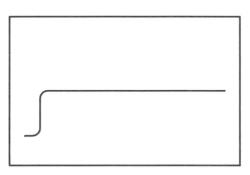

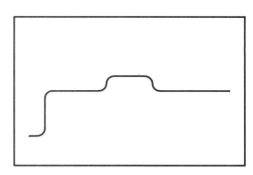

Cable terminated in resistive Load equal to cable impedance

Terminated cable having a section of higher impedance cable in the middle

Figure 1-4 Sample TDR presentations of good and bad coaxial cables

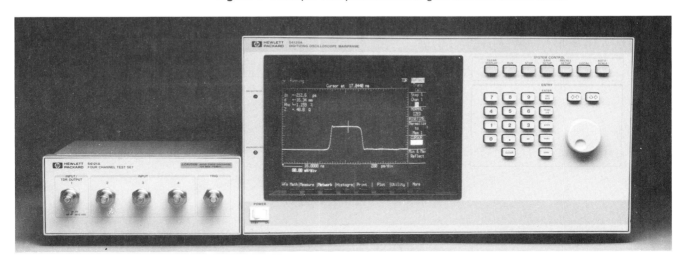

Figure 1-5 A TDR instrument. This instrument is valuable for the maintenance and troubleshooting of coaxial cables (Photo courtesy of Hewlett-Packard Company)

Testing coaxial cables with an RF source, wattmeter, and terminating load — An alternative way of testing coaxial cables is to supply one end of the cable with a few watts of radio frequency power and to put a wattmeter and proper terminating load at the far end of the cable. See Figure 1-6.

If the cable is "perfect," the wattmeter should read essentially the same power when connected between the far end of the cable and the load as it does when the wattmeter is connected between the RF source and the near end of the cable.

8 □ Finding the Defective Equipment

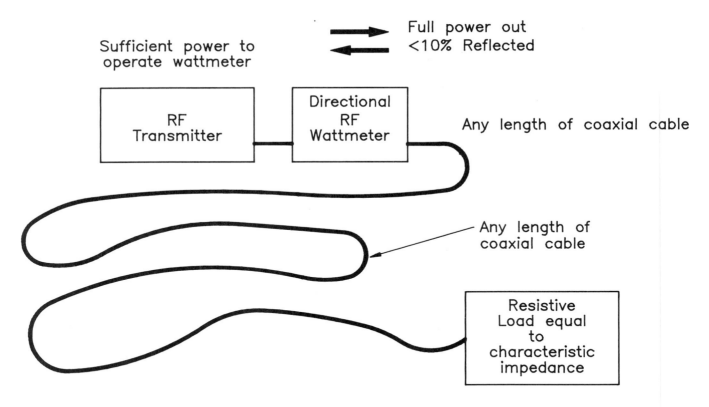

Figure 1-6 Wattmeter test indications for a good cable and for a shorted or open coaxial cable

The attenuation factor of the cable may have to be taken into account for this test. Attenuation will depend upon the specifications for the cable and the frequency used for the test. Generally speaking, the higher the frequency, the more the cable will look like a long resistor rather than a perfect cable.

A shorted cable will show little or no RF coming out the far end, while nearly all of the power supplied at the RF end is reflected back. Again, refer to Figure 1–6.

A cable that has less than a drastic failure, such as a short or open, will show varying degrees of reflected energy. The ideal is for all power to flow outward from the source with nothing reflected back to the source from discontinuities; however, crushed cables and damaged shielding can result in partial reflections, as shown in Figure 1–7.

The higher the frequency used for these tests, the more sensitive the tests will be to find small problems like small tears in the shielding. On the other hand, the losses of the cable may become predominant at higher frequencies.

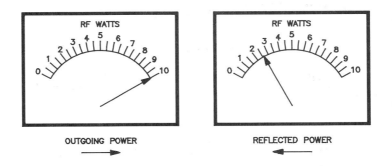

Meter indications of less than drastic failure of a coaxial line, such as crushing or bad shield.

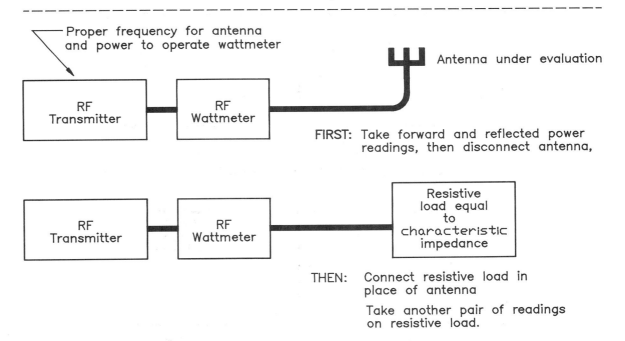

The closer the antenna approaches the readings of the resistive load, the better the antenna

Figure 1–7 Detecting a damaged coaxial cable, and a method of evaluating an antenna

1.1.2.3 ARE ALL INTERLOCKS CLOSED?

— An interlock is a switch designed to be automatically actuated when a cover or panel is removed. Some interlocks remove all power from the equipment and prevent everything from working. Other interlocks are more subtle in their operation. They might remove drive power to a particularly dangerous motor if the cover of the equipment is removed, leaving power applied to all other circuits. If power is obviously still applied to the equipment, partial failure might be due to an interlock. Look carefully around cover openings for concealed interlocks. An example of this kind of interlock is the safety interlock on a washing machine that quickly stops the rotation of the inner drum if the cover is lifted during the spin cycle. The machine might continue to pump water in and out (power is obviously still applied) but the spin cycle won't work. The interlock, sometimes cleverly concealed, prevents the hazardous spinning.

1.1.2.4 Close or Bypass ONLY if Necessary

— Bypassing an interlock is a dangerous act. The interlock is there to protect people, not the equipment. Bypassing a safety device should be a sensible and necessary act. Work with the interlocks functioning normally whenever possible. In those instances when it is absolutely necessary to bypass an interlock—to get a high voltage reading, for instance—turn off the power, connect the voltmeter, and get your hands out of the equipment before reapplying power.

Some interlocks are not meant to be bypassed. Others are designed so that the service technician can easily bypass them, yet the switches will return to their original safety operation when the panel or cover is installed again. See Figure 1–8.

1.1.2.5 IS THIS SYSTEM DRIVEN BY SOFTWARE?

— Equipment that is operated by software or "programs" is dealt with in a different manner because of the interaction of the software and the circuits it controls. If this is the case the reader is referred to Phase 8, "Troubleshooting Software-Driven Circuits."

1.1.2.6 IS THIS A DIGITAL SYSTEM?

— Not all digital systems are operated by software. A digital clock is an example of such circuitry. Some printers are "dumb" and operate with basic digital circuits without benefit of the flexibility and smartness of a programmed microprocessor. If there is a microprocessor or microcontroller chip in use, the equipment is software controlled. Without these chips the equipment is a simple digital circuit and troubleshooting continues here in this flowchart.

1.1.2.7 NOTES ON ANALOG SYSTEM TROUBLESHOOTING INSTRUMENTS

Built-in system indicators — Analog systems frequently have system indicators with which to localize a problem. One example is a radio broadcasting station. As the audio progresses from a turntable through the control console, the line amplifiers, and into the transmitter, its progress is constantly monitored by various meters and is controlled by various adjustments. A technician experienced at such an installation could locate an audio failure anywhere within the entire system within minutes, using built-in system indicators.

Specialized system test equipment — Complex systems often use specialized test equipment for system testing. An example might be the instrument landing system of an aircraft that involves the receiving antenna, the receiver, and an indicator. The input signals to this system consist of multiplexed signals that must be precisely generated to evaluate the operation of the aircraft system. A test instrument for this system is made up of a signal generator on specific frequencies, using special modulation waveforms.

The RF wattmeter as a system troubleshooter — The wattmeter is one of a few general purpose instruments that can be used as both a system and equipment test instrument. The wattmeter can be used with a dummy termination load resistor to test transmitting equipment, coaxial lines, and antennas.

Antennas can be compared to the "perfect" load of a termination resistor. The more nearly an antenna approximates the wattmeter readings of a resistive load, the better matched and more efficient is that antenna. See Figure 1–7.

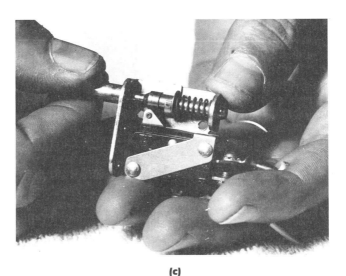
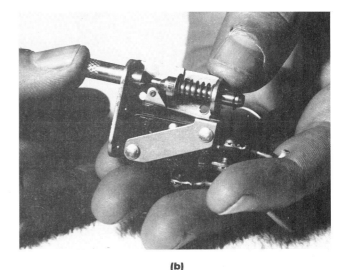

Figure 1-8 A typical interlock switch that removes circuit power when the plunger is allowed to come out under spring tension when the equipment cover is opened. Note the special "service" or bypassed condition in photo 1-8(c)

Transmitting equipment can be tested and tuned for proper power output using the wattmeter and terminating resistor. See Figure 1-9.

1.1.2.8 ARE THERE SEPARATE EQUIPMENTS WITHIN THE SYSTEM?

— It may be that the problem reported involves a single piece of equipment that is not part of a larger system. There is only a single unit. An example is a TV set. Here is a case of beginning troubleshooting at the equipment level rather than the system level, the subject of the next phase. The checks to be made as mentioned earlier in this phase apply to the equipment entry level, too. Continue from this point to Phase 2 if the equipment cannot be broken down further by substituting other major units of a system.

1.1.2.9 ARE OTHER EQUIPMENTS AVAILABLE TO SUBSTITUTE?

If there are two or more interconnected equipments that could be causing the reported problem, by all means substitute others, including the cables, to try to pin down the offending equipment. Each case will be different, but the important principle applies:

SUBSTITUTE TO ISOLATE THE PROBLEM!!

12 ☐ **Finding the Defective Equipment**

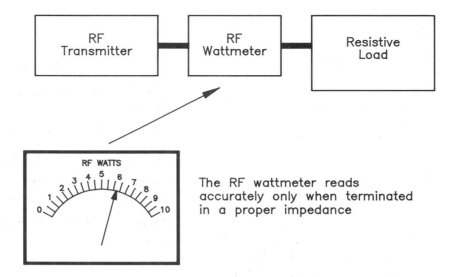

Figure 1-9 Testing a transmitter for proper power output using a wattmeter and terminating resistor

Substitution is the single most powerful troubleshooting technique the technician can employ. It applies to any system, to any PC board, to any component; it also applies to possibly defective test instruments. When in doubt, substitute a known good "whatever."

A special case warrants mentioning here. If a component is known to be bad, and if substituting a good one *makes* the good one go bad too, stop. The bad circuit could possibly be destroying new replacements faster than you can install them.

1.1.2.10 NOTES ON USING SYSTEM INDICATORS AND SYSTEM TROUBLESHOOTING CHARTS TO LOCALIZE PROBLEM TO ONE EQUIPMENT

A familiarity with the block diagram of the entire system, together with observation of the installed indicators, provides a very effective means of localizing the problem.

System indicators can take many forms. Common indicators are meters, indicators, and cathode-ray tube displays. Such indicators are always operating, providing instant information for trouble analysis.

Instruction manuals may include troubleshooting charts that help to localize a problem. These charts are available in one of two forms, tabular and flowchart. See Figures 1–10 and 1–11.

Troubleshooting charts are an aid in troubleshooting but are sometimes less than adequate for a specific failure. Once a technician is familiar with the block diagram of the system and the location and significance of the installed indicators, the troubleshooting chart will no longer be useful. These charts are written to substitute for system or equipment familiarity. They may be of use at first when the technician is initially learning the system.

1.1.2.11 Substitute One Piece of Equipment or One Cable — If the reported problem is still present, the offending unit of the system must be identified. This is done by substituting other known good working equipments into the system until the system runs properly. The last equipment removed must have been the bad one. One caution should be kept in mind when changing equipment and cables: it is easy to mistake a problem. If, for instance, a cable is disturbed during an equipment change, a defect in that cable may easily make a piece of equipment seem faulty when it is not. For this reason, *it is often a good idea to change BACK and put the suspected unit into the system again, to be sure that the problem re-appears as expected.* If it does not, the problem of an intermittent cable or circuit is strongly suggested.

Substitute known good equipment and cables until the defective equipment is definitely identified.

POWER

SYMPTOMS

1. Printer is completely inactive when the power switch is set to the ON position.

2. Power LED on front panel illuminates, but printhead does not move to home position. (This symptom can also indicate a problem with logic or hardware; refer to CARRIAGE MOTION of this section.

3. Printer operates in self test, serial interface will not function.

PROPOSED CHECKS

1.
   a. Check line voltage and power cord for proper connection.
   b. Check for blown fuse (F1).
   c. Check for defective power switch.
   d. Test for defective components in +5VDC power circuit (BR1, VR4, capacitors).

2.
   a. Check for +24VDC at the output of transistor Q3 in power supply.

3.
   a. Check for +12VDC at the output of VR2.

CARRIAGE MOTION

1. Power LED on front panel comes on, carriage does not move to home position.

2. Power LED on front panel illuminates, carriage does not move to home position and fault LED blinks.

3. Carriage stops during print operations
   or:
   Carriage slews to home and re-initializes during print operations
   or:
   Unusual or excessive noise during print operations. including print head hitting sides of print mech with excessive force.
   or:
   Characters not vertically aligned (every other line offset).

1.
   a. +24VDC not enabled via 24EN by microprocessor (defect on PCB).

2.
   a. Carriage drive cable disconnected.
   b. Obstruction in carriage path.
   c. Carriage motor winding open.
   d. Defective carriage motor drive components on PCB.

3.
   a. Duty Cycle on phasing circuits out of adjustment.
   b. Excessive slack in carriage drive cable.

Figure 1-10 Partial troubleshooting chart of the tabular type (Figure courtesy of Mannesmann Tally)

14 ☐ **Finding the Defective Equipment**

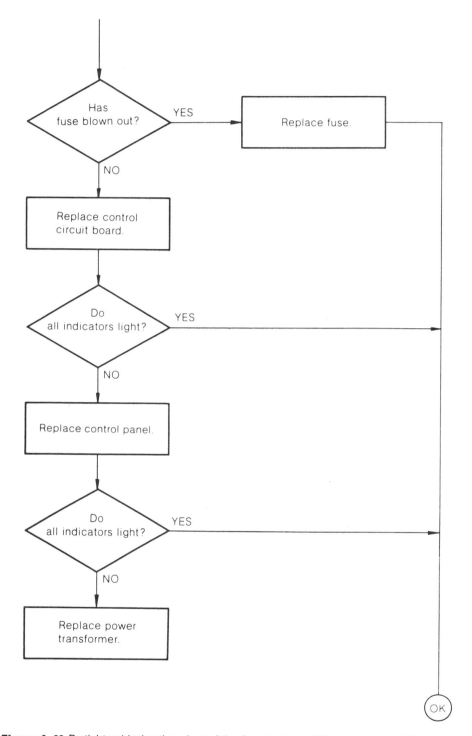

Figure 1-11 Partial troubleshooting chart of the flowchart type (Figure courtesy of Epson America Inc.)

On occasion, there may be just a shade of doubt that the problem is still really identified. In this case, it is an excellent idea to put the system back to its original non-working configuration, test for the problem, and then again change out the suspected equipment. Doing this once or twice will give the technician confidence that the problem was found.

1.1.2.12 Retest System — Each time a piece of equipment is substituted in the system, the system must be re-energized and the problem sought using the new setup. This procedure involves repeatedly turning off the system or equipment, putting in another known good one, turning on the system again, and looking to see if the original problem is still there. If it isn't, the most recently removed equipment is probably the bad one.

CAUTIONS: Don't be surprised if the insertion of a "good" piece of equipment introduces a new problem. "Known good" equipment can turn out to be bad, too.

There is also a small chance that a defect elsewhere in the equipment could damage good boards inserted for troubleshooting purposes. As mentioned before, if a known good board also tests bad, there is a possibility that a failure has occurred which will continue to damage every good board inserted!

To add still another variable, the substitution of a good piece of equipment might only partially cure the original problem. It is possible that the original problem was actually two or more problems.

1.1.3 IS POWER TURNED ON AND PLUGGED IN? — It is common for a servicing technician to find that the cause of a service call is simply that a plug has been dislodged from the wall outlet. Some plugs may appear to be plugged in well enough, but a solid push proves otherwise.

If there is any doubt as to whether the wall outlet is energized, another appliance such as a lamp can be substituted to prove the point. Of course you must make sure that the lamp itself is turned on! An AC voltmeter can also be used, of course.

Intermittent plugs may sometimes be cured, at least temporarily, by spreading the prongs of the plug apart, inward, or by putting a slight curve in them. Particularly troublesome outlet receptacles should be replaced by an electrician to prevent future problems.

Sometimes equipment power switches are located in very inconspicuous places. They might be located underneath the equipment, in the rear, or perhaps half way down the power cord. The equipment operator's manual, if available, may have to be consulted to find the switch in some equipment. See also Paragraph 1.1.2.1.

1.1.3.1 ARE ALL FUSES IN AND GOOD? — If power is still apparently not available to the equipment, a check of the fuses is the next logical step. The fuse is often mounted in a quarter-turn holder, removable from the rear panel.

In some equipment, the fuse may be located internally and it is necessary to remove the equipment cover to gain access to it.

In either case, the fuse should be carefully inspected. There may be a small break in the internal wire or the fuse may have blown violently, vaporizing the wire and depositing it as a silvery spot on the inside of the glass. See Figure 1–12.

Some fuses cannot be visually checked for an open. These fuses should be tested *out-of-circuit* using an ohmmeter on a low range. There are too many possibilities for paralleling paths if the fuse is left installed, not to mention the risk of damaging the ohmmeter if there is any power applied.

A very conclusive and simple method of testing a fuse is to leave the fuse installed and take an appropriate voltage measurement *across* the fuse *with switches on and power applied.* On 117 VAC circuits, the voltmeter should be on a high AC voltage scale. On low-voltage circuits, such as automotive lighting circuits, the 12- or 15-VDC scale would be appropriate to use in finding open fuses. Note that *all* power switches must be on (including any interlock switches) for this test to be valid. A fuse that is blown will have the full supply voltage across it. See Figure 1–13 for examples of this method of testing fuses in-circuit with power applied.

1.1.3.2 Analyze the Fuse Failure — If the blown fuse has been badly overloaded, which is indicated by the mirrored appearance of a hard fuse failure, refer to Phase 7, "Troubleshooting the Power Supply." There is nothing to be gained by trying another fuse as the new fuse would only blow immediately. A major failure is indicated, probably a short circuit.

16 □ **Finding the Defective Equipment**

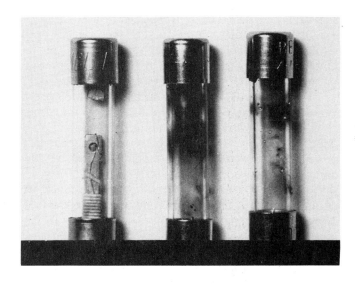

Figure 1-12 Photograph of the two types of fuse failures, the "soft" and "hard" blown fuses

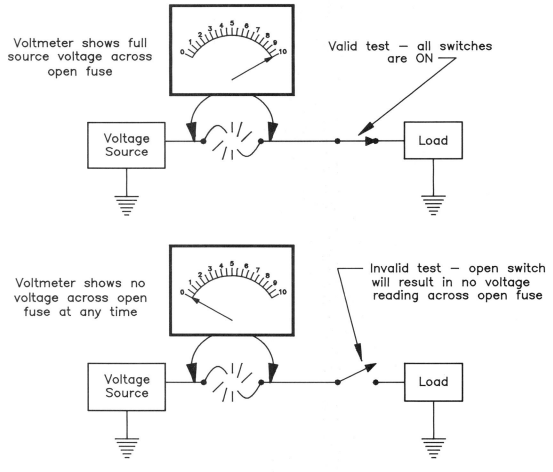

Figure 1-13 Valid and invalid circuits using a voltmeter to test for a blown fuse

On the other hand, if the fuse has "gently" blown by separating the internal wire, there is a good chance that replacing the fuse with another of the proper rating may solve the problem. In circuits that begin operation with a surge of current, the fuse is momentarily overloaded. This is the case with equipment that has a simple DC filter circuit with large value capacitors installed. The initial surge to charge these capacitors causes a momentary surge of current when the equipment is first turned on. Repeated overloads like this can cause the fuse to fatigue and open without there being a failure within the equipment.

Whenever a blown fuse is replaced, it is a good idea to be very sure that the fuse to be installed is of the proper current rating and type. It sometimes happens that a prior repair results in the wrong fuse being installed. Do not simply install the same rating that was installed before the failure. Look at the rating stamped on the chassis or consult the equipment service or operations manual for the proper size. The current rating of a fuse is usually stamped on the end ferrule. There may also be a voltage rating, which can be confusing. It is the *current* rating that must be matched first.

The type of fuse is also important. Fuses can be manufactured to withstand initial surges common to electronic equipment. These fuses are called *slow-blow* or *time-lag* fuses. They should be used only in equipment that specifies their use. Using a slow-blowing fuse in a circuit that is supposed to have a fast-blowing fuse may result in a great deal of damage to the circuit before the fuse finally blows. On the other hand, installing a fast-blowing fuse in a circuit that should have a slow-blowing fuse will probably result in the fuse immediately blowing, without damage to the circuit at all.

1.1.3.3 Replace Fuse and Reapply Power

— Put in the new fuse. During installation, note the amount of force required to install the fuse. Is it reasonable for the size of the holder? Fuses with high current ratings of perhaps three amperes or more are prone to heating of their holders. This is a common problem with automobile fuse holders where currents of 15 amperes or more flow continuously during the operation of the vehicle. If a fuse holder develops corrosion or is not clipped tightly to the fuse, resistance can build up at the ends of the fuse. This resistance heats the end of the fuse because of the normal high currents flowing. This heating at the ends of the fuse contributes to the melting of the fuse element. The result is to reduce the current rating at which the fuse will blow.

To afford a better contact with the fuse, the spring clips of the fuse holder can often be resprung closer together with a pair of pliers. Corrosion of the clips can be cleaned with emery cloth or sandpaper in an emergency, but a corroded or rusted holder should be replaced at the first opportunity.

Since we are considering that the original fuse was a "soft" failure, re-applying power may or may not cause the fuse to blow. If the fuse does blow, then the power supply should be investigated. Phase 7 should be consulted for further help. If the fuse seems to hold the load, place an ammeter *in series* with the fuse to monitor how much current is flowing. Normal currents for a fused circuit should be below the fuse rating by at least ten or 15 percent. Higher currents indicate an overload of the circuit, and the power supply and loads should be investigated as explained in Phase 7.

In any case where a fuse blows once, but a substitute fuse of the same rating holds the load, the equipment should be operated for an extended period of time. This procedure should reveal any problems that might occur and cause the overload after the equipment warms up. Another tip is to gently rap the equipment to see if there is a mechanical intermittent that might be causing the fuse to blow.

1.2 Problem May Be Intermittent

— Many electronic problems are simple to analyze and to repair. The equipment fails, and that's all there is to it. The complete failure is relatively easy to repair.

The other side of the coin is the equipment that sometimes works fine—and sometimes doesn't. Under Murphy's Law, the equipment always works fine when the technician arrives. Intermittent problems, however, can cause real misunderstandings between the technician and the operator. The general nature of intermittents, with or without Murphy's law, should be explained to the customer.

1.3 Attempt to Induce the Intermittent Problem

— Once the operator has been eliminated as the possible cause of a problem that is no longer evident, the technician should suspect an intermittent problem. This may require the removal of the equipment to the shop, because the removal of the equipment covers will probably be necessary.

Intermittent problems are of three types: the *thermal* intermittent, the *mechanical* intermittent, and the *erratic* intermittent. There is one key to repairing any intermittent problem: *The intermittent problem must be present long enough to effect a repair.* If equipment is working normally, you cannot fix it.

1.3.1 THE THERMAL INTERMITTENT — The thermal intermittent is a problem that becomes evident when the equipment warms up. It might occur more on hot days or when the room temperature increases for any reason. It may also occur with regularity as the equipment reaches normal operating temperature. A problem that is described as being fairly consistent with regard to operating time can often be confirmed as a thermal intermittent.

The reason for the thermal intermittent is that materials usually expand when heated and contract when cooled. On a microscopic scale, circuits can actually pull apart with the thermal stresses placed upon them. Integrated circuits sometimes develop these problems, as do poor wiring connections and discrete components, such as transistors, resistors, and capacitors. While intermittent wiring problems can often be solved by simply cleaning and tightening the connections, defective components must be replaced.

The thermal intermittent is best handled by letting the equipment cool down completely without power applied. In severe cases, the equipment may have to be refrigerated to make it work normally. After normal operation for a period of time it should, if this is a true thermal intermittent, produce the reported problem. Cycling once or twice between normal and abnormal operation under only the influence of changing temperature will verify that the problem is a thermal intermittent.

Stubborn cases may make it necessary to cool the equipment while it is in normal operation. A special product is made for this purpose, generally a spray can of freon. The freon, upon being released from its pressurized container, evaporates very rapidly and absorbs heat from whatever it contacts. With a thermal intermittent, cooling a circuit that has a problem will make it resume normal operation. Often the circuit will again show abnormal operation upon coming back up to normal operating temperature.

The offending component can be isolated by cooling progressively smaller areas of the circuit. Large areas can be cooled by removing the small directional pipe from the spray head of the can of freon. Individual components can be cooled by using the pipe and a lighter touch on the spray can's button.

1.3.2 THE MECHANICAL INTERMITTENT — The mechanical intermittent is familiar to every technician. This is the problem when equipment changes from an operating to nonoperating state or back again when tapped or thumped.

Tracking down the mechanical intermittent is fairly easy if the problem can be made to come and go at will. This might be done by flexing a circuit board a certain way or pushing it in a particular spot. *As the actual defect is approached, the amount of effort required to make the problem come and go should be less.* In other words, if you must push hard on the right side of the circuit board to make the problem appear but it requires much less push on the left side of the board to accomplish the same result, then the problem is somewhere on the left side of the board. It may take a great deal of playing with the problem to isolate the exact area or component that is defective.

1.3.3 THE ERRATIC INTERMITTENT — The erratic intermittent is the worst intermittent of all to repair. There appears to be no initial pattern to the failure, such as temperature or mechanical sensitivity. Many service calls have been made with nothing accomplished because of the erratic intermittent. Paragraph 1.5 gives some suggestions as to how this might be resolved.

A possibility that should occur to the servicing technician is that the reported problem might have been caused by a power line transient or a static discharge to the equipment. Proper grounding of the chassis to an earth ground, via the third wire of the power cord, is good insurance against this problem. Computer equipment is particularly prone to this kind of erratic behavior. This topic is covered more in paragraph 1.6.

1.4 DID THE PROBLEM RECUR?
— After having attempted to no avail to verify a thermal or a mechanical intermittent, it can be assumed that the problem is an erratic intermittent. Consult the operator or owner of the equipment and inquire about conditions at the site where the equipment is used. See paragraph 1.5 for inquiries to make.

1.5 Checking Other Possibilities
— Is there a possibility of the line voltage being too high or too low where the equipment is used? Must the customer change burned-out light bulbs frequently? This is an indication of higher than normal line voltage. Is the equipment operating in temperature extremes of hot or cold? Is there a chance that there may be transients that could affect digital equipment such as computers? Is there any transient protection on the site such as surge protectors or battery backup systems? See Phase 8 for more information on these topics, in regard to computer interference.

The answers to these questions may alert the technician to specific environmental factors that may be causing the problem.

In some cases, it may be necessary to deliberately stress the equipment under controlled conditions to attempt to duplicate the reported problem. Two ways to stress equipment are: 1) to apply higher (or lower) than normal line voltage, and 2) to place the equipment in a special enclosure where the temperature may be elevated.

1.5.1 Controlled Overvoltage Stressing
— The manual for the equipment under test should be examined closely to see what the line voltage limitations might be. In the absence of information to the contrary, from 110 percent to as much as 115 percent of normal voltage might be applied to the equipment. The actual voltage that should be applied is a matter of judgment, considering the cost of the equipment and the urgency to find a definite problem.

A variable transformer can be used to apply variable AC voltage to AC-operated equipment. In the absence of a transformer, the technician can make a buck/boost arrangement that will add or subtract a small voltage from the AC line. This arrangement is shown schematically in Figure 1–14.

Using this circuit, an inexpensive 12 VAC transformer can produce a voltage of either 129 VAC (117 + 12) or 105 VAC (117 – 12). The current that can be safely drawn from this circuit is the current rating of the transformer *secondary*.

When using this overvoltage technique it would be a good idea to leave the equipment operating at the abnormal voltages for some time to allow the equipment temperatures to stabilize and thus show a problem that would not be evident if operated for only a few minutes.

Problems may also surface if the equipment is operated at less than normal line voltage. This is sometimes a problem in remote areas where lines must transport power long distances into the country. Operation at approximately 80 percent of normal line voltage is probably the minimum at which most equipment could be expected to operate properly.

1.5.2 Controlled Overtemperature Stressing
— One of the causes of the erratic intermittent might be a sensitivity of the equipment to higher temperatures than those that exist in the technician's work area. In order to get the equipment to an elevated temperature such as might be encountered at the installation location, a special "hot box" can be made from inexpensive parts. Just mount a few 100-watt light bulbs (the heaters) inside a box along with a common household thermostat to control them. When the heat accumulates in the box it turns off the lamps, thus regulating the temperature to a safe maximum. See Figure 1–15.

Be sure never to leave a hot box operating unattended as it could be a fire hazard!

1.6 NOTES ON USING THE STORAGE OSCILLOSCOPE

Power line transients — Power transients are very short-term fluctuations of the line voltage, substantially above or below normal levels. These can account for some very

20 ◻ **Finding the Defective Equipment**

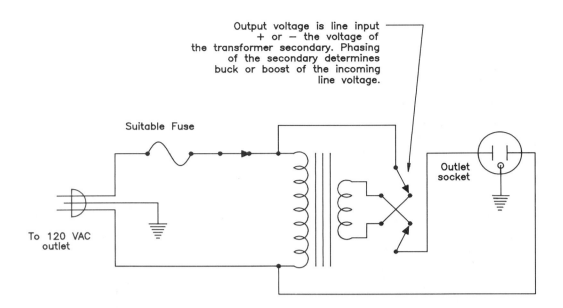

Figure 1–14 Normal AC line voltage can be boosted or bucked by the voltage generated across the secondary of this transformer

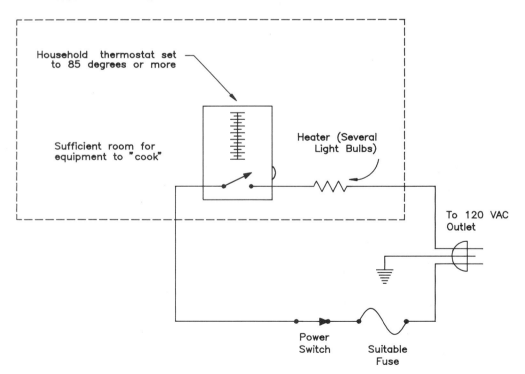

Figure 1–15 Schematic for a "hot box," useful in tracking erratic intermittents

mysterious problems in electronic equipment, particularly computers and other digital circuits. Spikes of high voltage can, in extreme cases, cause the failure of integrated circuits. Moderate protection from transient damage and the resulting equipment operational failures can be obtained by installing surge suppression components such as those in Figure 1–16. These components absorb the energy of the transient and keep line voltage at a safe value.

Figure 1–16 Two components used for protection against line voltage transients; the Zener Diode and the Metal Oxide Varistor (MOV)

Outlet strips with these components installed are marketed under a variety of brand names. The best ones have three of these components installed, one between each of the three wires in the power cord.

Digital memory and analog storage oscilloscopes are particularly well suited for capturing "pictures" of voltage spikes and short-term events that occur infrequently. The triggering circuits of the instruments can be adjusted so that the instrument is not sweeping but is waiting for an unusually high voltage spike to initiate the trace.

Once the trace occurs the instrument is able to hold the pattern for a relatively long time. The pattern can be observed and the transient voltage spike positively identified as the cause of an erratic intermittent.

Digital equipment and computers are sensitive to transient voltage spikes on the power supply lines. A voltage spike can be mistakenly interpreted as a digital signal where there should have been none, or, if high enough in amplitude, can cause failure of the digital hardware. Chips can be damaged by voltage transients and not fail until a much later time, months or even years.

1.7 NOTES ON USING THE RECORDING VOLTMETER — The recording voltmeter is very useful for graphing line voltages during times when most people want to sleep. The recording voltmeter can be used to monitor the 120 VAC or 240 VAC line voltage and record the voltage on a graph at specified time intervals. This instrument will show line voltage fluctuations over relatively long intervals. A short spike or a few cycles of unusually low or high voltage will not be shown. Only sustained variations will show. While some recording voltmeters are designed for use only on 120 or 240 VAC lines, others are available with full volt-ohm-milliammeter and multiple range capability. The recording voltmeter might be called for when equipment runs fine during the day and develops problems at night when no one is around.

1.8 The Stalemate: Return Equipment for On-Site Testing

A very small percentage of problems will not emerge in spite of the technician's best efforts. The only recourse in this case is to tell the customer that efforts have been made to make the problem recur, all to no avail. This is a stalemate situation that has been responsible for some equipment staying in the shop for months on end. Move it back to the customer, with a request to put it back into service and let you know if it still gives problems. In the worst case, it may be necessary for you to go to the installation site to evaluate the problem there rather than have the equipment brought in to the shop.

☐ Phase 1 Summary

This phase has described some of the common problems that a technician may observe in the first investigations of a repair. A surprising number of problems can be quickly resolved by the application of these tips: Be sure power is applied, be sure the operator has neither caused the problem nor called something a problem when it really is not, and be able to recognize an intermittent problem when it occurs. Appendix A, the "Frustration List," may help in stubborn cases of troubleshooting.

At this point we should have something like a defective unit or piece of equipment that can be put on the bench for further work.

☐ Review Questions for Phase 1

1. Name three indications you might get to prove that power is available to a piece of equipment.
2. Name three "simple things" that should be checked before equipment is assumed to have a defect.
3. A particular inter-unit cable is often kicked, moved, or otherwise abused. A problem develops. What is the 100 percent sure way to test the cable quickly?
4. Discuss in detail the testing of a multiple-wire cable having 12 wires within it plus an outer shield. The cable is a city block long, and the technicians cannot get within 100 feet of each other, but can communicate by talking loudly. Only one has an ohmmeter. The cable must be tested for shorts and for opens. Include use of the shield as a return for the ohmmeter when testing for opens and logical progression through the wires for both tests.
5. Three circuit points must be connected together with wires. Each point has a single wire end in sight. What is wrong?
6. What method of testing coaxial cables will give information on whether the cable is open or shorted, and the location of the problem?
7. Draw the pattern you would expect for a 50 ohm coaxial cable terminated in 600 ohms if tested with a TDR instrument.
8. When does the interlock perform its primary function?
9. When troubleshooting a large system, what would the technician use to initially troubleshoot a problem?
10. What instrument would be most valuable for testing coaxial cables *and* antennas?
11. What is the most powerful method of troubleshooting that is useful from system troubleshooting all the way to the component level?
12. When will troubleshooting charts be the most useful to troubleshoot a system?

13. A suspected unit has been replaced. The problem is no longer present. What can be done to verify that the removed unit is really the problem?
14. You wish to test fuses with the voltmeter. What conditions must be met? What are the indications of a good fuse?
15. Explain the conditions that may lead to the "gentle" blowing of a fuse.
16. Name the three kinds of intermittent problems.
17. What is the secret in repairing intermittents?
18. Name two methods of troubleshooting an erratic intermittent.
19. Describe the stalemate problem.

PHASE 2
ISOLATING THE DEFECTIVE CARD

☐ PHASE 2 OVERVIEW

Now that the defective equipment has been isolated from an entire system, the technician must further narrow down the problem. Most equipment is made up of individual circuits, many of which are separate cards within the equipment. We shall now consider the decisions to be made along the way and the alternatives that should be considered as the problem is isolated to a single defective card or area in the equipment.

2.0 NOTES ON GENERAL SAFETY AND ELECTROSTATIC DISCHARGE

Grounding — Most high-quality equipment used today is operated with the chassis at earth ground potential and connected to the third wire of the power cord, which is eventually connected to earth. See Figure 2–1.

Carefully observe the wiring within the equipment. It is possible there are one or more green wires with a yellow tracer connecting various pieces of chassis sheet metal together. These are grounding wires that should be kept intact to reduce the danger of electrical shock. These wires also reduce electromagnetic interference by bonding all the chassis components together.

Less expensive equipment may not have the safety feature of this third wire, depending upon the insulation of a plastic case to protect the operator from dangerous shock hazards of an ungrounded chassis. Television sets and stereos often fall into this class of equipment. Servicing such items can involve serious shock hazards for the technician. See Figure 2–3 (page 25).

Some manufacturers have attempted to ensure that the chassis is at least connected to the neutral side of the power line. This is done by connecting the chassis to the same wire that is connected to the wider of the two blades, which is *supposed* to be the neutral side of the power line. Bypassing this safety feature by trimming the wide blade of the plug or by nonstandard wiring of the wall receptacle will make the chassis "hot" and hazardous to service. See Figure 2–2.

Whenever equipment using a power cord with only two wires is serviced, there is a possibility that the chassis of that equipment could be dangerous. Because of this, it is a good idea to use an *isolation transformer* between the power line and the equipment being

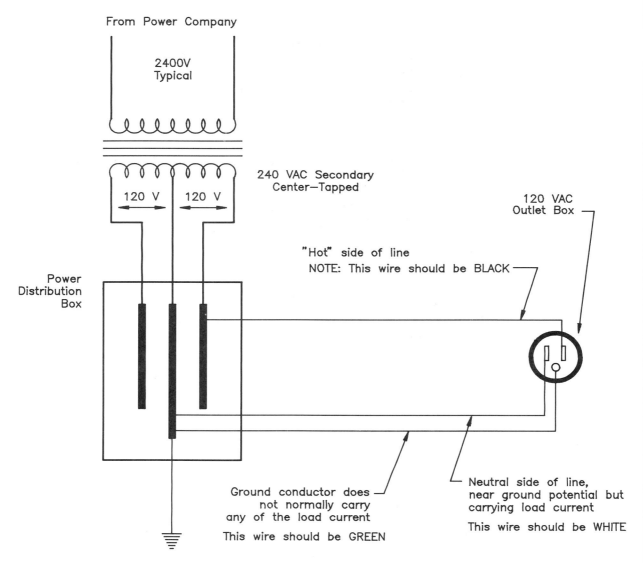

Figure 2-1 The third wire of a three-wire cord is normally connected to an earth ground through the power distribution panel

Figure 2-2 Polarizing of the power plug can be accomplished by making one of the blades wider, thus making it necessary to insert only one way into the wall receptacle. The wider blade should be connected to the neutral side (white wire) of the AC line.

serviced. Once the power line is electrically isolated from the chassis with one of these transformers, the chassis can then be safely grounded. See the schematic of this situation in Figure 2-3.

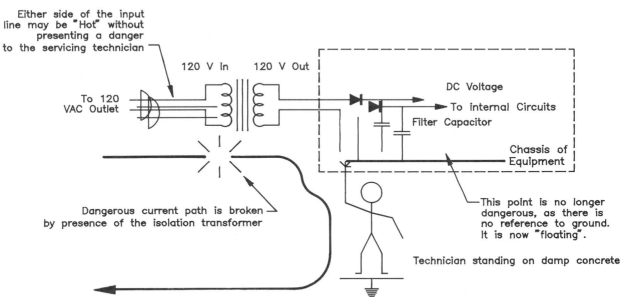

Figure 2-3 Schematic of a servicing setup using an isolation transformer

There also can be a safety hazard when coaxial cable is improperly providing an AC ground for defective equipment. A technician can get a nasty shock by disconnecting two coaxial cables from one another by becoming part of the AC ground path for a defective piece of test equipment. To avoid this potentially dangerous situation, separating two coaxial cables at a connector, loosen the connectors, then slide a least one hand back onto the insulation of the cable before pulling them apart.

Safety when making voltage measurements — If voltage measurements must be taken on high-voltage circuits of perhaps more than 60 VDC or 30 VAC, a series of steps should be observed to prevent electrical shock.

1. Turn off all power.
2. Discharge capacitors.
3. Attach test leads.
4. Turn on power.
5. Take reading.
6. Turn off all power.
7. Discharge capacitors.
8. Remove test leads.
9. Proceed with other tests as necessary.

Equipment using vacuum tubes — Although most modern circuits using semiconductors and operating on less than 12 volts above or below ground potential are quite safe to service without special care, this does not include all of the circuitry that must be serviced.

Vacuum tubes are still in use in some high-power and high-frequency applications. Examples include the magnetron used in radars and microwave ovens, cathode-ray tubes used in television receivers, computer monitors and terminals, and the high-power radio frequency amplifiers in radio broadcasting stations.

The safety precautions covered in the next section apply to most of these high-voltage applications. Usually, high-voltage capacitors are used in the power supplies. Each of these circuits has additional danger points that the servicing technician should be aware of. Some of the dangerously high potentials are shown in general terms in Figure 2–4.

Equipment using high-voltage capacitors — If the circuit uses high-voltage capacitors (over about 25 volts), high-capacity capacitors (over about 1000 ufd), has any vacuum tubes including cathode-ray tubes (CRTs), or contains a laser, the technician must take certain precautions to avoid electrical shock.

Capacitors with more than a 25-volt rating may still be charged, sometimes long after the equipment is turned off. These capacitors should have bleeder resistors provided to discharge the capacitors after power is removed. See Figure 2–5.

Electrolytic capacitors will eventually discharge themselves, but this could take a long time. Oil-filled capacitors, on the other hand, can seem to hold a charge forever.

Even after discharging, some large-capacity, high-voltage capacitors may regain a partial charge over a period of time.

The dead man stick — The shorting stick, or "dead man" stick as it is sometimes called, is a safety device that can provide peace of mind for the technician. See Figure 2–6.

Use of the dead man stick is simple enough. Turn the equipment *all the way off,* including circuit breakers on the wall, if applicable. (It would be easy to short the incoming power mains by accident.) Clip the dead man clamp to a solid ground somewhere on the chassis of the equipment. A solid ground is one that attaches directly to the metal of the frame, a place that has no paint or corrosion to prevent a good connection. The hooked end of the stick is then touched to any potential high-voltage points in the circuit. Any "snap" or "bang" heard is confirmation that a shock has been avoided and possibly that the bleeder resistor is not doing its job.

Once the circuit is rendered safe to touch, it is a good idea to leave the stick hanging right on the high-voltage supply bus during any maintenance. This accomplishes two purposes: it prevents further accumulation of charge in the capacitors, and it prevents personal danger due to accidental application of power to the circuits when the technician is "into" the equipment. A technician can feel considerably safer while working on dangerous circuitry when the shorting stick is used this way.

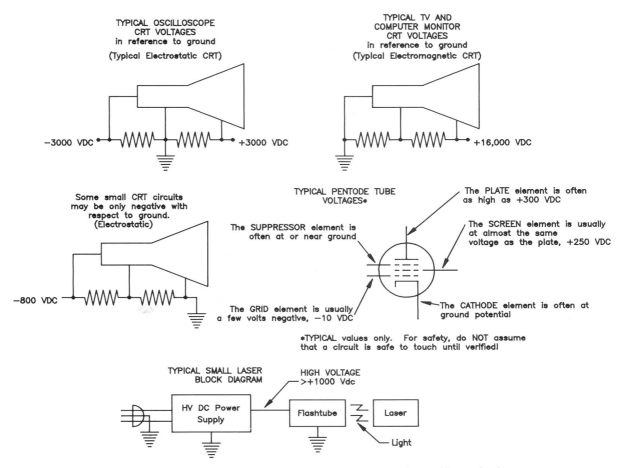

Figure 2-4 Danger points in cathode-ray tube, vacuum tube, and laser circuits

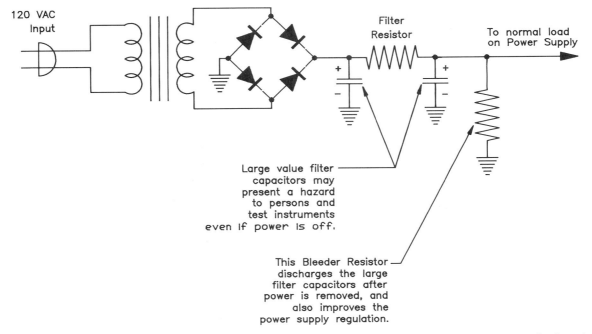

Figure 2-5 Circuit of a power supply that has a bleeder resistor installed for improved regulation and personnel safety after equipment shutdown

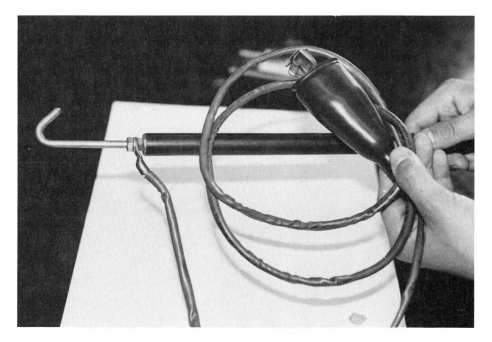

Figure 2-6 The "dead man" stick, used to discharge capacitors that present a potential danger to the technician

Be sure to remove the shorting stick after the repair and before applying power.
The only way to discharge a capacitor that is not referenced to ground is to bleed off the charge by a short, or, in the case of large capacitors, it is better to drain the charge through a resistor of a hundred ohms or so. Smaller capacitors can be discharged by shorting directly across the terminals of the capacitor with a screwdriver or other insulated tool. On very high-voltage, floating circuits, the technician could use one dead man stick to ground one end of the capacitor, then another stick to short the remaining capacitor terminal to ground.

Never work on hazardous circuitry alone — In the painful event of electrical shock, having someone available to turn off the power and perhaps perform cardiopulmonary resuscitation (CPR) can save a life. It is better to delay a repair than to work on dangerous equipment alone. Every technician working on high-voltage equipment should be well versed in CPR procedures, a subject taught by several health organizations. Consult your local Red Cross or industrial safety and health organizations for information on classes.

Electrostatic discharge precautions — As technology advances, the size of circuitry diminishes. We are now dealing with circuits within chips that have dimensions measured in micro-inches. With these spacings, it doesn't take much voltage to arc across from one point to another, making small welds, thus damaging the circuit. Some of these failures will result in the immediate destruction of the chip, while others may "weaken" the chip, causing it to fail prematurely later while in operation. Much research has gone into the investigation of this static electricity discharge problem. The result has been new preventive products and the recommendation of new procedures to minimize hazards to the chips.

Some of the most important precautions are the use of commercial items to make a static-free technicians workstation. This includes a grounded surface upon which to work and a way of grounding the operator. See Figure 2-7, and note the use of a wrist strap that keeps the operator from building up any static charge. The wrist strap and the conductive mat upon which the work is to be done are grounded to earth ground, often the ground connector of a power outlet. No dangerous current can flow through the technician because of a series resistor of about a megohm installed in the grounding wire.

The technician should observe precautions other than the use of a static-free workstation, including the use of component packing for static protection. See Figure 2-8.

Figure 2-7 A typical static free technician's workstation with its various components

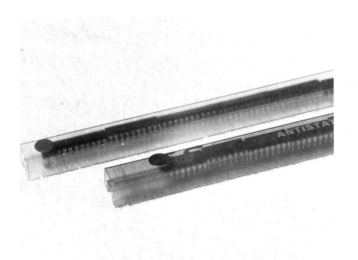

Figure 2-8 Chips are protected during shipment by these special anti-static tubes

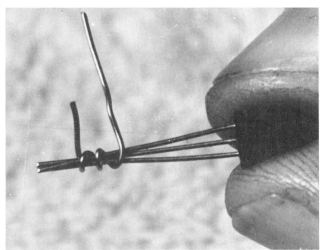

Figure 2-9 Keep the leads of all FETs shorted together until installed

Complete circuits should be shipped and handled while encased in anti-static bags. These bags help avoid static damage to the circuit board and also keep out dust and grime.

All types of field effect transistors (FETs) should have all their leads shorted together with a small piece of wire, until inserted into a circuit. This will prevent accumulation of charges on one lead with respect to another, which could cause failure of the device. Once installed, the normal circuitry takes over this function to a large extent. See Figure 2-9.

Observance of these ESD precautions will help significantly in the reliability of repairs and in preventing new problems during repairs.

2.1 Remove Equipment Covers — Before removing the covers of a piece of equipment, it is wise to make a careful check of the connectors that may attach to the outside of the unit. A bad connector often causes a problem, making the removal of the covers unnecessary. Be certain that they are all firmly in their sockets and inspect the cables for possible lack of cable support to the body of the connector and the resulting breakage of the wiring within the connectors. If no problem is evident with the cabling, the covers will have to come off.

There is usually no particular trick to removing the screws that keep the covers on equipment. It is a trick, however, to keep from losing them. Don't fall into the habit of laying the screws on the bench top. Those screws will likely be knocked off the bench onto the floor and will be lost. The technician should make a habit of replacing screws into the nuts provided for them, often found on one side or the other of the cabinet. This way, regardless of how long it may be before the cabinet is re-assembled, the screws will be there in the right places when they are needed.

After all of the screws are removed, be careful when removing the cabinet. Be sure that all internal wiring is clear and won't be pulled off with the lid. If the cabinet is tight, get someone to help remove the interior if necessary, to avoid possibly dropping the unit if it should come apart suddenly. Look over the interior of the equipment and heed any warning signs that may be posted. They have been put there for the technician's safety and must be observed. See Figure 2–10.

Figure 2–10 This symbol means "READ THE INSTRUCTION BOOK"!

2.2 Check for Loose Connections — Loose cabling and inter-card connections account for a high percentage of equipment failures. Cabling and connectors on PC boards within the unit should be carefully inspected. This is particularly important to check on new equipment that may have had rough handling during shipping. Look for such things as PC boards partially out of their sockets. Look also for loose hardware, such as small bolts that may have been dislodged from their proper places and may be rolling around within the equipment. Turn the equipment upside down carefully and listen for any rattling of loose parts.

If any loose connections are found, they should be corrected and the equipment tested again for the reported problem.

2.2.1 Make Necessary Repair — All internal connections should be clean and tight. If they are not, fix them. The cure should be obvious for a specific job. Don't be hesitant about dismantling a suspected connection and looking at it more carefully. A bit of corrosion can make the difference between normal operation and erratic or no operation at all.

Terminal boards and connectors that carry substantial current, say in excess of three amperes, are typical places to verify for good connections. A small connector carrying as little as three amperes can, over a period of time, get hot, corrode, and cause major problems in the form of abnormal voltage drop.

2.3 Make a Detailed Visual Inspection — This is a good time to carefully inspect the entire interior of the equipment, looking for situations that could create problems.

Broken wires account for a large percentage of problems. Wires break most often when subjected to constant movement. Cables between units of a system, particularly cables leading to a human operator, are subject to failure. An example is the cord leading from a computer keyboard to the computer. The constant shuffling back and forth of the keyboard is a strain on the conductors within the cord. The cable from a computer to a printer, on the other hand, is seldom moved and therefore is less likely to cause problems.

The technician is another source of broken wires. When removing or replacing the cover on a piece of equipment, the internal wiring may be stressed needlessly if the technician is in too much of a hurry. Removal of covers is a particularly sensitive action, since the internal wiring is often hidden. *NEVER* force the cover off or on to something! Find out what is binding the cover rather than applying more effort. The resulting damage often can be more disastrous than a mere broken wire.

Burned resistors are sometimes discolored, favoring a brown to black color in the case of carbon composition resistors. Burned resistors also are often far out of resistance tolerance because of their overheating. A small resistor of perhaps one-fourth watt or less should not be discolored in any case, as these small resistors almost always are operated far below their wattage ratings. A burned resistor indicates that a problem exists somewhere else in the circuit. See Phase 4, "Identifying the Defective Component."

Integrated circuits (ICs) will sometimes overheat when they are shorted internally, causing a small dimple or discoloration in the center of the IC. This is a good item to look for if an overload on the power supply is indicated.

The circuit board itself may have been damaged by heat from resistors. Repairing the cause of the resistor overheating problem should prevent further damage to the board.

The insulation on wiring can become brittle and even flake off, exposing the conductors of the wire. Wiring routed against high-powered resistors is often affected this way. Reroute wiring away from sources of heat. Replace any wiring that has brittle insulation.

The observant technician might note that the small dabs of paint on calibration and tuning adjustments have been cracked or broken, indicating that those adjustments have been changed from the factory settings. This is often a problem on audio or CB equipment that belongs to people who like to tinker.

Any connectors should be carefully inspected to be certain that they are properly engaged. Sometimes, these connectors can be turned around end-for-end and plugged in backwards. Such errors will make the equipment act strangely, at the least, and could cause severe damage.

If you are fortunate enough to be able to spot the problem with a careful inspection, the answer to the next question will determine what to do about the problem. More often than not, though, you must proceed to the problem of applying power to the equipment to troubleshoot any further.

2.3.1 CARD UNDER WARRANTY?

— If the card is under a manufacturer's warranty, it would be foolish to attempt component-level troubleshooting and thus void that warranty, unless it was of unusual importance that the card be put back into service immediately. If this is the case, the reader is referred to Phase 4, "Identifying the Defective Component." Otherwise the card should be carefully packed and sent to the manufacturer or dealer. Some manufacturers require that advance notice be given and authorization received for the return before the card is mailed. It should also be kept in mind that some of the internal cards used in equipment may carry a separate warranty from other parts of the equipment.

2.4 CAN POWER BE APPLIED?

— A very important question must now be answered. This will determine the approach to be taken in repairing the equipment to the component level. If for any reason power cannot be applied to the equipment, further troubleshooting must be restricted to the techniques covered in Phase 6, "Dead Circuit Troubleshooting." Note that dead circuit troubleshooting is done on a component or possibly a card level. Being unable to apply power is a particularly severe hindrance to troubleshooting success at the equipment level.

Power application might not be possible because of the lack of any markings or documentation that tell how much voltage to apply, which polarity, and where to apply it. Multiple voltage inputs are required on some cards. Another reason might be the lack of the power sources required, such as one or more bench power supplies.

2.5 Apply Power and Check *ALL* Power Supply Outlets —

While the application of power to a piece of equipment often means simply plugging it in, sometimes a bench power supply must be used. This is covered in more detail in paragraph 3.2 of the next phase. The primary input voltage to the equipment should be checked to be sure it is proper. For instance, the 12-volt source for mobile equipment may be low. It is better to make a check than to assume primary power input is adequate.

The technician should then use a voltmeter to test for the proper output voltages from the equipment's internal power supply. Consult the schematic for the equipment and test for the proper voltage outputs as appropriate. Power supplies often have more than a single output voltage. Each voltage must be checked using the correct reference point, or "common," for that circuit. *The chassis of the equipment is not necessarily the proper point to connect the negative lead of the voltmeter.*

If a circuit board is dismounted from equipment during testing, it is very important to prevent the board from contacting ground or any other sources of voltage. Place insulating material under the board before reapplying power.

Cathode-ray tube circuits, as used in televisions, computer monitors, and terminals, have an additional supply to check for proper output—the DC-to-DC converter that produces the very high voltage required for the CRT. This very high voltage power supply is similar to the switching power supply of Phase 7B. The input signal to the supply's power switching transistor comes from the horizontal oscillator. The transformer used in the supply is called a flyback transformer. One of the windings on the output side is rectified and filtered to obtain the necessary DC voltage. See Figure 2–11.

If power supply output voltages are normal, proceed to paragraph 2.6. If not, go to Phase 7, "Troubleshooting the Power Supply."

2.6 ALL POWER SUPPLY OUTPUTS OK? —

Power supplies frequently have more than one output voltage. Additional outputs can come from separate transformer windings or from simple voltage dividers across a primary output voltage. In either case, be sure to check all output voltages. The lack of a single voltage or a voltage that is out of tolerance can cause numerous problems within the equipment.

2.7 Smell and Feel Components —

In addition to sight, two other senses are important for troubleshooting: smell and feel.

An experienced technician will recognize the smell of a burned transformer. It is similar to the smell of an overloaded and hot electric motor. The smell alone can tell the technician that there is a current overload of that component. This is often caused by the combination of an overload on the output of the transformer and the installation of a fuse that is too large, thus removing the normal circuit protection.

Another familiar smell is that of the burned carbon resistor. In many cases, this is the resistor used in the filter application in the power supply. It is often the first one to burn up when there is an overload on the power supply. Such a smell indicates the same thing as that of the overloaded transformer—an overload of current through the resistor.

When feeling components for possible overheating, be mindful of two cautions: electrical shock and thermal burns. Both cautions can be eliminated by using a temperature probe. See Figure 2–12.

When a temperature probe is not available, the technician can use his or her fingers, but must do so with care. It is easy to get burned electrically and thermally if not very careful. An abnormal circuit (that *is* the reason we're working on this thing, isn't it?) can produce both abnormal temperatures and unexpected high voltages where they wouldn't be present in a normal circuit.

NOTE: An oscilloscope is useful in measuring *some* voltage waveforms in this circuit. The voltages employed in some of the output windings will damage the oscilloscope probe and/or input circuits, so do not use a scope on any circuits with DC or AC voltages in excess of about 500 V.

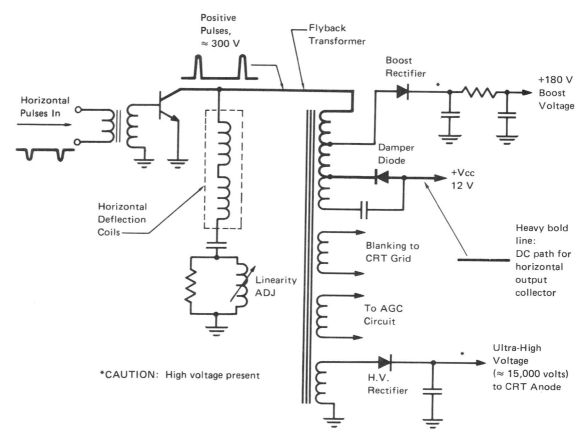

Figure 2-11 Basic flyback power supply schematic

When the technician wants to get an indication of small changes in temperature, the area under the nose, on the upper lip, is very sensitive to small changes of temperature. Just press a finger firmly on the suspect component for a moment and then quickly put that finger to the upper lip. Very small differences in temperature can be compared this way.

2.8 NOTES ON USING TROUBLESHOOTING CHARTS

— Paragraph 1.1.2.10 has already covered the use of the troubleshooting charts provided in the equipment manuals at the system level. When using these charts to troubleshoot within a piece of equipment, however, the technician may find that they are of little use because of the many failure possibilities inside the equipment. These troubleshooting charts are occasionally a source of help for the new technician, who has little experience with the equipment, or knowledge of its circuits. After gaining experience and understanding of the equipment, however, the troubleshooting chart will slip into disuse.

2.9 DOES THIS EQUIPMENT CONSIST OF SEPARATE CARDS?

Finding a problem within equipment that is modular in construction is easier than trying to find a problem on a single, complex circuit board. Individual inputs and outputs are easier to identify and isolate. An even greater benefit of having separate circuit cards is the ease of using substitution to isolate the card that has the problem. This gives rise to the next consideration.

Figure 2–12 Using the temperature probe offers safety and precise measurement techniques for troubleshooting (Photo courtesy of Fluke Co.)

2.9.1 ARE OTHER GOOD CARDS AVAILABLE? — The simple availability of a few known good cards can cut troubleshooting time by a factor of 10 or more. Substitution of good cards is the most effective way to both troubleshoot to the board level and to get the equipment back up and on-line in the shortest possible time.

2.9.2 Substitute One Good Card for an Old One — If there are any known good cards available to substitute in the equipment, substitute good cards one at a time until the equipment runs normally. The last card removed before the problem is corrected is the suspect card. Be sure to re-install the suspect card at least once to verify that the problem reappears and this is not an intermittent problem. Putting the suspect card back into the equipment should cause the original problem to recur. If it does not, go to Point A of Phase 1's flowchart and look for the information there on intermittents. When the defective card has been identified, go to paragraph 2.3.1.

The substitution of good cards is based on the assumption that all of the cards used are identical to those in the equipment, and that all of the possible jumpers and switches on the new cards are in the same positions as those on the original cards.

2.9.3 Problem Is in Backplane or Mechanical Portions — It is quite possible that replacing all of the cards, one at a time, has not cured the problem. This means that what has not been replaced is the problem. Depending upon what the equipment actually is, the direction of troubleshooting will vary. As an example, consider the case of a computer data printer that doesn't want to work. Replacing all of the circuit cards still does not result in operation of the printer. Obviously, the problem is still there. The only things that have not been replaced are the printing head and a couple of servo motors. The technician should now replace one of these items, beginning with the one most likely to cause the problem, or possibly with the one that is most easily replaced. Remember that mechanical things wear out, and the things that move most often are those most prone to failure. This includes switches and relays. Items generating heat also are more apt to fail than those operating at room temperatures.

Another possibility is that the board into which other boards must plug is the cause of the problem. This board may be called the system board, backplane, or mother board. It may be practical to try a replacement of this board to eliminate it as one of the causes.

2.9.4 Repair Mechanical and Trace Circuits as Necessary —
Mechanical repairs are often obvious: things get broken and you simply replace them or put them back together. Nuts and bolts come loose, and are not difficult to ascertain, but more subtle failures will not be obvious. Tension adjustments, such as those found in some tape recording mechanisms, are a good example. A little too much or not enough tension could result in improper operation. This usually happens when the mechanism is dropped. Mechanical repair is possible if the proper mechanical alignment procedures are available, but repair is difficult, if not impossible, without these procedures. Special tools may be required for the repair, such as tools to handle springs in tight confines, sensitive spring scales, and so forth.

It may be necessary to trace wiring from the backplane PC board to the mechanical portions of the equipment. The ohmmeter is often the best instrument to use, utilizing the continuity function while the circuit has no power applied.

2.10 WILL INTERNAL SIGNALS SUFFICE? —
The need to identify the defective card in the equipment may make it necessary to inject a proper signal and trace it through the equipment. In the case of an audio amplifier, for instance, a signal generator could be used. A stereo amplifier could conveniently use a tuner input for the same purpose.

2.11 NOTES ON USING SIGNAL GENERATORS —
When it is necessary to use a signal generator to produce an external signal, keep several things in mind:

1. Apply only sufficient signal to accomplish the purpose. Do not overload the circuitry by forcing too high an amplitude into it. This can cause problems such as severe audio distortion or poor tuning response in RF amplifier stages.
2. If there is a possibility of DC being present in the circuitry at the point to which the signal is to be applied, place a DC blocking capacitor in series with the signal generator output. See paragraph 3.5 for details of this procedure.
3. Provide the proper frequency. Some applications are not critical, such as audio circuits. Others, such as radio frequency circuits, may be very sharply tuned and will require very precise frequency setting of the signal generator.

2.12 IS DOCUMENTATION AVAILABLE? —
Paragraph 3.6 discusses documentation in some detail. The determination to make now is whether or not there is a block diagram of the equipment available to help determine which of the cards might be malfunctioning.

2.13 Review the Equipment Block Diagram —
If substitute cards are not available, other methods must be used to localize the problem to a card within the equipment. Modular construction is not a troubleshooting advantage without substitute good cards to use. Without them, the technician will have to resort to more time-consuming means to isolate the bad card in the equipment. At this point, *the block diagram for circuitry in the equipment should be studied to get the overall picture of signal flow through the equipment.* This information will be needed during the next step in localizing the problem.

2.14 ANY INDICATORS OR CONTROLS TO HELP FIND THE BAD CARD? —
Front panel controls and indicators, such as LEDs, can be used to great advantage to locate the area of the equipment causing the problem. Compare the normal result of manipulating a control to what actually happens. A malfunctioning control is prior to the location of the problem. *The problem is near the latest malfunctioning control in the block diagram.* See Figure 2–13.

38 ☐ **Isolating the Defective Card**

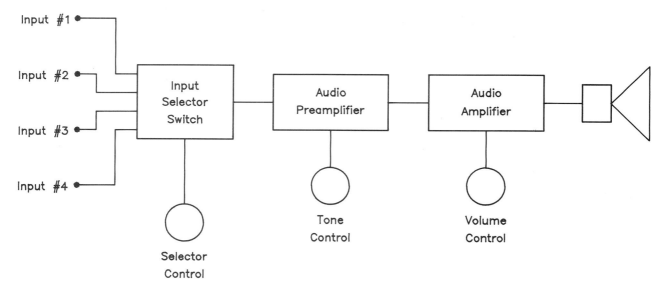

Figure 2-13 Demonstration of the principle "The problem is near the latest malfunctioning control"

2.14.1 Use Them to Isolate the Defective Card — The block diagram for a stereo might show, for example, that either the tuner or the tape deck are selected by a switch that then feeds the audio preamplifier. If the tape deck works normally but the tuner does not, simple logic would indicate that the problem lies within the tuner circuits. See Figure 2-14.

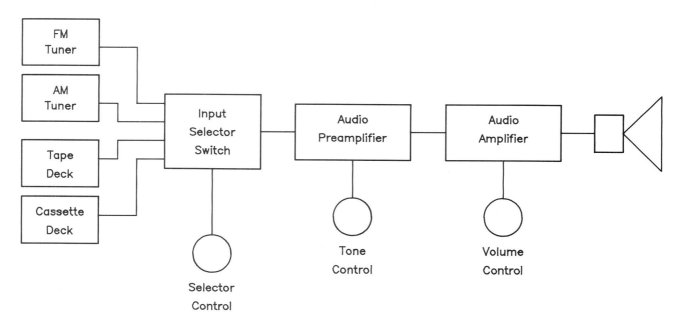

Figure 2-14 If the tape deck works normally but the tuner does not, the problem lies within or prior to the selector switch

2.15 ARE DESIGNATED TEST POINTS AND DOCUMENTATION AVAILABLE?

— The combination of designated test points and information about what these test points should have in the way of waveforms and/or voltages is important for troubleshooting. These test points have been chosen by the manufacturer as being good points at which to conveniently determine where a problem might lie. See Figure 2–15.

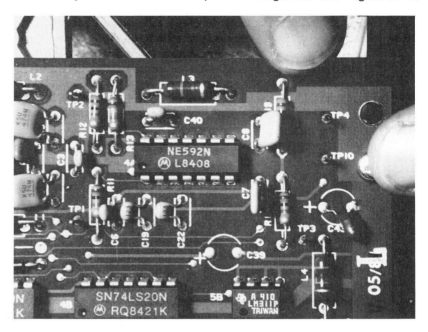

Figure 2–15 A circuit board with test points provided by the manufacturer

Test points have little or no value without documentation that tells what to expect at the various points.

2.16 NOTES ON USING AN OSCILLOSCOPE FOR SIGNAL TRACING

— It is easy to forget to attach the probe ground when using the oscilloscope probe. At low frequencies, such as audio, a long, common ground attached between the oscilloscope and the circuit under test might be acceptable. It is better to form the habit of using the short ground lead provided on a standard attenuator probe. Forgetting the ground connection can result in the display of false signals superimposed on the actual signal under test.

A technician who cannot operate an oscilloscope properly reveals this in adjusting the instrument. It is tempting and natural to twist knobs until the trace seems normal on the scope. In contrast, a technician who really knows how to use the oscilloscope will make very few adjustments compared with the newcomer. Interpretation of what is seen versus what is desired is made and the appropriate control is all that is changed. The new technician should spend considerable time with this instrument, simply thinking about what the instrument shows and the interaction of the various controls.

New technicians may make the mistake of operating the instrument with the intensity set too high. This has a tendency to burn the trace on the oscilloscope, and it is especially damaging if the spot on the CRT is not moving at all.

The oscilloscope probe — The oscilloscope should only be used with a proper oscilloscope probe. See Figure 2–16.

Using common test leads for connecting an oscilloscope to the circuit under test is not recommended. While test leads used this way may make little difference while testing common audio or DC circuits, their use at high frequencies and with digital signals will produce erroneous results. See Figure 2–17.

Using a proper probe, on the other hand, will show the waveform properly on the screen. A schematic of such a probe is shown in Figure 2–18.

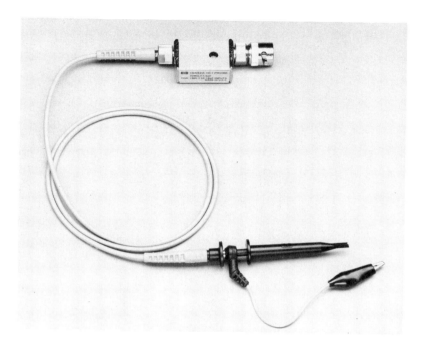

Figure 2–16 The oscilloscope attenuator probe causes very little circuit loading, shields against stray signal pickup, and is compensated for a wide band of frequencies (Photo courtesy of Hewlett-Packard Company)

A digital signal as it should appear when using a properly compensated oscilloscope probe.

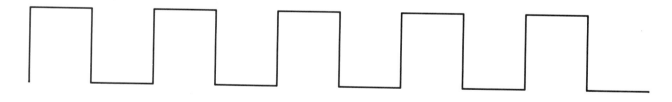

Same signal as above, but viewed with common test leads instead of a proper oscilloscope probe.

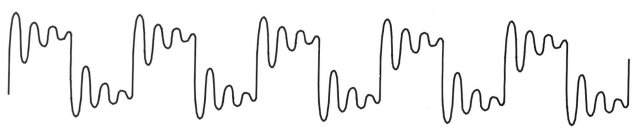

Figure 2–17 A digital waveform as it really is and as it might appear if using test leads to connect an oscilloscope

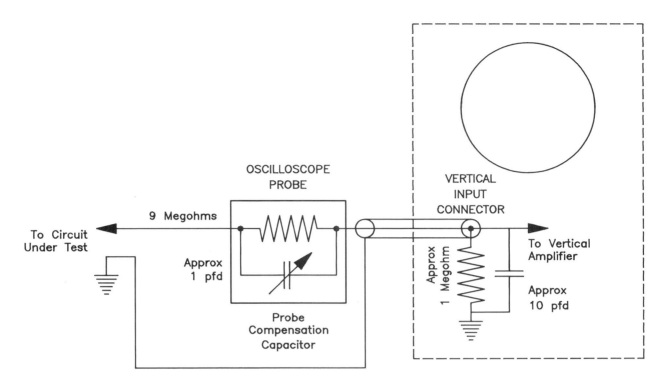

Figure 2-18 A high-quality attenuating probe schematic diagram

The high value of series resistance used in the probe, about 9 megohms, results in a negligible loading effect on most circuits. The probe resistance is in series with the 1 megohm resistance of the oscilloscope itself, making about 10 megohms total loading effect.

Since the cable from the probe to the oscilloscope cannot avoid a small amount of capacity, this can be compensated for at the probe. Provided with the DC resistance divider is a capacitive divider of the same ratio. This results in all frequencies being attenuated by the same amount. As a result, the digital waveform has no overshoot or undershoot, and shows no ringing.

The probe is compensated using a squarewave provided on the front of the instrument. The probe trimmer capacitor is adjusted for the "perfect" squarewave shape.

The AC/DC coupling control —
This switch, located next to the vertical input connector of the instrument, switches a series DC blocking capacitor into or out of the circuit. When tracing audio, pulse, and RF signals through a circuit, it is probably better to use the AC position. When tracing DC signals, the capacitor must be removed from the circuit. See Figure 2-19.

Looking for missing or unusually low amplitude signals —
Analog signals are often traced using an oscilloscope. The common problem of missing signals is easily detected using this instrument. Signals that are lower than normal amplitude are more difficult to trace because of the necessity of interpreting the oscilloscope amplitude readings and deciding if the signal at a given point is within normal limits or not.

While using the oscilloscope to identify and analyze the card-to-card signals, don't fall into the trap of continually varying the oscilloscope vertical gain to see the screen better, ignoring what the scope is revealing. In other words, a low-amplitude signal may be just as important in troubleshooting as the lack of a signal altogether. Turning up the sensitivity can make the signal seem normal visually unless the significance of having to turn up the sensitivity is also carefully considered.

42 □ Isolating the Defective Card

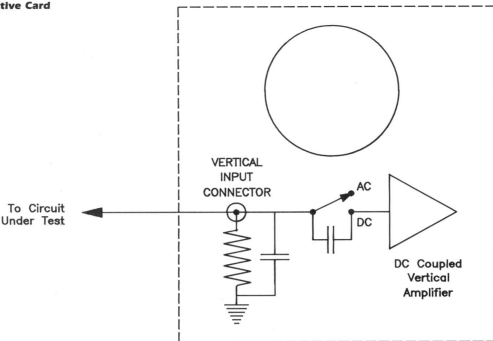

Figure 2-19 What the AC/DC switch on an oscilloscope does inside the instrument

Tracing distortion — Tracing distortion is another use for the oscilloscope. Large percentages of distortion are easily detected by this instrument. See Figure 2-20.

The technician should be aware that the voltage waveforms observed at some points in the circuitry may not appear as expected, particularly if there is a transistor base involved. See Figure 2-21 for an example.

Gross distortions of signal voltages in amplifiers are usually an indication of a biasing problem. See paragraph 3.15.2.1 for more information on the subject of audio distortion.

2.17 Signal Trace to Find Defective Card
— Signal tracing to identify the bad card in a piece of equipment will mean that *only the signals passed from card to card are of interest.* The number of signals that must be considered is not as great as might first be anticipated. The specific equipment at hand will determine where the signals must be monitored. The block diagram and possibly the schematic will be used to determine where to monitor the signals as they flow from one card to another. The signals in question are those passed from one card to the next, not to those signals on the cards themselves.

2.18 IS CARD WORTH TIME TO FIX?
— As the cost of labor increases, printed circuit boards are becoming more and more disposable. It doesn't make much sense to spend an hour of a technician's time at $35 an hour to repair a circuit card that costs only $25. Of course, there may be special circumstances that will demand otherwise, regardless of the monetary value of the card. (Then, too, perhaps the technician has nothing more important to do for a while.) If the card is worth repairing, then we can proceed to the next phase now that we know which card is causing the problem. Phase 3, "Locating the Defective Stage" is the next step.

□ Phase 2 Summary

At the conclusion of this phase, the technician has been led to a more specific area in localizing the problem. Instead of an entire piece of equipment with a problem, the technician now knows if a particular module or circuit card is causing the problem. On nonmodular, large boards, the general area that is causing the problem should now be known. Many repair jobs will require no further work beyond this point, since many PC cards are inexpensive and therefore disposable. In the case of expensive or special equipment, the problem may be pursued further in the next phase, progressing to repair to the component level.

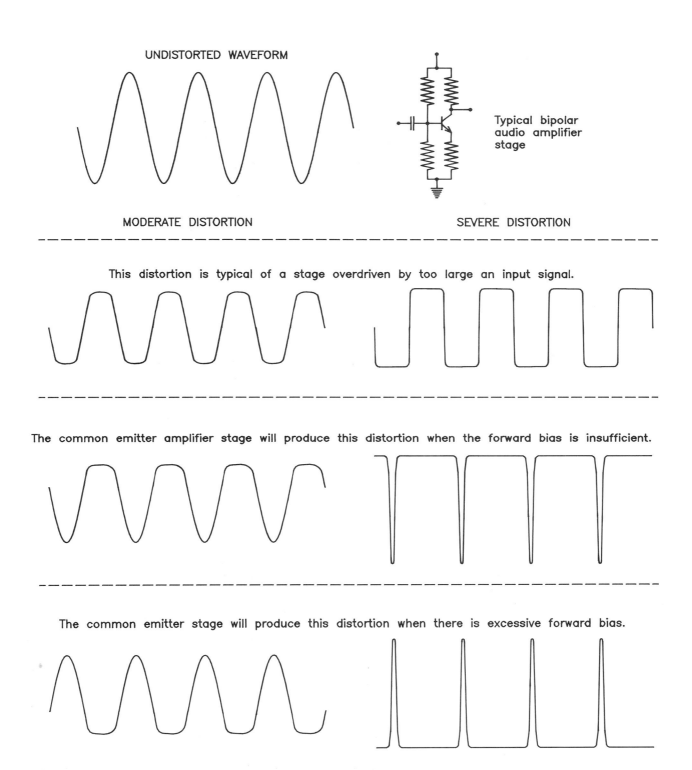

Figure 2-20 Examples of good and distorted audio waveforms

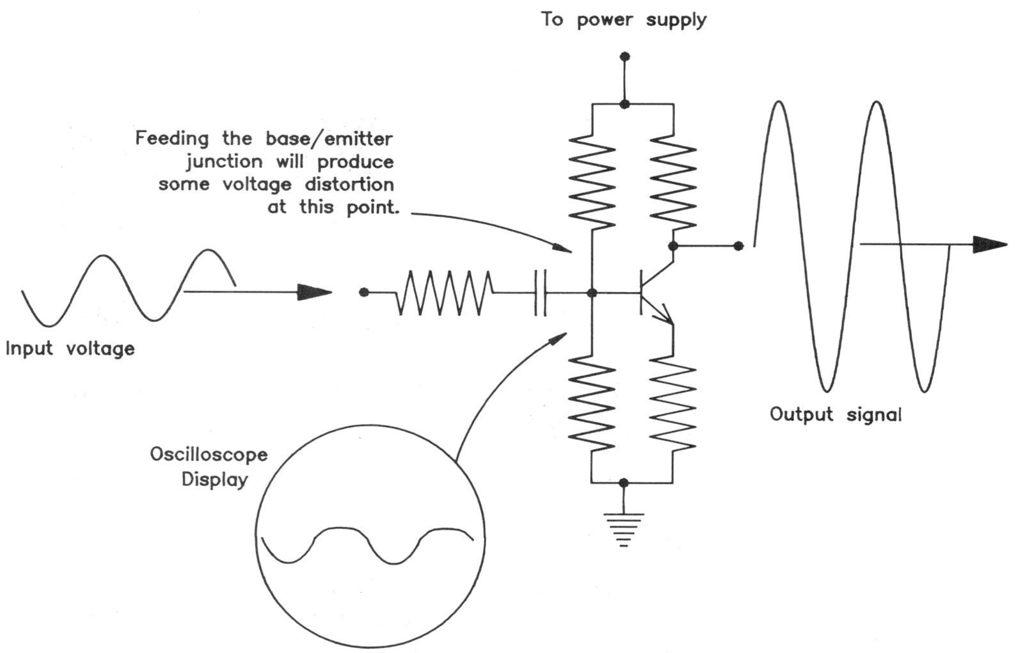

Figure 2-21 An example of an apparently distorted waveform at the base of a transistor that could be misleading

☐ Review Questions for Phase 2

1. Discuss methods of preventing damage due to electrostatic discharge (ESD).
2. What is the significance of green wires with yellow tracers as used in equipment?
3. Name three methods of preventing electrical shock caused by touching the chassis of equipment.
4. What is the sequence of actions to follow in taking the measure of a very high voltage?
5. What is the principal danger of working on vacuum tube equipment as compared to solid state circuits?
6. What components of a television might be dangerous even if the set is turned off?
7. What piece of equipment can provide peace of mind for a technician working on high voltage circuits with the power off?
8. What is the symbol for "Read the Book!"?
9. You have just removed the cover of a piece of defective equipment. What is the next logical step in troubleshooting?
10. Discuss some of the things that might be visually detected as the cause or relating to a problem upon removal of the equipment cover.
11. The defective equipment is open and nothing is obvious as to the cause of a problem. Name the next two steps in troubleshooting.
12. Name the three cautions in using a signal generator.
13. Discuss why the block diagram is used before referring to the schematic drawing.
14. Discuss the principle "The problem is near the latest malfunctioning control." Use Figure 2-13 and simulate various stage failures.
15. What is the significance of test points?
16. How can an inexperienced technician be spotted by the way an oscilloscope is used?
17. What are the advantages of using a proper oscilloscope probe over simple test leads?
18. Demonstrate the proper compensation of an oscilloscope probe.
19. Discuss the economics of repair of inexpensive boards.

PHASE 3
LOCATING THE DEFECTIVE STAGE

☐ PHASE 3 OVERVIEW

Some equipment consists of a single printed circuit board while other equipment has individual cards. While it is often sufficient to troubleshoot to this board or card level and discard the faulty board, the technician may be required to troubleshoot to the component level. It is a fact that the deeper into the circuit you must go, the more electronics you must know. In this case, the next step is to further narrow the problem to a single stage within the circuitry. In accomplishing this step, the signals present *between* the stages will be of primary concern to indicate which stage it is that we seek. This phase will include ways to narrow the problem and will point out facts that should be known, the decisions to be made, and the pitfalls along the way.

3.0 ONLY A FEW, SOCKETED CHIPS IN QUESTION? —
Now that the area of the problem is generally defined, it is sometimes advisable to take a shortcut at this point. If there are only a few chips or components that might be causing the problem, and if they are installed in sockets, the repair might be accomplished quickly by component substitution. This is a good alternative only when the substituting can be done quickly, otherwise the problem will need to be traced in a more laborious fashion.

3.0.1 Try for a Quick Fix: Replace a Few Socketed Components —
When replacing IC chips, keep the new ones in anti-static packing tubes or inserted into conductive foam until just before installation. Change the chips, either in a batch to determine if this approach is going to work, or one at a time, retesting each time to see if the problem has been resolved. If the batch method is successful in determining that the problem can be quickly repaired, replace with the original chips one at a time until the problem again appears. Discard the last chip installed and replace all of the remaining chips of the batch with the originals.

3.1 TRACKER AVAILABLE? —
Another shortcut that may be successful at this point in the repair job is to make some quick tests with a Huntron Tracker or other solid state tester. If the general area of the problem is known, the tracker can be used, testing the circuit without power applied, to identify the problem quickly. Phase 6 describes the steps to use.

3.2 NOTES ON USING A BENCH POWER SUPPLY

— If these two shortcuts are not used or are not successful, it will be necessary to apply power to the board. If application of power is not possible because of lack of information or any other reason, there is little else that can be done other than proceeding to Phase 6, "Dead Circuit Troubleshooting."

There are two possibilities for applying power to a circuit: a) Using a general purpose bench-type power supply and b) using the power supply within the equipment to supply all of the necessary operating voltages. The choice of the method will depend upon several factors. Will there be sufficient room inside the equipment to have access to the circuit components and PC board traces during troubleshooting? Paragraph 3.3, Using Extender Boards, gives some tips on their use. If space in the equipment prevents access to both sides of the card it may be necessary to remove the card from the equipment and power it from a bench power supply.

There are several kinds of test bench power supplies, classified according to their output features. Of course, voltage and current limitations are of primary concern. Besides these basic requirements, bench power supplies are available with additional features that should be thoroughly understood by the technician.

Voltage regulation—As more current is drawn from a power supply by a progressively heavier load, there is a tendency for the output voltage of the power supply to decrease because of the unavoidable internal resistance inside the power supply. See the upper half of Figure 3–1.

If there is no manual adjustment or special circuitry to minimize this voltage "drooping" tendency, the output voltage will depend to some extent upon the amount of current demanded by the load. If a supply shows considerable drop in output voltage from an open circuit to a normal load, this could be called a "soft" power supply.

We could add an adjustable power resistor to the output of the power supply and manually vary this resistance to maintain a given output voltage. See the lower half of Figure 3–1.

Since many circuits require the voltages supplied to them to be within narrow limits at all times in spite of varying current demands, a soft power supply would not be satisfactory. It would be better to use a supply that would automatically maintain the output voltage at the correct level for currents up to the maximum current rating of the power supply. For these applications the *voltage-regulated* power supply should be selected. The voltage-regulated power supply internally generates more than sufficient voltage to supply the load. By sensing the actual voltage at the output terminals, a variable resistance (actually the collector to emitter resistance of a power transistor) in series with the circuit is automatically decreased, thereby holding the output voltage relatively constant. See the upper half of Figure 3–2.

Bench power supplies for digital circuits commonly have a voltage regulated output of 5 V, intended to be used with TTL (transistor-transistor logic) integrated circuits. The regulated output voltage can sometimes be reset to a new regulated value by varying the output voltage sample with a potentiometer. Once changed, the power supply will hold the new output voltage level constant with a varying load. This is a *variable voltage-regulated* power supply. See the lower half of Figure 3–2.

Remote voltage sensing—Applications requiring heavy current at low voltages, at some distance from the power supply, will require additional circuitry to maintain proper voltage at the remote load. See Figure 3–3.

The use of sensing leads makes it possible to maintain regulated voltage at the load in spite of varying voltage losses in the main current lines caused by changing current demands.

Current limiting—The next feature to consider concerns what happens when too much current is demanded of the power supply. Simple power supplies will attempt to maintain a voltage output with overcurrent loads until the fuse blows. This kind of power supply might be called a "brute force" or "hard" or "stiff" power supply. Two refined current protection schemes are the linear current-limited and the fold-back power supplies.

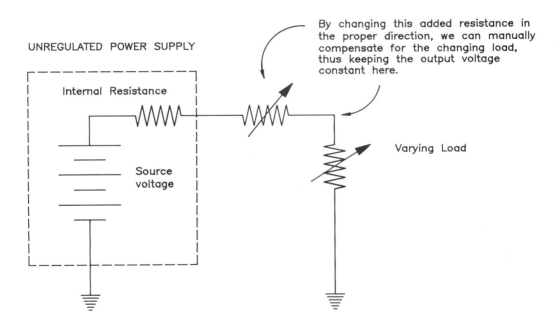

Figure 3-1 Correcting for the varying output voltage of an unregulated power supply

The current-limited power supply is usually also voltage regulated. When the output current exceeds a set amount, the output voltage of the power supply is then allowed to decrease so as not to exceed the specified maximum current. In other words, the power supply changes from a constant voltage to a constant current supply at this "overload" point. See Figure 3-4.

48 ☐ Locating the Defective Stage

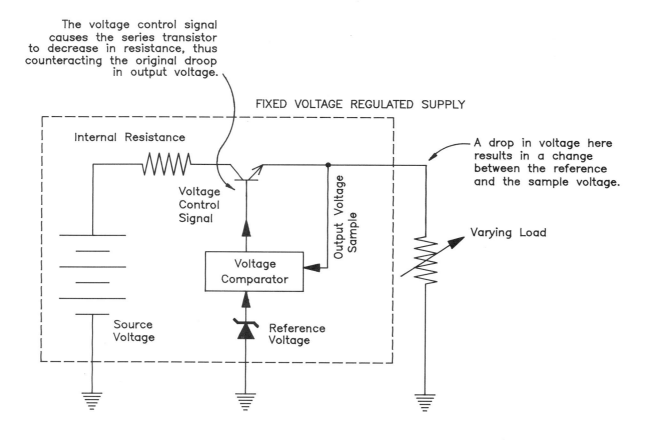

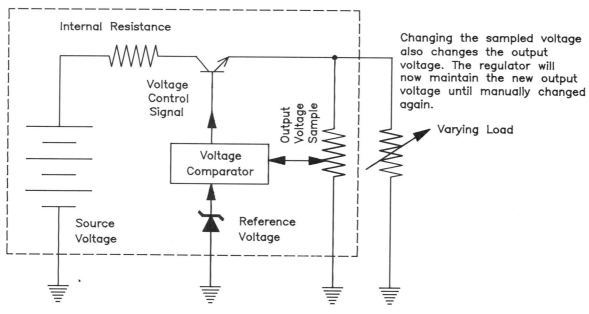

Figure 3-2 How the fixed and variable voltage regulated power supplies work

Electronic Troubleshooting □ 49

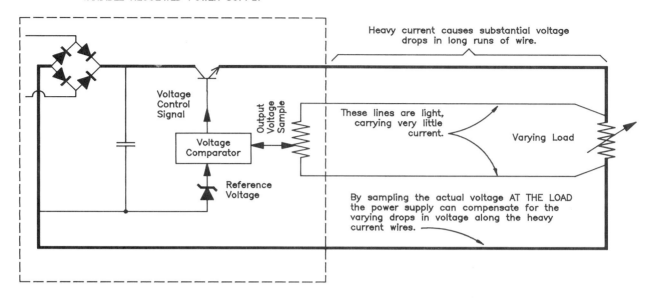

Figure 3-3 Sensing leads are used to "feel" the voltage at a remote load and compensate for voltage loss on long runs

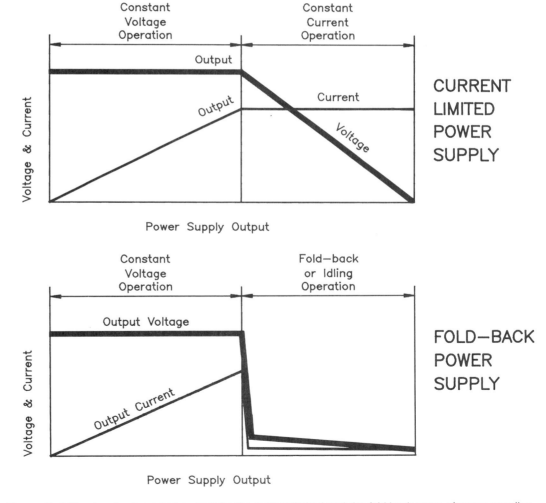

Figure 3-4 Graphs of voltage and current for the current-limited and the fold-back types of power supplies

More sophisticated power supplies include not only a variable voltage regulated output but also a means of varying the set point for current limiting. This is a desirable feature when the technician wishes to limit the maximum amount of current flowing, such as when testing random loads. An example of this situation would be if the technician is asked to test several identical cards, some of which may have shorts from Vcc to ground. If a good card draws 1.2 amperes of current, the set point of the power supply could be set to about 1.3 amperes, a few percent higher than normal. Connection of a shorted card would cause the power supply to have a very low voltage output, thus protecting the power supply, the bad card, and indicating the problem by the lack of normal output voltage from the supply.

The fold-back power supply has a different approach to the problem of overcurrent demands. When the power supply senses an overload, the output voltage drops immediately to a very low value. After the overload is removed, some fold-back power supplies require a manual reset to obtain normal output voltage. A different type of fold-back circuit requires only that the load resistance increase sufficiently to turn the supply on again. This type of fold-back supply senses the continuing overload or the overload removal with the very low voltage output after fold-back occurs. See Figure 3-4 again.

A note of caution: While the bench power supply is versatile, it should be used to charge batteries only when absolutely necessary. The output circuitry of these instruments will be damaged if connected to a reversed voltage source such as a partially discharged battery. There may easily be enough charge remaining in a discharged battery to ruin the output circuits of the supply.

Power supplies with multiple outputs—Some bench supplies have more than one output voltage. Additional output voltage sources may or may not share a common output terminal. The output ground connections of a power supply should be investigated to determine their status. Floating grounds—those not referenced to any other circuit in any way—are provided on a bench power supply so that an output voltage can be either positive or negative with respect to the common of the circuit under test. Some power supplies connect one side of the output voltage—usually the negative—to the chassis ground of the power supply. The technician must be aware of these subtle differences to use the supplies without harm to either the power supplies or the loads that might be connected to them.

Some bench power supplies have a ground connection at the front panel that is connected to the chassis of the power supply. Remember that this connection is the one to which the ground lead of the oscilloscope is always connected by means of the third wire of the power cord.

3.3 **NOTES ON USING EXTENDER BOARDS** — The power supply installed in the equipment can be used to supply the required voltages to a card during stage-by-stage troubleshooting if both sides of the bad card are exposed enough to allow the connecting of test leads and probes. It is terribly frustrating to try to reach between two cards fully installed in equipment to try to take readings. Some of the important test points will invariably be completely inaccessible. Attempting to troubleshoot under these conditions also carries a great risk of not only taking the wrong reading because of reduced visibility of the circuit, but also of shorting something on the circuit with a poorly placed test lead.

Extender boards are a solution to this problem. These boards allow the card to be operated up out of the equipment and in the clear. See Figure 3-5.

There are several cautions to be observed when using extender boards. First, be sure that the extender is placed into the equipment socket in the proper orientation. Some extenders can be installed "backward." In a similar manner, be sure the PC card is then placed into the extender properly. It is sometimes possible when using an extender to effectively reverse the card in the circuit. This could be very hazardous to both the card and the power supply. To avoid this, some extenders have a keying arrangement to prevent improper installation.

Another option is the use of flexible extenders. These are cables with appropriate male/female connectors at the ends. They may even allow the card under test to be laid flat on the workbench or easily turned over during troubleshooting.

Electronic Troubleshooting □ 51

Figure 3-5 The extender board allows work on compact card installations

It may be worthwhile to obtain or fabricate the extenders necessary for equipment that is often serviced. Where plug-in, stiff extender boards may require considerable work to build, flexible extender cables may be easier to fabricate, given the necessary connectors.

Another good idea is the marking of the extender board to help trace signals right on the extender traces. See Figure 3-6.

Figure 3-6 How an extender board might be marked to help trace signals

3.4 NOTES ON TROUBLESHOOTING CLOSED-LOOP CIRCUITS

— Before getting deeply into the techniques for isolating a defective stage, the technician should be aware of the unique problems of troubleshooting closed-looped circuits. Such circuits are called closed-loop because they depend upon their outputs for an input. A problem anywhere within this loop will cause an apparent malfunction of all of the stages involved. *The key to troubleshooting a looped circuit is to break the loop at an appropriate point, thus causing it to become just another series of cascaded stages.* Under these conditions the circuit will not do what it is supposed to, but by varying a *dummy input* to the loop, each stage should be operating sufficiently normal to troubleshoot it as an independent stage. See Figure 3–7 for an example of a closed-loop DC circuit, a DC voltage regulator. The regulator varies the resistance of the series transistor to maintain a specified DC output voltage. If there is any disturbance in the output voltage, the feedback loop back through the amplifiers to the base of the series transistor counteracts the change at the output.

A problem in this particular closed-loop circuit could be easier to trace if the feedback voltage were supplied, not by the output voltage, but by a separate voltage input from a variable voltage supply. Then, by varying this external input voltage, the circuit becomes a series of ordinary DC amplifiers. See Figure 3–7.

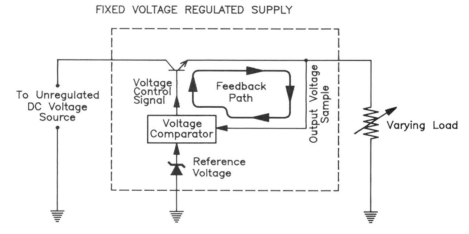

A DC voltage regulation circuit, an example of a looped, or feedback, circuit.

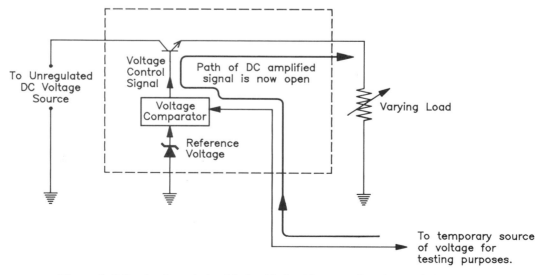

Figure 3–7 Break a looped circuit to troubleshoot it as a series of cascaded stages

The specific closed-loop circuit under test will have to be carefully examined for the signals normally present in various portions of the circuit. Break the loop at a point where external dummy signals can be introduced with the greatest ease and simplicity. Keep in mind that DC levels and AC signals can be simultaneously present in a normal working circuit, and that breaking the circuit at such a point will require similarly superimposed signals to operate it as an open-loop circuit. In Figure 3–7, note that the break point chosen had only a simple DC voltage input requirement with reference to ground.

There are some limitations to this technique of breaking a closed-loop circuit that should be understood. The overall gain of many looped circuits is very high. The voltage regulator in our previous example is extremely sensitive to variations of the dummy voltage input. The output voltage of the regulator can go from one extreme to the other with a very small adjustment of the dummy input voltage. This is normal, but even with these symptoms the circuit can be analyzed effectively for a malfunction.

Some closed-loop circuits serve a less radical purpose; they reduce signal distortion produced in the stages within the loop. Such feedback is out of phase with the input signal, causing a canceling effect on the input signal. See Figure 3–8.

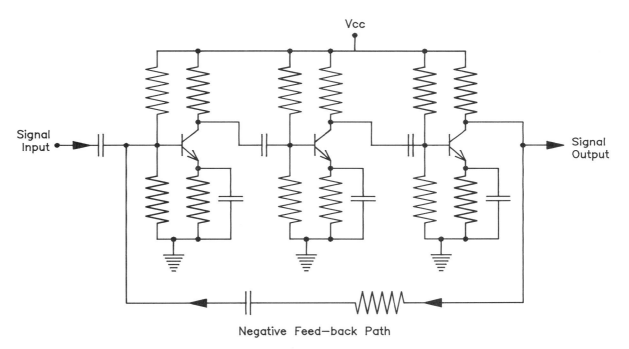

Figure 3–8 A sample discrete circuit using negative feedback to reduce distortion produced within the stage

Breaking such a feedback path will not only cause the circuit to produce more distortion, but the lack of negative feedback may cause the gain of the stage to increase dramatically.

3.5 Apply Input Power and Signals If Applicable

Application of power—When power is applied to a card outside its normal plug-in environment, it is particularly easy to damage the card. Great care must be exercised to prevent the accidental application of a higher-than-normal voltage, the application of a voltage where it doesn't belong, or worse yet, the application of the wrong polarity of voltage.

All power supplies used should be *OFF* while connecting them to circuits under test. Before applying power, be *absolutely sure* that the polarity and voltage are correct. Be sure that there is no metal under the circuit board, such as wire scraps or stray tools. Work on an insulated surface. Watch for the likelihood of shorts across the power supply outputs caused by using uninsulated test clips. When possible, work on circuit boards on a piece of anti-static rubber. The padded surface helps cushion the board during handling and provides suitable ESD protection. See Figure 3-9.

Figure 3-9 A circuit card repair workstation (Photo courtesy of Hewlett-Packard Company)

Circuit cards are very susceptible to damage from static electricity discharge while work is being done on them. The servicing technician must take every precaution to prevent damage from this source. Even though a mishandled card might still work, static damage can reduce the service life of a card, causing early failure. These precautions will become even more important as technology continues to produce ever smaller circuitry inside the chips we use.

Connections from the circuit under test to the power supply must be reliable. Use banana plugs that fit snugly, for instance. A poor connection to the power supply can cause intermittent problems that can be frustrating to trace.

High-current applications of more than about an ampere may require the use of heavier wiring to the power supply. The size of wire used will depend upon the amount of current and the affordable voltage loss from the power supply to the load. TTL digital circuits, for instance, require current of this magnitude and larger power leads should be used than are necessary for the low-current complementary metal oxide semiconductor (CMOS) family of ICs. If in doubt about the size of wire in use, measure the voltage at the inputs to the circuit under test. If the voltage there is OK, the wires to the power supply are fine.

Applying input signals—Depending upon the kind of problem and the circuit being tested, it may be necessary to apply signals to the card for the signal-tracing technique to follow. These signals might be DC levels, audio frequencies (AF), RF frequencies, or digital signals. Although techniques of signal tracing begin with paragraph 3.6.1, some notes are in order in regard to the connection of an input signal to a circuit.

If the card has a normal signal input jack, chances are that the signal generator can be directly connected to that jack. The output voltage level of the signal generator will generally decrease as the number of amplifying stages in use increases. In other words, the more amplification, the less input voltage required.

Application of a signal at other than a normal injection point involves a probable hazard for the signal source. These intermediate circuit points will often have a DC component that must not be allowed to flow back through the output circuits of an AF or RF signal generator. See Figure 3-10.

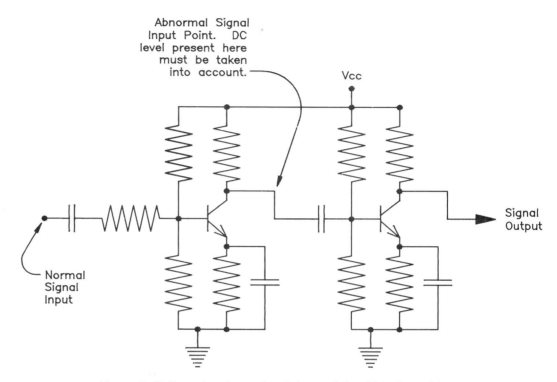

Figure 3-10 Examples of normal and abnormal signal injection points

A capacitor in series with an AF or RF signal generator will prevent the flow of DC back through the attenuator circuit of the signal generator. The size of the capacitor chosen to block DC should be suitable for the frequencies involved. Examples are as follows:

DC Blocking Capacitors
- ☐ Low frequency audio 1 to 10 ufd electrolytic with appropriate polarity.
- ☐ General audio 0.1 to 1 ufd electrolytic with appropriate polarity.
- ☐ Low frequency RF (to about 1 MHz) 0.01 to 0.1 ufd disk
- ☐ High frequency RF (to about 100 MHz) 0.001 to 0.01 ufd ceramic or Mylar

Some AF or RF frequency generators will have the DC blocking capacitor installed within the generator while others will require the technician to be aware of the requirement and to supply a capacitor external to the generator. Note that an electrolytic capacitor is not effective at radio frequencies. This is due to the unusually large value of internal series inductance inherent in an electrolytic capacitor that would be introduced into the circuit. Similarly, the smaller capacitors are not useful at the lower frequencies because of their high reactance at low frequencies.

Apply the signal generator input at the correct input level for the circuit under test. Overdriving a circuit will cause distortion, and in severe cases can damage sensitive input components. Even if the input voltage is not grossly overloading the circuit, it may still introduce a great degree of distortion, thus severely upsetting signal amplitude measurements further down the line.

The mockup—If the technician is expected to repair a quantity of boards, all of the same kind and configuration, it may be worthwhile to build a mockup. A mockup is an arrangement of connectors, power supplies, and input signals necessary to signal trace through a specific kind of card. Some mockups are able to test several different cards, and bear a functional resemblance to the actual equipment in which the cards are normally operated.

Dedicated test instruments—Some circuits are very popular and because of this, industry has produced specialized test instruments to meet the needs of the technician servicing these circuits. The floppy disk drives of a microcomputer are an example of a specialized unit that requires very sophisticated signals to control it. The outputs of the unit are also relatively complex, requiring a "smart" interpretation. Special test equipment is available to generate the input signals and analyze the outputs from these drives. See Figure 3–11.

Figure 3–11 The DYSAN PAT-2+ disk drive exerciser is an excellent example of a dedicated test instrument. This instrument produces input signals and also analyzes complex test results.

3.6 IS DOCUMENTATION AVAILABLE?
— One of the most important questions of all now becomes pertinent: How much detailed documentation is available for the circuit under test? Generally speaking, *the lack of documentation is a severe handicap to successful troubleshooting.* There are a few alternatives without it, however, beginning with paragraph 3.5. For now, consider that complete documentation is available. This will include:

1. *Schematics*
   Schematic diagrams of the circuits under consideration. See Figure 3–12.
2. *Component Layout Diagrams*
   Diagrams of the physical layout of the components, keyed to the schematic with reference designations, such as U5 (the fifth IC of the circuit) R12 (resistor number 12 of the circuit) used to relate schematic symbols to actual components. See Figure 3–13.
3. *Circuit Block Diagram*
   Block diagram of the circuits under test with the major signal flows indicated. This is useful in determining which stage follows which and gives clues as to the expected signals between stages. See Figure 3–14.

Electronic Troubleshooting ◻ 57

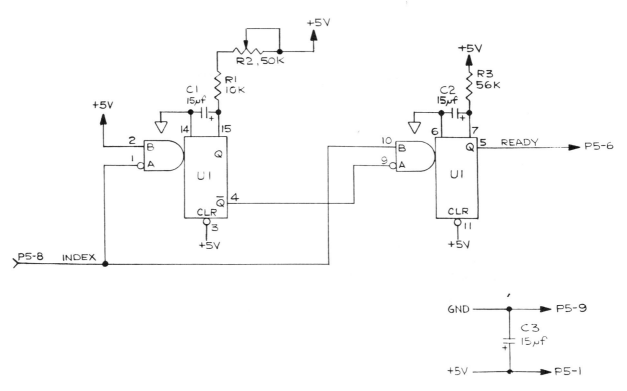

Figure 3-12 Example of a circuit Schematic (Figure courtesy of Micro Peripherals, Inc., Chatsworth, CA)

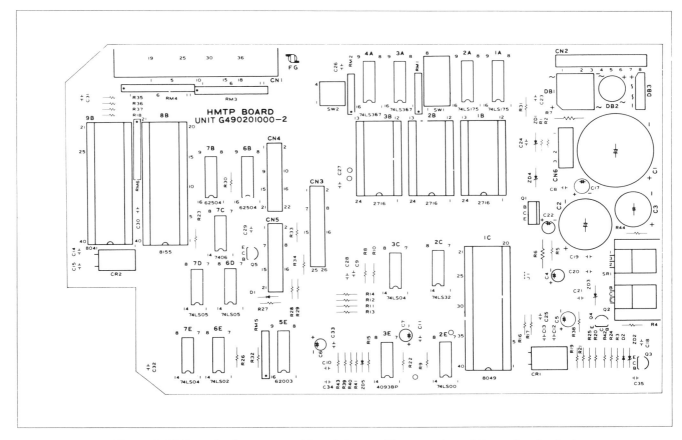

Figure 3-13 Example of a component layout diagram (Figure courtesy of Epson America Inc.)

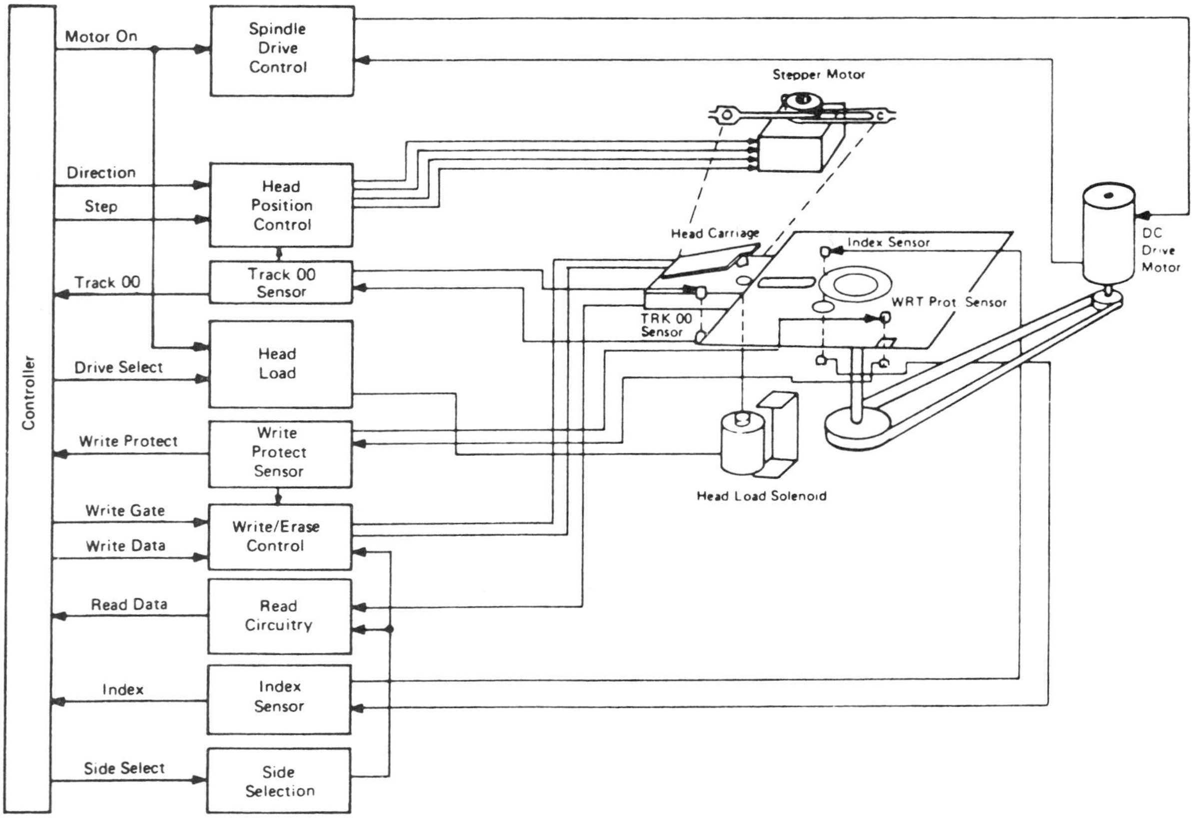

Figure 3-14 Example of a circuit block diagram (Figure courtesy of Micro Peripherals, Inc., Chatsworth, CA)

4. *Theory of Circuit Operation*
   Especially important for digital circuits, this section of the documentation is a detailed account of the signals as they flow through each stage. This theory often explains the necessity for most, if not all, of the components used in the circuit. Extremely helpful for the new technician faced with a complex component-level repair. See Figure 3-15.

5. *Parts List*
   Additional help in determining the detailed specifications for components, such as the power dissipation rating of resistors, designation of transistors (e.g., Q5 = 2N2222), and integrated circuits (e.g., U5 = 7406 integrated circuit). See Figure 3-16.

The lack of any of these items will make the repair job more difficult. It is often wise to put a job aside until the documentation can be obtained.

The documentation should be checked to be sure that it applies to the particular make, model, *and revision* under consideration. Accessories for the equipment may require additional supplements to the basic instruction/maintenance manual.

3.6.1 NOTES ON GENERAL SIGNAL TRACING —
The schematic and other documentation used for a repair job should not be placed on the workbench. This makes them prone to tearing with tools or a sharp chassis and getting them very dirty with solder splashings and other grime. Schematics should be clamped to a small easel so they are readily accessible but out of the primary work area.

With the power applied and the required signals present at the input to the defective card, it is time to consider what must be done to find the defective stage. The details will vary with the specific circuit being considered, but there are some guidelines. The input signal is assumed to progress through the stages of the circuit. This is the usual case with audio amplifiers, radio frequency circuits, and some digital circuits. Differences in this technique for more complex digital circuits are considered beginning with paragraph 3.10.

The voltages lowered by the power transformer are converted into required voltages by the rectifier and constant-voltage circuit. Fig. 2.5 shows the connection diagram of the rectifier and constant-voltage circuit.

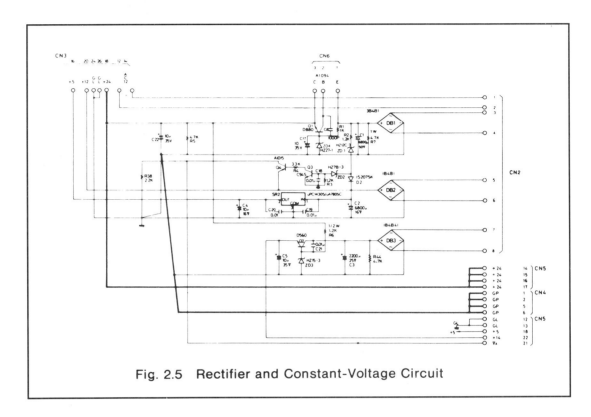

Fig. 2.5 Rectifier and Constant-Voltage Circuit

In this circuit, the following voltages are produced.
+24V DC ... Power supply for driving print head solenoid and stepper motors
+5V DC ... Power supply for TTL logic
+12V DC ... Power supply for buzzer, LED's, etc.
+14V DC ... For stepper motor holding current
+12V AC ... Power supply for serial interface (option)

Should the supply voltage become abnormal, this circuit detects such condition and causes a reset signal to be generated in the control circuit so as to return the control circuit to its original state, and at the same time, inhibits voltage from being applied to the print head solenoid driving transistors, thus assuring proper energization of the print head solenoid.

In this circuit, the voltage level at the zener diode ZD1 connected to the rectified output of the +24V line is normally 12V and the voltage level at the zener diode ZD2 connected to the base of transistor Q3 is normally 7V. When a total of these two voltages exceeds 19V, a current is supplied to the base of transistor Q3, causing transistor Q4 to turn on to permit the supply of +12V to the head driver circuit. Should the voltage of the 24V line reduce to below 19V for some reason or other, transistor Q4 is caused to turn off, thus prohibiting the supply of +12V to the base of the head driver transistor. The print head solenoid is thus inactivated to prevent the dot wires of the print head from being damaged.

+24V is produced from the 23V supplied from the power transformer into a current of 1.5A on the average at 24V by a series regulator consisting mainly of zener diode ZD4 and transistor Q4, after being processed through a bridge rectifier. The +5V supplied to the logic circuit is obtained by producing a current of approx. 1.5A with three-terminal regulator SR2.

Figure 3-15 Example of theory of circuit operation (Figure courtesy of Epson America Inc.)

ITEM NUMBER	COMPONENT	*CATALOG PART NUMBER	*NON-CATALOG PART NUMBER
R53, R54	RESISTOR 510 1/4W 5%		400072-90
R4	RESISTOR 560 1/4W 5%		400072-91 1/4W
R10, R16, R17 R19, R20	RESISTOR 1K 1/4W 10%		400071-01
R24	RESISTOR 2K 1/4W 1%		402285-130
R40, R41, R42	RESISTOR 3.3K 1.4W 5%		400073-13
R26, R27, R45	RESISTOR 4.7K 1/4W 10%		400071-09
R25	RESISTOR 3.4K 1/4W 1%		402285-152
R22, R23, R47, R48, R49, R50, R52, R58	RESISTOR 10K 1/4W 10%		400071-13
R44	RESISTOR 5.1K 1/4W 5%		400073-18
R31 THRU R39	RESISTOR 0.5K 1W 2%		401511-01
R3	RESISTOR 22 5W		400475-35
R56	RESISTOR .27 2W		402276-02
R6	RESISTOR 3.3K 1/2W 10%		400051-07
R5	RESISTOR 47K 1/4W 10%		400071-21
R1	RESISTOR 6.8K 1/2W 10%		400051-11
C9, C10	CAPACITOR 470 40V		401815-02
C17, C18	CAPACITOR 47 PED 50V		402371-05
C11, C12	CAPACITOR 100 PFD 50V		402371-08
C2	CAPACITOR 10000 PFD 500V		401004-71
C5, C6, C14, C15, C16, C19, C20, C23, C24, C25, C26	CAPACITOR .1 MFD 50V		402371-20
C21	CAPACITOR .001 MFD 50V		402371-14
C13	CAPACITOR .01 MFD 50V		402371-17
C7, C8	CAPACITOR 10 MFD 16V		402172-72
C22, C27, C28	CAPACITOR 33 MFD 16V		402172-74
C4	CAPACITOR 220 MFD 35V		402584-91
C1	CAPACITOR 4700 MFD 16V		402504-02
C3	CAPACITOR 6800 MFD 50V		402504-45
C29	CAPACITOR .22 MFD		400955-01
CR2, CR4, CR5, CR6, CR8, CR13 THRU CR18	DIODE IN4001		401150-01
CR1	DIODE IN4148		400376-01
CR3	DIODE ZENER 10V IN 4740		400664
CR7	DIODE ZENER 13V IN4743A		400664-53
BR1, BR2	BRIDGE RECTIFIER 4.0A		401841-02
Q14 THRU Q18	TRANSISTOR T1P120		401297-01
J5	CONNECTOR 20 PIN		402242-02
J6	CONNECTOR 25 PIN		402575-21
J7	CONNECTOR 36 PIN		402387-03
CR9 - CR12	DIODE IN4935		402749-03

*Catalog items are routinely available. Non-catalog items are available on a request-for-price-and-delivery basis only.

Figure 3-16 Example of a parts list (Figure courtesy of Mannesmann Tally)

Some of the better-engineered circuits will provide special test points on the board to help quickly determine the source of a problem. These test points are usually marked with a TP designation and number printed on the circuit board. To use the test points properly will require the service manual for the equipment. The manual will tell what signals should be at the test points, such as signal amplitude and/or DC voltage readings. See Figure 3-17.

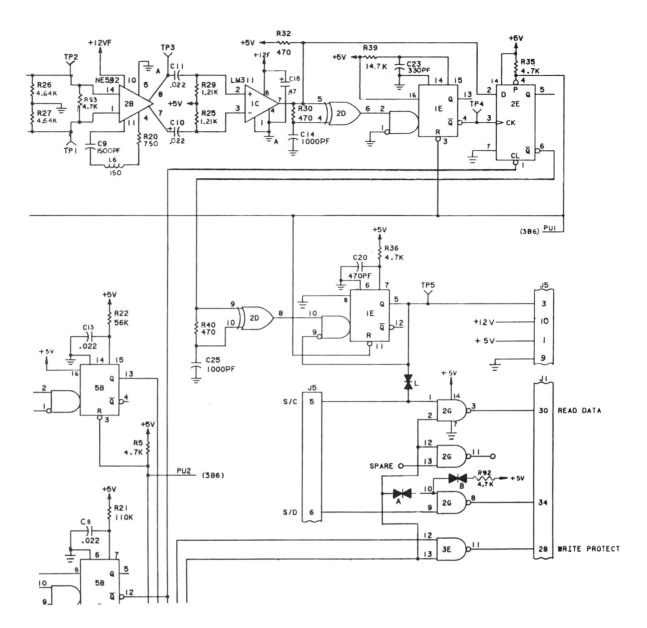

Figure 3-17 Schematic of a circuit that includes established test points for convenient troubleshooting of important functions (Figure courtesy of Micro Peripherals, Inc., Chatsworth, CA)

Without the benefit of established test points, five steps are necessary to take a specific stage signal reading:

1. Identify on the block diagram where the signal is to be sampled.
2. Identify the stages on the schematic diagram on either side of the intended measurement point.
3. Identify the line where the signal flows between the selected stages for a measurement point.
4. Identify a component on that line that will afford the intended test point.
5. Locate the component and test point on the card and make the measurement there.

Let us follow this summary with an explanation of each step in taking one signal tracing reading.

3.6.2 USING THE BLOCK DIAGRAM

Identify on the block diagram where the signal is to be sampled — This step often involves using the half-split, or "divide-and-conquer" method. Once the input signal is applied to the circuit, assuming the output is not normal (often missing altogether), we begin by looking for the signal in the middle of the circuitry, according to the block diagram. If there are a great many stages involved this can sometimes save a considerable amount of time. If only a few stages are involved, this step is less important. In that case the signal can be traced through sequentially from either end. See Figure 3–18.

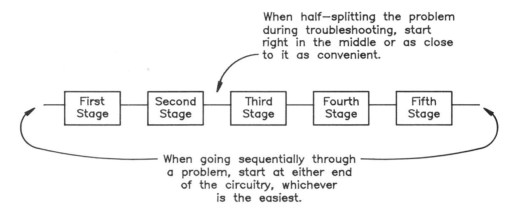

Figure 3–18 When using the half-split method, look in the middle of the suspect chain of stages for a missing signal. The sequential method starts at one end and proceeds to the other end stage by stage.

The block diagram is most useful at this stage of troubleshooting. Pick a point between two functional blocks. Now we progress to the next step:

3.6.3 USING THE SCHEMATIC DIAGRAM

Identify the stages on the schematic diagram on either side of the intended measurement point — Identify the two stages on the schematic that represent the two blocks on either side of the point we selected on the block diagram. While some block diagrams will indicate the circuit symbols of the stages (usually the transistors or ICs involved) on each block, it will often be up to the technician to interpret which stage corresponds to which block. Though not difficult, experience will help in executing this step quickly. See Figure 3–19 for an example.

Identify the line where the signal flows between the selected stages for a measurement point— After the two stages are identified, find exactly where the signal should be present between these two stages. See Figure 3–20 for an example.

The presence of added lines or additional circuitry between the two stages in question can sometimes be confusing. The schematic may have to be studied carefully to be sure of the proper signal path.

Identify a component connected to the line that will provide access to the signal — Once the signal path is identified, it is time to find a convenient point on the circuit card where that signal may be observed.

Transistors provide convenient test points in the circuitry for signal tracing. The leads are reasonably easy to identify and the transistors are likely to be identified on the circuit board as a further aid in finding well-identified points.

The ends of resistors are also common points at which to sample. Resistors, however, have two ends! If you should get the wrong end you will probably get an erroneous reading.

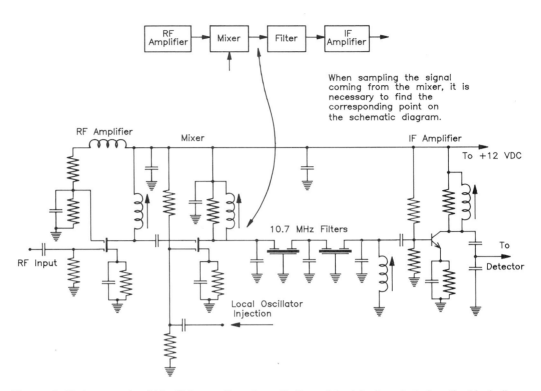

Figure 3-19 An example of identifying on the schematic the point originally selected on the block diagram

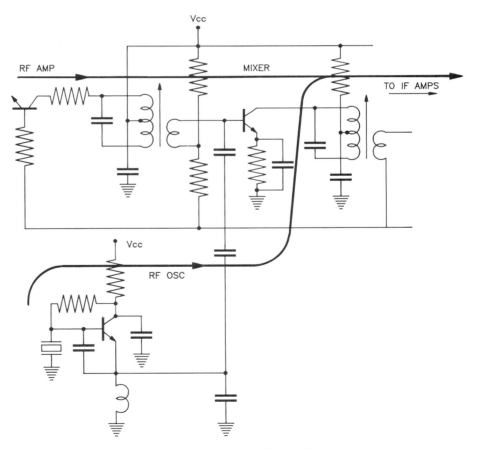

Figure 3-20 An example of identifying signal paths through several stages

Therefore, it will be necessary to verify you have selected the proper end of the resistor. Figure 3-21 shows one way to do this.

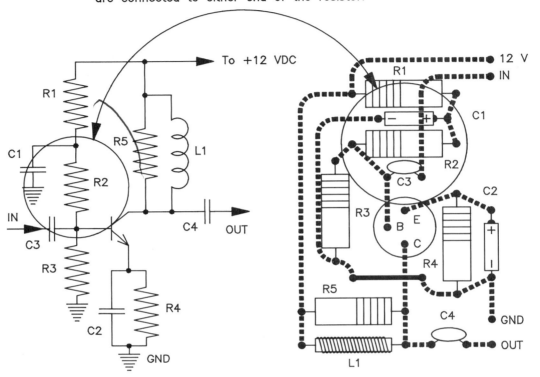

Figure 3-21 It is necessary to carefully note what is connected to either end of a component to determine which end to use for a signal or voltage reading

An alternative to finding the proper end of a resistor is to read the signal at either end of the series coupling capacitor, if any. The only difference from one end of the coupling capacitor to the other is that there is a DC component at the input end, which we can ignore by using the AC input mode of an oscilloscope. The AC signal amplitude should be essentially the same at either end of the capacitor.

When tracing DC signals, capacitors are "opens" and inductors are "shorts." This principle is covered in detail in paragraph 4.7. When tracing AC signals, things get a bit more sticky. Depending upon the reactances involved, capacitors can approach either shorts or opens, and so can inductors. In RF circuits, capacitors and inductors together in a circuit, in series or in parallel, can act even stranger. These subjects are covered in Phase 4.

Signal tracing in RF circuits requires, as will be explained in paragraph 3.15.3, that the technician take into account the relatively small values of unavoidable capacitance of the test probes. Radio frequency circuits can be detuned by the attachment of instrument probes and therefore exhibit erroneous results.

3.6.4 USING THE PARTS LAYOUT DIAGRAM

Locate the test point on the card and make the measurement there — When the most advantageous component for the test point is selected, that component must be identified physically on the card. Some circuit cards are labeled with component designations such as "R12" or "Q23." These cards are easier to work on than cards that have no identifications. If the card is not printed with circuit designations, the parts layout diagram in the documentation must be used to locate the component. See Figure 3-22(a) and 3-22(b) for examples of cards with and without component designations.

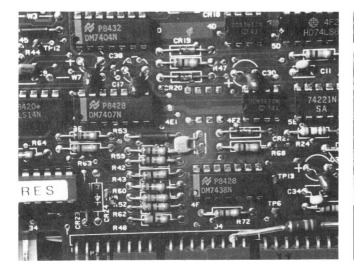
(a)

(b)

Figure 3-22 Cards printed with component designations are easier to work with than those without. The card shown in 3-22(a) has component designations; the card in 3-22(b) does not.

Once the selected component is located and the proper end of that component chosen, the signal tracing instrument may be attached to the component at that point and the reading taken.

3.7 IS ANOTHER GOOD CARD OR CIRCUIT AVAILABLE? —

In the absence of any documentation it may still be possible to find the more drastic problems, such as lack of power supply voltages, in certain parts of the circuitry. Compare voltage readings between two cards or two equipments, each operating under identical conditions of power supplied and inputs, one of them working normally and the other being the defective one.

A classic example of this technique is troubleshooting one of two audio amplifiers in a stereo unit. It is both quick and convenient to compare readings between the two channels.

3.7.1 Compare Interstage Signals Between Good and Bad Cards —

Compare readings from one circuit to another, looking for differences. Note that you are looking for *differences,* not necessarily just missing signals. Missing signals are only one form of signal difference that we might seek when troubleshooting. It is quite possible that the bad card may have a substantial signal at a point where the good card does not. Even this would still be a good indication that you are near the cause of the failure.

Keep in mind that at this point we are looking for *interstage* signals, not voltages inside a particular stage. Therefore, concentrate on looking where such signals would most likely appear, such as the collectors of transistors and at interstage capacitors or transformer windings.

Sometimes the application of a test probe to a circuit produces an unexpected result such as sudden and proper operation of the circuit or a loud thump from a stereo speaker or other major change. A drastic change such as this would indicate one of two possibilities: Either the probed point of the circuit has an intermittent connection or there is an open circuit, making the input side of the open circuit very sensitive to any disturbance. This is a particularly applicable tip when troubleshooting CMOS and FET circuits.

3.7.2 FIND A DIFFERENCE BETWEEN THE TWO? — If a substantial difference is found between the interstage signals of the working and nonworking cards, there is a good chance that the problem is within one stage, either toward the input or toward the output of this point. Move to Phase 4, "Identifying the Defective Component."

If no differences in the interstage signals are found between the good and the bad cards, it is time to verify that the suspected card is still defective.

3.7.3 IS THE CARD STILL BAD? — It is quite possible that the problem has "cured itself" sometime during the troubleshooting session. Test the card under the original conditions that showed a problem. If the problem is now gone, consider that the card may have an intermittent problem. If so, go to paragraph 1.3 of Phase 1.

On the other hand, if comparison testing still hasn't helped, the only course of action left at this point is to go to Phase 6, "Dead Circuit Troubleshooting."

3.8 IS SIMILAR DOCUMENTATION AVAILABLE? — It may be possible to use documentation for a circuit that is not identical, but is at least similar to the one that has failed. Revisions of cards from the same manufacturer sometimes involve minor changes and these changes may have been made in circuits that will not be considered during your troubleshooting. Be careful when applying the similar documentation alternative to your problem, however, and always keep in mind that the information is *not* as exact as you would like. Use the similar schematic as a guide rather than an absolute road map. This alternative is often sufficient for finding the defective stage within a card, particularly when the two schematics were issued by the same manufacturer on similar models. You could even compare *two* similar schematics to the circuit in question to try to find the best combination of information.

3.8.1 Use It, with Allowances for Differences Between Cards
Some differences between similar documentation and an actual circuit may be obvious. For instance, the components on the card may be numbered differently from those on the schematic; Resistor values may be different; Or, there may be more or fewer components on the schematic. However, there may be surprising consistency in the signal flow paths. The general functions of the schematic and the card under test must be nearly the same to get any real benefit from a similar schematic.

When using the information suggested by the similar schematic, troubleshooting should proceed beginning with paragraph 3.10.

3.9 IN A PANIC TO REPAIR CARD? — Without a good working card to compare or even a similar schematic to use, the next question to consider is the importance of repairing the card in spite of these obstacles. If there is a real need for immediate repair, the card might be repairable using the techniques of Phase 6, "Dead Circuit Troubleshooting," or by drawing your own schematic, as discussed in paragraph 3.9.2.

3.9.1 Order Manual from the Manufacturer — If time allows, an instruction book from the manufacturer might be available. Consult the nameplate on the equipment in question to get the name and address of the manufacturer. If the address is not available, the library may be able to assist you in getting the company's address. Make a telephone call to the manufacturer to inquire if an instruction book is available. Be sure to ask for the *technical manual* for the equipment. Otherwise, you might get only an operator's manual that tells you how to operate the equipment and nothing more.

Many manufacturers will not give out technical information at all, reserving all technical repairs for themselves or their designated "service centers." Don't be surprised if there is a charge for a technical manual as some manuals are large and expensive to print.

Televisions, stereos, other consumer equipment, and some personal computers are largely supported by the Howard SAMS company for technical information. Libraries sometimes stock SAMS folders and make them available to their users.

The best time to obtain technical information is at the time of the equipment purchase. Since equipment will be expensive to repair without documentation, it is a good idea to purchase only equipment for which information is readily available.

If the technical information is forthcoming, the equipment should be set aside and tagged with a status "awaiting instruction book." This tag is a professional touch, tying up one of the loose ends of daily work, and making it that much easier to keep track of equipment awaiting repairs.

3.9.1.1 Call Manufacturer for Help

Some manufacturers offer *technical support* for their products. All you need do is call the company and they will assist you in answering questions and sometimes will suggest repairs to solve the problem. As an example, a certain computer monitor was behaving erratically. Two attempts had been made to cure the problem in the vertical sweep circuit. The monitor would work fine for days, then begin to give trouble. The manufacturer was consulted. The technician was told that they had had similar problems before, caused by a defective batch of Mylar capacitors. One of the capacitors was in that circuit. It was replaced and the problem never recurred.

A few manufacturers are possessive of their schematics and their information. They may tell you that the information you seek is available only to their official representatives or authorized service centers. Documentation for these equipments may be all but impossible to procure through normal channels.

3.9.2 IS IT WORTH YOUR TIME TO DRAW A SCHEMATIC? —

It is possible, though difficult at first, to draw a schematic from the physical circuit. It is time-consuming, however. As a very rough estimate, a technician familiar with the procedure could probably produce a smooth schematic drawing of a circuit consisting of a dozen discrete components on a single-sided card in perhaps half an hour. (Single-sided means that there are circuit traces on only one side of the board.) If the circuit involves traces on both sides of the board, the time required may quadruple to perhaps two hours. Obviously, it would be financially unwise to draw the schematic if a whole new card costs less than the technician's estimated time to draw it. On the other hand, if the technician's time is not a factor, such as when the technician wants to repair something for himself as a "labor of love," or the board is a critical part of a system that is no longer being manufactured, this remains a viable alternative.

3.9.3 NOTES ON DRAWING A SCHEMATIC AND LAYOUT FROM AN ACTUAL CIRCUIT —

Drawing a schematic from an actual circuit is a skill that is quickly sharpened by practice. The first few times it is attempted, though, it can be frustrating. Be sure to practice on simple circuits first.

Drawing the schematic involves several steps:

- ☐ Sketch the component layout of the board.
- ☐ Assign arbitrary circuit symbols if not already provided.
- ☐ Identify probable Vcc, ground, inputs and outputs.
- ☐ Start tracing with probable signal input.
- ☐ Complete intermediate drawing.
- ☐ Reduce, redraw, and straighten drawing in stages.
- ☐ Draw final schematic with inputs on left, and so forth.

Now let's expand on each of these steps. Taking them one at a time and being thorough in their execution should make it possible to draw a circuit in the least possible time. Be forewarned: missing or skimping on a step will make it harder to do the next step.

Sketch the component layout of the board — This step is necessary to positively identify the same component as you look back and forth between your actual circuit and the drawing you are attempting to make. While this may not be such an important item on a simple circuit, it is positively necessary on more complex circuits. Making the layout is also important when it comes time to troubleshoot. You will be working principally from the schematic, but when the schematic refers to a particular component, the layout comes into play again and allows you to find the correct component with a minimum of additional trouble.

Assign arbitrary circuit symbols if not already provided — Some circuits will already have component designations silk-screened on the board. Others will have a few or no markings at all. It doesn't matter which component has the first, second, or last designation, as long as all components have a unique number. Follow the conventions of the table to mark the components to make your schematic readable by any other technician.

C = Capacitor	Q = Transistor
D or CR = Diode	R = Resistor or potentiometer
F or X = Fuse	S = Switch, Socket
I = Indicator (LED)	T = Transformer
J = Jack (receptacle)	U = Integrated Circuit
K = Relay	TB = Terminal Board
L = Inductor	Z or ZD = Zener Diode
M = Meter	VR = Voltage Regulator (or Variable Resistor)
P = Plug	

See Figure 3–23 for an example of a layout sketch.

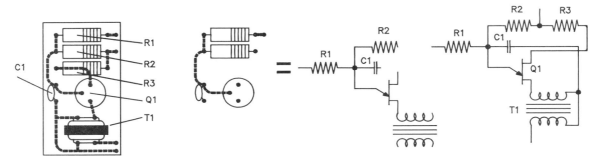

One of the first steps is to make a component layout sketch, complete with arbitrary circuit designations if not already provided.

Beginning preferably at an input, begin drawing a schematic with the first component encountered. Make a dot if more that one component is connected to a particular PC trace.

Completed intermediate drawing.

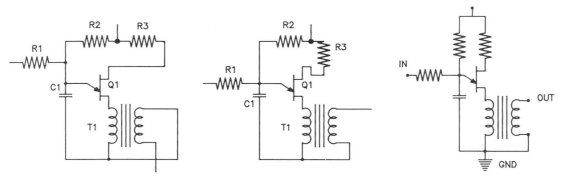

Two intermediate drawings. Try to avoid crossing lines and attempt to put all suspected inputs on the left, outputs on the right, grounds to the bottom, and power supply inputs at the top.

The finished schematic, drawn from an actual circuit

Figure 3–23 Examples of the steps necessary to generate a schematic from an actual circuit

Label the components with consecutive circuit designations. For instance, label the first resistor you encounter R1, the second R2, and so forth. The same would apply to other components, the first capacitor would be C1 and the second C2.

Identify probable Vcc, ground, inputs and outputs — The ground connection of a PC board is usually the widest trace on the board. It frequently surrounds the mounting holes and may depend upon the mounting bolts to make a solid connection from the ground trace to the metal chassis of the equipment. This is probably the easiest and therefore the first lead that should be identified. If a black power supply wire is attached to the circuit board, there is a good chance that this is the ground lead. Sometimes there are relatively large electrolytic capacitors on the board that are clearly marked with + and − symbols. While some audio amplifiers use electrolytics to couple between stages, most large-value electrolytic capacitors are connected directly across the Vcc and ground busses. An easy way to identify ground on boards that have TTL digital ICs is to use the continuity function of a digital ohmmeter and probe around until you find the traces that have continuity to pin #7 of 14-pin ICs or to pin #8 of 16-pin ICs. Before considering this conclusive evidence of a ground, however, verify with an IC databook that the pin #7 or #8 of the specific chip you are using is really ground. While these two pins are *usually* ground, the rule applies in only about 95 percent of the cases. There may be more than one ground on the card, too, depending on mounting bolts to a metal chassis to tie them all together.

The Vcc or positive supply voltage lead may be, but is not necessarily, a red lead. Vcc leads have even less color significance than the black of ground leads. Any further color coding suggestions cease here. Any color wire besides black and red is "up for grabs" as far as guessing what purpose they may serve. The Vcc lead for TTL digital ICs is almost always tied to the last lead of the IC, pin #14 or #16. Be sure to check this against a databook to be sure in a specific case.

PC boards that are densely packed with ICs will sometimes have multiple layers of conductors sandwiched within the board. These circuit boards most often have an internal layer for a ground. Holes are punched through it where necessary to pass signals through the board. See Figure 3–24.

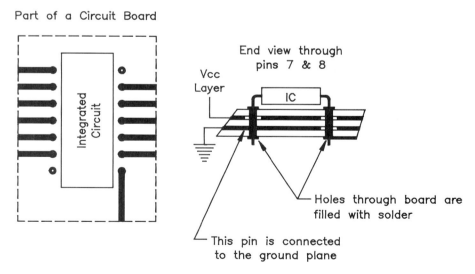

Figure 3–24 Cross section of a multi-layer PC board. Note how the ground in the middle of the board is cut back to allow signals to pass through the ground plane without shorting to it

Another plane could be added for the Vcc bus, too. This kind of construction makes a compact PC board with excellent immunity to its own electrical noise pulses.

There may be additional supply voltages applied to some printed circuits, but in our example we will assume only one supply voltage is required.

Input signals to a PC board may be DC, AC, digital, or RF signals. When drawing a schematic of a discrete semiconductor circuit, look for the input to connect to the base of a transistor or an operational amplifier input. AC and low-frequency circuits usually use a series coupling capacitor. RF input circuits almost always have series capacitors or RF transformer, either tuned or untuned. On digital circuits, the inputs are relatively easy to identify with the aid of the right databook for the chips involved.

If information on the semiconductors and chips is not available, it may be impossible at first to say for certain which is the input lead, the ground, or the output lead. In this case, all that can be done is to make that particular component a "black box" on the schematic drawing without assuming anything at all about inputs and outputs until the drawing is finished. When the drawing is nearing completion, the inputs and outputs may then be obvious.

Output signals from circuits using discrete components come from operational amplifier outputs or from transistor collectors or emitters. AC low frequency (audio) circuits usually use an output transformer winding or medium- to fairly large-value series electrolytic capacitors. RF circuits use RF transformers or relatively small capacitors for output components. Digital circuit outputs are identified with the aid of a databook.

Start tracing with probable signal input
— Once the most likely input (or one of them if multiple inputs) is identified, the actual drawing of the circuit can begin. Starting from this point, follow the circuit trace until the first component is encountered. Draw a horizontal line and the schematic symbol for that component. If the circuit trace continues on to some other component, don't be concerned with it at this point. Just put a large black dot at the input component and a small line coming from it for each additional component connected to this point. This will serve as a reminder to yourself when you check this point later to be sure you have accounted for all of the connections to this point. See Figure 3–23.

Label the circuit designation adjacent to the first component. Now follow the circuit *through* the first component. If the opposite end of this component is connected to more than one other component make another dot with a small line for each additional component that is connected to that point.

Remember that all points tied together by a trace are at the same voltage and may be drawn together with lines in any way you choose.

Complete the intermediate drawing
— Follow through each component on the board in a like manner. Every time there is a crossroads, make the dot and short line. When all components are drawn, go back and find out where each of the short lines connects. You will probably find that most of the circuits loop back on themselves as in Figure 3–23.

Once all of the short lines are accounted for, this drawing should be an accurate, if confusing, circuit diagram.

Reduce, redraw, and straighten drawing in stages
— Schematics are supposed to begin with the inputs on the left and follow horizontally across the page to the right. Beginning with your assumed input line, begin redrawing your circuit into something that looks a bit cleaner. Complete a drawing all of the way through before beginning another.

Try to keep inputs on the left, outputs on the right. Any connection to ground should go down and connections to Vcc go up. Grounds do not have to connect together, but should end in a ground symbol. See Figure 3–23 for an example of such a circuit.

It may help to have handy a typical circuit that does the same job as the one in question. As an example, if you are drawing a circuit that you know is an audio amplifier, it may help to have a schematic of a similar audio amplifier available to help get the final drawing laid out properly.

Draw final schematic with inputs on left, outputs on the right, Vcc lines upward, and grounds terminating in several downward symbols
— The final schematic should look something like the last drawing of Figure 3–23. Note that this drawing is much easier to understand than any of the intermediate drawings.

3.A Locating the Defective Stage

3.10 IS THIS A DIGITAL CIRCUIT? — At this point we should have a circuit card that we wish to troubleshoot to the component level, and for which we have full information by way of schematics, layout, and block diagrams. Signal tracing techniques from here onward depend upon one major distinction: Is this a digital-type circuit or is it analog? Digital signals are either at a high logic level or a low and include critical timing as a third factor. Analog signals, on the other hand, include variable amplitude and the widely varying effects of sinusoidal and non-sinusoidal frequencies as factors to consider. The timing of signals is seldom important in RF circuits. Pulse circuits often use precise timing as a factor. Let us first consider signal tracing with digital signals.

3.10.1 Be Sure Digital References are Available — Having suitable component reference materials on hand is important for the servicing of digital circuits. While the operation of an analog circuit might be inferred from the components used, the digital schematic diagram often provides only the generic number of the integrated circuits. The function of a digital circuit is seldom given. Without references, the symbols on the diagram are mysteries. See Figure 3–25. Reference books also help in defining obscure terms and special purpose abbreviations.

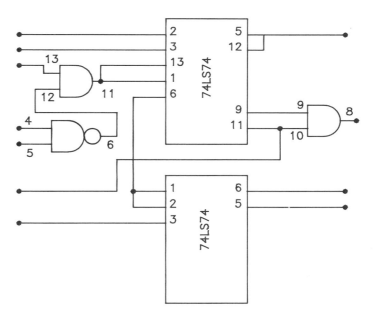

Figure 3–25 Example of a digital circuit that, without references for the chips involved, would be all but impossible to understand

There are two families of integrated circuits in common use: the TTL and CMOS families. The technician should at least have databooks for each of these families. They are available through local electronic parts distributors and from the manufacturers of ICs. Information on voltage regulators and operational amplifiers, often used with digital circuits, is found in the linear databook, also published by the same sources.

Digital chips other than generic gates and simple ICs may be encountered and will have to be researched in a variety of references. A particular serial data transmitter/receiver chip might be found in a manual that deals only with data communications chips. Memory chips would be in a memory devices manual. Microprocessors would be found in a microprocessor manual. Input/output chips would be found in a peripheral chip manual, and so on. Add to this the confusion of manufacturers often instituting their own chip numbering systems and it becomes evident that to have information on all the chips you use may require many books from many different sources. Depending upon the chip information needed, the appropriate manuals will have to be obtained.

If for some reason information is not available for a particular chip, it is often possible to get enough information for the job at hand from an instruction manual for equipment that uses that chip. What the schematic may not tell, the parts list might. Unrelated schematics showing the chip in which you are interested are particularly helpful in answering questions on the connections and/or use of specific pins.

Another possibility for getting scarce information is to visit an electronic parts distributor. See Figure 3-26. Many such businesses make a single copy of references available at the counter for their customers.

Figure 3-26 An electronic parts distributor may be able to assist in finding parts, getting part specifications, and finding substitute parts (Photo courtesy of Radar Electric, Seattle, WA)

If your troubleshooting inquiry can be answered simply, consider calling your friendly parts distributor and asking for the information. This will require that you have cultivated some degree of rapport. Remember that the distributor doesn't memorize the specification books any more than you do. It will take time to look up the information for you and it is a courtesy.

3.10.2 NOTES ON PREPARING DIGITAL SCHEMATICS FOR USE — The average digital schematic will show resistors, connectors, and switches as any other schematic might. Since the integrated circuit is actually several transistorized stages in itself, it is customary to show the entire chip as a single block. Trouble is, these blocks are seldom labeled as to what they do in the circuit. A single chip having several independent gates may be shown, scattered, where needed on the schematic. To use the digital schematic efficiently it is wise

to customize it, especially if the schematic will be used a great deal. Here are some suggestions for making digital schematics more "friendly":

- *Color code your schematics.* This topic will also be covered later in this chapter, paragraph 3.11. Color coding is applicable to digital circuits. Vcc and ground should be colored standard colors, red and black respectively.

  Signal flow lines are not quite the same for digital circuits as they are for analog. Rather than show signal flow it may be better to color code groups of similar lines the same color. For instance, a microprocessor schematic might have all the data lines one color and the address lines another. There are generally too many different signals going every which way in a large, complex digital circuit for the term "signal flow" to make much sense. Of course, a color coding key should be provided on the schematic.

- *Mark the Vcc and ground connections,* if any, on the schematic. It is standard practice with digital circuits to eliminate these connections. If you feel that it might be a help to you, put these connections on the schematic along with the appropriate IC pin numbers and color code them too.

- *Solid-state tester patterns* are very helpful on both digital and analog PC cards. The typical patterns may be shown for many points in the circuit. Use a ground for the common reference point unless the waveform is specifically labeled otherwise.

- *Marking normal indications of a logic probe* on a schematic is a very good idea. With the circuit in full, normal operation, the schematic should be marked with a simple code at all appropriate places indicating one of the following:

  1. A (there is normally an active signal here)
  2. L (this point normally at a low logic level)
  3. H (this point normally at a high logic level)
  4. B (this point normally at a "bad," "floating" logic level)

More information on the logic probe is given in Notes on Using the Logic Probe, paragraph 3.10.8.

If the circuit you are dealing with is software-driven, remember that you must have the same program running during the testing as when the readings were originally recorded as "normal." The software will have complete control of what is running and at what logic levels.

3.10.3 NOTES ON DIGITAL CIRCUIT TERMINOLOGY AND SYMBOLOGY

Terminology — Some of the terms used in dealing with digital circuits are unique. Here are some of the common ones and what they mean:

ACTIVE. Active simply means "what satisfies the circuit." The AND gate, for instance, is satisfied and acts as an AND gate when the input #1 AND input #2 are both high. The gate is satisfied that the conditions for an AND gate are met and the output goes ACTIVE. In this case, the output is high.

If this is confusing, consider the fact that logic circuits can have positive logic, where a logic "0" is false, and "1" is true. This is "right-side-up" logic, and easiest to follow. On the other hand, we have negative logic, too. Using this "upside-down" logic, the "1" is now false and the "0" is true. With these reversals, logic levels are confusing to keep track of while troubleshooting. It is easier to think of an active high or active low than to be concerned about whether the logic is a "1" or a "0."

CLEAR. This means to change to, or make sure that a logic level remains, a "0." Often used complementary to a PRESET input, the CLEAR input will cause an immediate change of state of a flip-flop to the "0" state, regardless of the clock signal.

CMOS. Complementary Metal Oxide Semiconductor. A family of logic chips characterized by very low power requirements and high sensitivity to static damage.

DYNAMIC. Moving, in operation. Used to describe circuit operation wherein the voltage levels are constantly changing, as during normal circuit operation. The opposite of STATIC.

PRESET. An input to a digital flip-flop which will cause an immediate change of state to the SET or "1" state, regardless of the clock signal.

RESET. To change to, or make sure that a logic level remains, a "0."
SET. To change to, or make sure that a logic level remains, a "1."
STATIC. Without motion. Often used to describe circuit operation wherein the voltage levels remain the same until manually tripped or incremented by the troubleshooting technician. The opposite of DYNAMIC.
STROBE. A logic change from one state to the other and back again rapidly. Used as a signal to tell another circuit when to operate. A commonly used term with circuits that LATCH data.
TTL. Transistor-transistor logic. Integrated circuits that depend mainly on interconnected bipolar transistors to perform circuit tasks. Characterized by ruggedness, high speed, and operation at exactly 5 VDC.

Symbology — Some of the symbology used on digital schematics needs explanation:

$\overline{CE}$ A bar close over the top of the letter designation of a pin means that input is satisfied or active when that pin is at a logic low. Lack of a bar over a pin designation *probably* means that the pin is active high. There are a few exceptions to this on some schematics, so it is always a good idea to check a suspicious schematic designation against a databook for verification.

—o   A small "o" at the input to a chip, like the overbar, means that this line is active low. On the other hand, the lack of an "o" *probably* means that this is an active high pin. It could also mean someone was inattentive when drawing the schematic.

—▷   A small triangle just inside the IC symbol indicates that this input is edge-sensitive. The circuit will act upon receiving the specified polarity of *transition* on this pin. Without a small "o," this circuit would be one that would act upon a positive-going signal transition. In other words, at the instant this input went from a logic low level to a logic high, the circuit would act. A transition later from a logic high to a logic low would not affect the circuit. Adding an "o" at the input would designate this as a negative-going signal input. While it is often used at the clock input of a variety of chips, this symbol does *not* by itself mean this is the clock input pin, a common misinterpretation.

Equivalent gates — When a circuit is designed and a particular chip is used, there are often times when not all of the internal gates of the chip are used in the circuit. If a need arises for another gate of a different type, it is a good idea to see if something could be done to use the "extra" gates on the board instead of adding an extra chip. For instance, if three NAND gates are used of the four available in a 7400 chip, one is left over. Elsewhere in the design it is found that an inverter is needed. It is a simple matter to tie the two inputs of the surplus NAND gate together and use it just as though it were an inverter. See Figure 3-27.

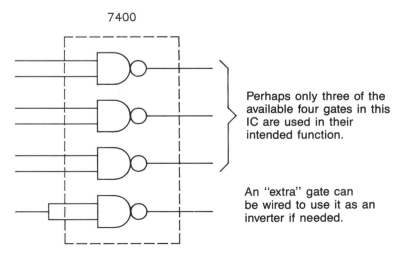

Figure 3-27 Using the "extra" NAND gate of a 7400 IC as an inverter instead of adding another IC to the board

The substitution of one kind of gate for another is often done in digital circuits. One of the most frequent examples has just been discussed, but other combinations are possible, too. Some of them can be a bit extreme, but will work. See Figure 3-28.

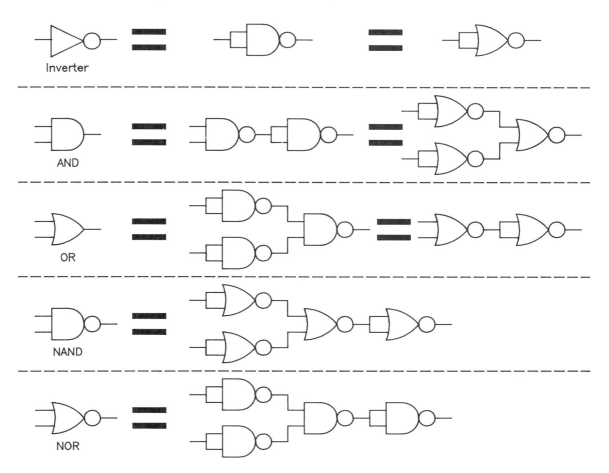

Figure 3-28 Summary of common substitutions of one type of gate for another

One of the things that can be confusing about digital schematics is that there is no standard way of showing these equivalent circuits. Using the NAND as an inverter, for instance, might be indicated on a schematic as the NAND with the inputs tied together as it actually exists, or it may be drawn as the equivalent inverter. Beware of this potentially confusing information.

Delay Circuits — Another circuit that might cause confusion is that of Figure 3-29.

Figure 3-29 Two cascaded inverters can serve an important, though not obvious function: time delay

This circuit does apparently nothing since the signal comes out with the same logic level as went in. There is a change between the input and output signals, though. The output signal transitions are delayed *in time* from those at the input due to the delay in getting through both gates. Technically speaking, this circuit depends upon the *propagation delay* of the two inverters to delay the signal. This delay, incidentally, might be on the order of 50 nanoseconds.

Another possible reason for using two inverters in this manner is to provide more noninverting driving power so that it can operate additional chip inputs (increased fan-out). One gate is used for the increase in driving power, and the other to re-invert the signal to its original active state.

Non-standard gate symbols — Some manufacturers make an attempt to draw their schematics so that it is easier to see how they operate. In using nonstandard gate symbols, they indicate what is wanted at the inputs and the function to be performed with those inputs. See Figure 3-30 for an example.

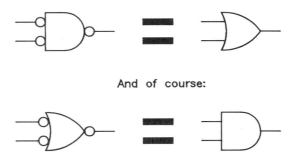

Figure 3-30 Occasionally, a circuit may be easier to understand if the logical equivalent for the gate is used instead of the actual symbol for the gate used

Another example of such a gate is the NOR gate with both inputs indicating an active low. This circuit is the same as a standard AND gate.

If these nonstandard symbols are encountered, don't bother to try to look them up by function. They don't exist in any databook. Pay attention to the IC identification number and don't be surprised when the gate in the book is entirely different from the schematic symbol. *The truth tables will be identical for each.* An example of this kind of digital notation is given in the partial schematic of Figure 3-31.

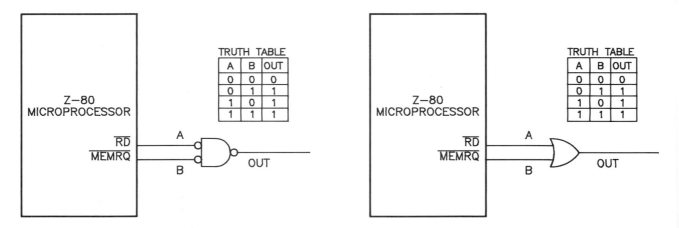

Figure 3-31 If we wish to have a line go LOW when both RD and MEMRQ lines go low, we could draw the circuit either of two ways

3.10.4 NOTES ON USING DIGITAL SCHEMATICS

IC Output Configurations — Most IC output stages pull the output line up or pull it down, using two separate transistors in the output stage. See Figure 3-32.

The common TTL totem-pole output will drive as many as 10 other TTL inputs. This type of signal flow might be considered a *"Type I" data bus - One source IC and one or more receiving ICs.*

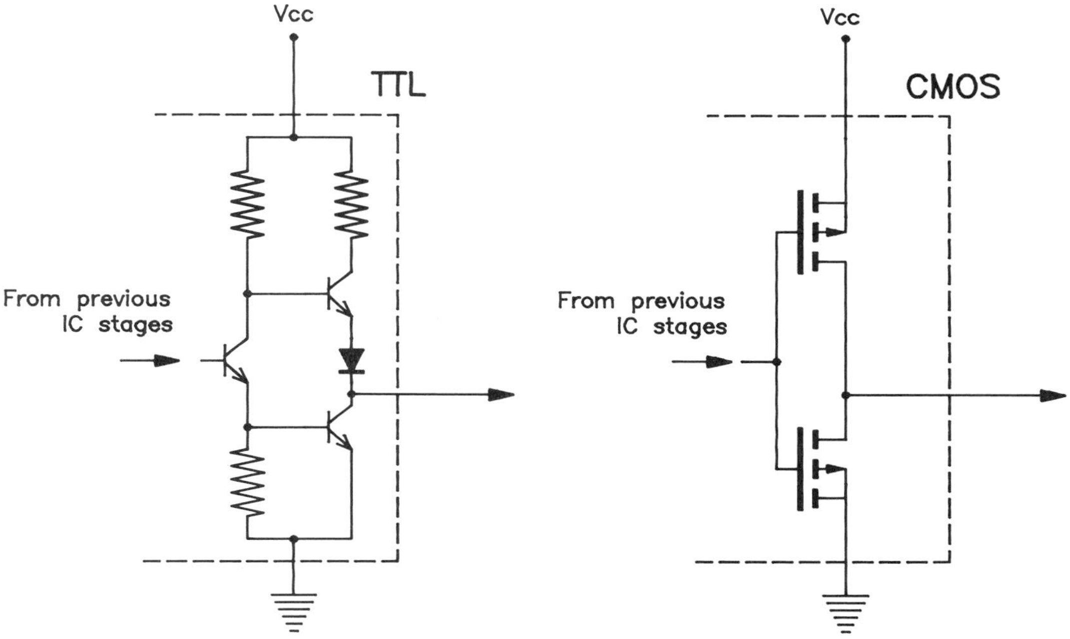

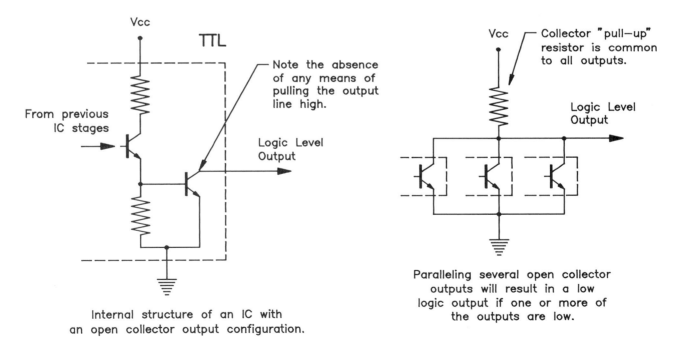

Figure 3-32 Basic totem-pole output stages, TTL and CMOS

A schematic might also have an IC output with a resistor to Vcc. This would strongly indicate that the IC does not have the totem-pole output, but rather has only the pull-down transistor of that circuit. See Figure 3-33.

Figure 3-33 The open collector IC output stage and how it can be used in a wired NOR configuration

The resistor connected from the IC output to Vcc is necessary to pull the output line high when the transistor in the output stage is not conducting.

Where only one totem-pole output can be connected to a given PC trace because of possible conflicting output logic levels, the open collector IC can be paralleled with other open collector chips. If any one of the chips causes a logic low on the output line, the other chips may either agree and help hold the line at the logic low or they may remain off, thus making them "transparent" and of no consequence to the low output logic level. This kind of bus can be thought of as *a "Type II" data bus, with more than one source IC and one or more receiving ICs.* It is restricted in use, however, because it is wired to be a NOR function at the output.

The third type of IC output stage is the three-state or floating output. This configuration allows two or more source ICs to be connected together. Such a configuration is common in microprocessor circuits, where many chips must be able to produce signals on a single line. Only one of these chips can be allowed to "talk" at any given time, however. This is the *"Type III" data bus, the three-state bus.* See Figure 3–34.

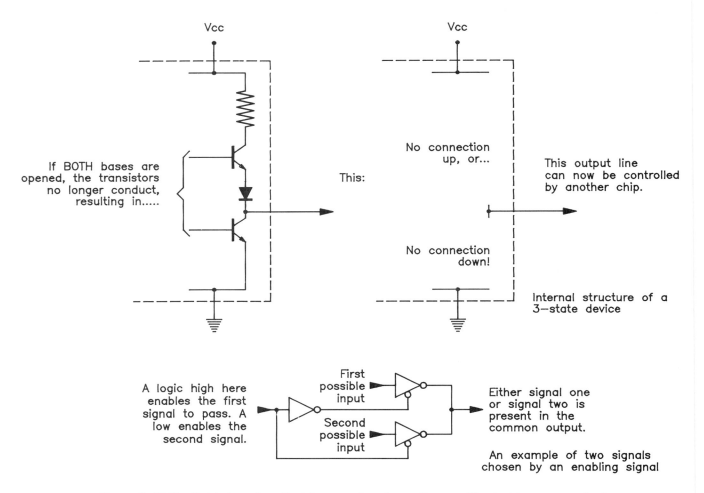

Figure 3-34 The 3-state bus allows the interconnection of more than one IC output to a common load

Digital timing — Where analog schematics usually have only a few signals to deal with, digital circuits often have many. A radio receiver may have two RF signals and an audio signal for the technician to contend with. A digital circuit, on the other hand, can have dozens of signals entirely different from each other—but they don't *appear* different! The signals are either at a logic high or a logic low. Those are the only acceptable active levels. The key to understanding the interaction of the signals in a digital circuit is to visualize the *timing* between these signals that is so important.

Why is timing such an important part of digital circuitry? Take the ordinary AND gate, for instance. Without timing information there is no way to tell what the output of the circuit might be. Sure, we could quote what it is supposed to do according to its truth table, but that has little to do with knowing what the instantaneous relationships of the two actual input signals might be. Here is an example: What is the output of the AND gate of Figure 3-35?

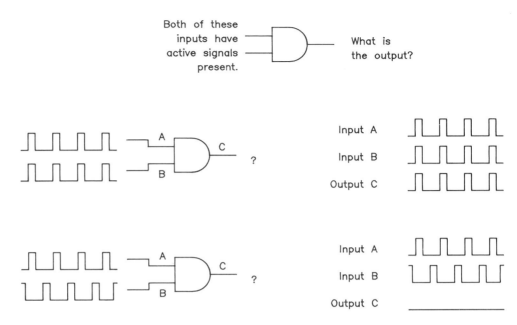

Figure 3-35 The presence of a signal is only part of the digital story. Of equal importance is the timing involved

A little thought will make the point clear that *digital circuits have a timing factor that is not evident on the schematics!*

Now let's see what adding the dimension of timing will do toward understanding this circuit. Take the same circuit as above but provide two identical squarewave inputs. These squarewaves, by their physical relationship above and below each other, infer that they go high and low together. In this case, the output of the AND gate simply follows them both. See Figure 3-35.

Now let's slip the *timing* of one of the signals so that it arrives later, exactly one-half of the cycle timing. See what happens to the output. See Figure 3-35.

To understand these differences for any circuit, plot the high or low logic levels along a common time base. In the above example, copy the input signals to a graph and name it *timing diagram.*

While it is necessary to know the functions of basic logic gates, flip-flops, registers, and other typical IC circuits, it is extremely important to know in a particular circuit exactly how the signals are arriving *with respect to one another.*

Interpreting Timing Diagrams — Timing diagrams are merely representations of what happens first, second, third, and so forth. The horizontal axis represents time, just like an oscilloscope. See Figure 3-36.

The theory of operation — Fortunate indeed is the servicing technician who has access to a complete theory of operation for the equipment to be repaired. Too many digital circuits are in use today without good documentation to back them up during service. The theory of operation is a step-by-step description of exactly what is happening at all of the important points in the digital circuit. Digital signals are explained with respect to each other. The theory of operation of a circuit will probably be very difficult to read through and to make much sense of the first time or two through it. But, *understanding the theory of operation is the fastest way to learn the complete operation of a digital circuit.* Some companies offer schools on their specific models of equipment. These sources often spend a great deal of time on the theory of operation, making it the core of their course.

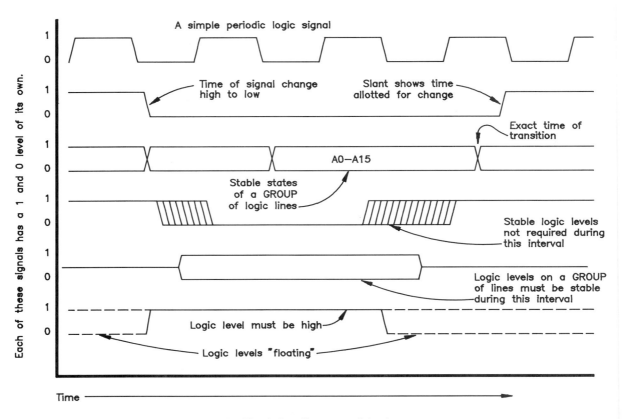

Figure 3-36 The timing diagram explained

The proper way to use the theory of operation is to somehow remove the schematic (a copying machine can allow you to "remove" the schematic without disturbing the book) and put it right beside the theory of operation. As the various signals are explained in the theory, follow them on the schematic. At the same time try to visualize the timing that must be in effect for the circuit to operate as described. Draw a timing diagram to summarize the operation of the circuit if it will help to visualize the timing sequences. Doing so will complete your understanding of the circuit. The timing diagram is a very good item to place on the schematic, up in a corner somewhere.

Reverse engineering — All too frequently, the technician is given only a schematic and a bad circuit, and from this alone is expected to make the repair. While some digital circuits are not too complex, others can be very complicated to figure out or imposible without the theory of operation. Figuring out a schematic and reasoning out exactly how it works without benefit of the theory of operation is called *reverse engineering*. The technician is left to deduce what the design engineer had in mind.

Reverse engineering can take a great deal of time. Even circuits containing as few as two or three integrated circuits can be very confusing.

Let's "reverse engineer" a sample circuit. See Figure 3–37.

For reverse engineering, go backward. Begin with the desired output and determine the logic levels necessary at the IC inputs until the board inputs are reached. The logic levels along the way can be noted on the schematic with 0s and 1s See Figure 3–38.

The operation of the circuit can be deduced from the indicated inputs. Digital reference books are necessary to identify the use of the various input lines and to give clues as to what kinds of signals can be expected.

And then there is the situation where the technician is given a bad board...and that's it. He or she has neither a schematic nor any theory of how it works. Assume a difficulty factor of 4 for this situation over the last situation. First, a schematic should be drawn for the

Input #1
Input #2
Input #3
Output #1
Output #2
Output #3
Output #4

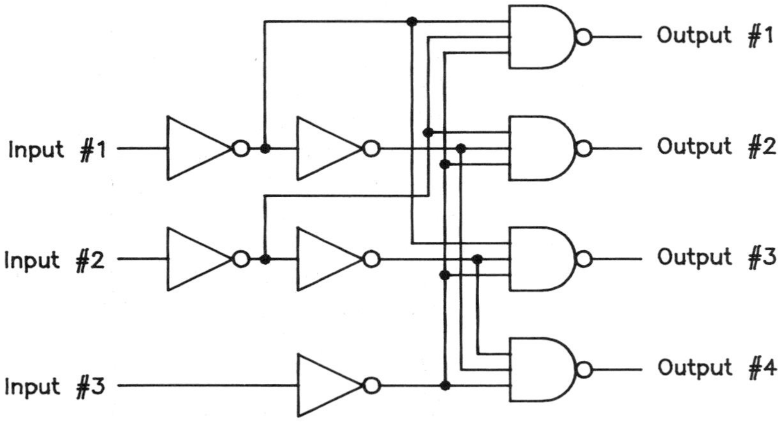

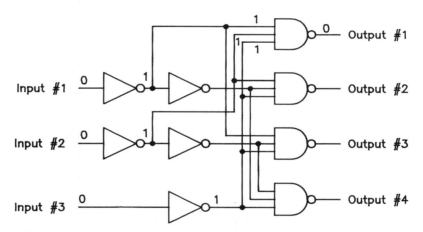

Figure 3–37 Circuit to be reverse-engineered

No output at #1? Reverse–engineer this circuit by beginning with the desired active low output at #1. Determine logic levels required at all previous stages.
Input #1
Input #2
Input #3
Output #1
Output #2
Output #3
Output #4
By working backward, we can determine that all "0" levels would be required for the "0" output desired. With this information, the problem can be found and the circuit repaired.

Figure 3–38 Schematic of Figure 3–37 with logic levels added

circuit, providing the circuit is not too complicated and that there is time to do it. See paragraphs 3.9.2 and 3.9.3 for help in drawing the schematic. After the schematic is drawn, the circuit can then be reverse engineered to find out what the inputs should be.

All of this assumes that the circuit in question has static inputs. The inputs to a digital chip can also be a squarewave, a waveform whose duty cycle changes and short pulses. Unless more is known about special inputs, the circuit may be frustrating to understand.

A limited amount of signal analysis can be inferred from the application of the circuit. The use of the circuit can provide an indication if the inputs are digital or analog, for instance.

Reverse engineering becomes easier as the technician gains experience with each individual chip and common circuits.

3.10.4.1 NOTES ON USING CORRELATION DATA — A logic analyzer can record blocks of binary information actually present in a normal digital circuit. This information can be easily stored by an analyzer on floppy diskettes or other media. The binary record of operation of a partially operable digital circuit can then be compared, bit by bit, to this stored record. The point at which the defective digital circuit logic levels depart from the recorded operation of a good circuit provides an excellent clue as to where the trouble lies. Specific details of the collection and storing of correlation information will depend upon the specific logic analyzer in use.

3.10.4.2 Identify Bad Stage Using Correlation Data — The logic analyzer itself can be used to automatically compare, byte for byte, the digital levels and timing of a normal circuit with a defective one that is being analyzed. The computer can do this quickly, providing the exact location in time where information does not exactly match. Again, the exact method of setting up a logic analyzer for this use will vary with the logic analyzer used.

The technique of comparing defective operating levels with normal levels is fast and efficient—providing normal circuit logic levels are available in a correlation file. Obviously, the programs or operating inputs for the digital circuits must be identical for both good and bad circuits.

3.10.5 IS THE CIRCUIT CONTROLLED BY ANY FORM OF SOFTWARE? — This is an important determination. If the circuit is controlled by software there is another dimension of complexity to consider: software dictates all the digital signals. Signals can be periodic squarewaves, always low, always high, usually low spiking high or vice versa, or they can even be "bad" levels. In addition, software-driven circuits usually have a third logic level involved part of the time—the open circuit (floating) state. Timing is extremely important for software-driven circuits.

If the circuit is controlled by software the reader is referred to Phase 8, "Troubleshooting Software-Driven Circuits."

3.10.6 NOTES ON DISCRETE DIGITAL CIRCUITS — The technician should at this point be working with a circuit made up of discrete logic gates. No microprocessor is involved. The circuit is designed to accomplish a single purpose and is not reprogrammable for a different use. It may have a clock circuit to keep all of the digital circuits in step with one another or it may be *unclocked* logic. Unclocked logic means that each IC responds immediately to its inputs without synchronizing to any master timing signal.

Some of these discrete digital circuits cannot be easily single-stepped to find problems in the logic circuits. Troubleshooting may have to be done using dynamic signals while the circuit is operating at full speed.

3.B Locating the Defective Stage

3.10.7 NOTES ON USING THE LOGIC CLIP AND SINGLE-STEPPING CIRCUITS — A few digital circuits run very slowly, so slow that you could attach a voltmeter to the pins of the ICs and watch the levels toggle back and forth. This would require signals that changed less than perhaps once a second. Other circuits can be *forced* to run very slowly or even be completely stopped for detailed troubleshooting. Some digital circuits that use a clock signal can be run one step at a time for troubleshooting purposes. All the technician must do is stop the clock completely. Then, by inserting a single pulse instead of the normal clock, the circuit will trip to the next state. This is called *single-stepping* a circuit.

Using this technique of single-stepping, the individual high and low signals at each pin of the ICs can be monitored. While a logic probe can help monitor the pins of an IC, it can watch only one at a time. What is needed for single-stepping is some way to watch all of the pins of an IC at once. This is done with the *logic clip*, sometimes called a logic monitor. See Figure 3–39.

Figure 3–39 A logic clip (Photo courtesy of Hewlett-Packard Company)

The clock pulses driving a circuit can be stopped completely by removing the clock IC if installed in a socket, or, if TTL ICs are used, by grounding the clock IC output stage. Figure 3–40 shows why this does not harm the TTL output stage.

While it is not hard to remove a chip or ground the clock driving a circuit, it is a bit more difficult to cause the circuit to advance one and only one step. A common mechanical switch will have a considerable amount of *switch bounce* and will inject many pulses into the circuit before settling down. What is needed is some way of putting a single pulse into the circuit. The circuit of Figure 3–41 is called a bounceless switch. When the switch is changed from one position to the other, the output of the circuit will toggle only once to the new logic state.

To use the bounceless switch just apply ground and Vcc to it and attach the output of the NAND gate to the circuit to be toggled. Each flip of the switch will change the circuit output to a new logic state.

Another possibility is to use a *logic pulser* as a signal source. This small instrument could be described as a digital signal generator of sorts. See Figure 3–42.

This instrument is constructed so that, when its control button is pressed, a single pulse (the logic level changes, then changes back very quickly) is injected into the circuit. It is designed in such a way that it will pulse a high logic level low or a low logic level high. In other words, the technician need not be concerned whether the original logic level is high or low, it will automatically be pulsed to the opposite state.

84 □ Locating the Defective Stage

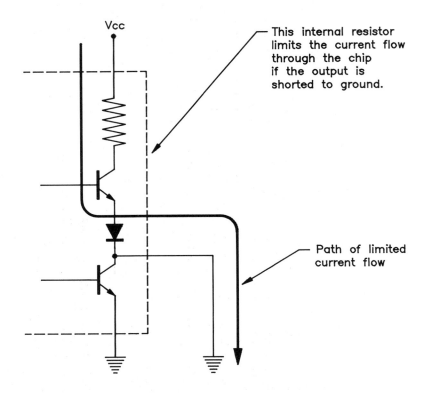

Figure 3-40 Why the output of a TTL stage can be tied to ground without damaging the chip output drives

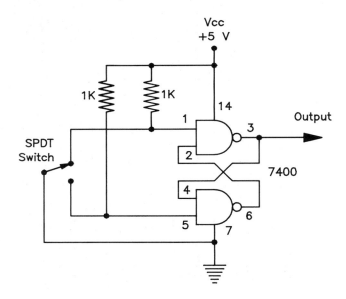

Figure 3-41 Circuit for a bounceless switch

Using a combination of a logic clip as a signal indicator and a logic pulser as a signal generator, digital logic circuits can be traced from one end to the other, pin by pin. This technique is called *static troubleshooting* of a circuit, static meaning that the signals do not change (unless you make them change).

Static troubleshooting using a logic pulser and a logic clip is probably a method of last resort. A more effective method for troubleshooting digital circuits follows.

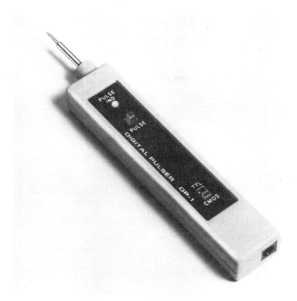

Figure 3-42 A logic pulser

3.10.8 NOTES ON USING THE LOGIC PROBE

While it is necessary to understand the circuit timing to understand how a digital circuit operates, *it is often quite sufficient for troubleshooting purposes to know merely if there is digital activity present or not.*

About 75 percent of digital IC failures involve an open circuit, either at the input or the output of the IC. The remaining 25 percent involve shorted inputs or outputs. Herein lies an important point: *Open circuits are found with voltage-based test instruments and short circuits are traced with current-based instruments.* Some thought will show why this is true: An open circuit cannot generate any current, and the short circuit cannot develop a voltage across the short. The 25 percent of digital problems that involve shorted digital circuitry is covered in paragraph 4.3.5.1.1.

For about 75 percent of the troubles that ICs cause, all that is needed is a means of looking for the presence or absence of a signal with a voltage-sensing instrument. The most convenient instrument is the *logic probe*. See Figure 3-43.

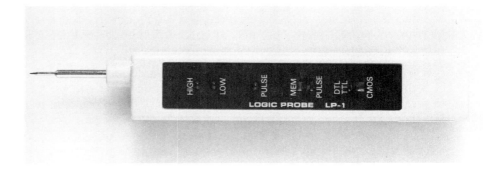

Figure 3-43 A logic probe

This little instrument is handy to carry and very easy to use. It can tell the technician enough about most situations to narrow a problem down to two of all of the ICs on the board. If a signal is not present between two ICs it could be either the transmitting or the receiving IC that is shorted and causing the symptoms. More is explained about the logic probe in paragraph 4.3.5.1.1.

Logic probe indications — The instrument has four basic indications with variations upon these. These basic indications are:

CIRCUIT ACTIVITY — One of the handiest features of most logic probes is the "pulse" or activity light. This LED will flash relatively slowly whenever the tip detects logic activity going on between logic levels. It also flashes once if the logic level changes. One of the logic probe's strongest points is that it will *stretch* a short pulse lasting for as little as 50 nanoseconds (0.00000005 second!) and create a long flash out of it, long enough for the operator to detect.

LOGIC HIGH — Another LED will light when the probe is in contact with a logic high. This LED lights with an intensity that varies with the duty cycle and the frequency of the sampled signal. This LED and the one for logic low often have a much lower capability of responding to high-frequency or short-duration signals. As a square wave goes higher than about 100 kHz in frequency, the logic probe high and low level LEDs may become progressively dimmer. The pulse LED, however, should continue to report the presence of a high-frequency signal far beyond this frequency. One probe manufacturer cites a frequency of 1.5 MHz for the pulse LED of one probe model, 50 MHz for another. The higher the frequency of response, the better the probe.

LOGIC LOW — This LED is identical to the one for the high logic level, but lights when the logic level being sampled is low.

INDETERMINATE LEVELS — When none of the LEDs of the logic light, this is an indication of a floating or disconnected line.

Important Note: If none of the 3 logic probes light, this is an indication of an open input line to this IC, between the source IC and the probe tip. The "floating" input is reported as a "bad" logic level without activity indicated.

While these indications may suffice for most trouble-shooting requirements, there are additional features sometimes offered on a logic probe such as:

IC FAMILY SWITCHING — This switch on a logic probe allows the use of a single probe on both the popular IC logic families, TTL and CMOS. This switch changes the high and low voltage levels that the probe reports via the LEDs. When switched to TTL, the probe conforms to standard TTL logic level definitions. TTL operates from a fixed voltage of 5 V. Voltages above 2.4 V are shown as logic highs and voltages less than 0.8 V are shown as logic lows.

On CMOS circuits the voltage levels are interpreted differently. CMOS circuits may operate on a supply voltage from 3 V to as much as 18 V. The logic high and low levels are interpreted as a *percentage* of the voltage applied to the probe. The probe must be using the same voltage as the circuit under test. Under these conditions, the logic high will be shown for any voltage sampled that is more than 70 percent of the Vcc, and a logic low shown for voltages less than 30 percent of the Vcc.

PULSE MEMORY — This option allows the logic probe to act as a digital "watchdog." If the probe tip is applied to the circuit under test and then the "memory" switch thrown, the pulse LED assumes a new function. The pulse LED should be extinguished under these circumstances and will remain so until a logic change occurs at the sampled point. This could be a complete, one-time change, the appearance of a normal on and off signal, or a very short pulse of only a few nanoseconds. Any change from the original logic level, whether originally high or low, will cause the memory LED to come on and remain lit. Think of the LED as a telltale sign, informing you that, even if you weren't looking, at least one logic change did occur. The pulse memory must be manually reset to indicate a second change.

SOUND EFFECTS — Some logic probes provide a low-toned beeping sound when the probe is connected to a logic low and a higher sound when connected to a logic high. Some technicians may find this helpful.

Using the logic probe — Using the logic probe is very simple. Connect the power cord of the digital probe to the 5 V supply for TTL or the 3 V to 18 V supply for CMOS. (This applies only if your probe can be used on either IC family. If not, be careful not to apply excessive CMOS voltages to a strictly TTL probe.) Provide a short ground nearby for the probe, using the lead provided. This avoids sensing the logic level through the relatively long power cord ground. High-frequency circuitry can give misleading or erratic indications without the short ground lead. Not using the short ground lead at the probe may be acceptable for slow digital circuits.

Every technician who works with digital circuits should have their own personal digital probe or one for their exclusive use. This rugged little instrument will find the majority of digital circuit problems and will prove to be an extremely handy instrument for all digital troubleshooting.

3.10.9 NOTES ON USING THE OSCILLOSCOPE ON DIGITAL CIRCUITS —

Before using an oscilloscope to troubleshoot signals through digital circuits, it may be a good idea to review paragraph 2.16.

The oscilloscope is one of the most common analog electronic test instruments, but it is often less than adequate for generic digital troubleshooting. It does not do a good job of showing the nonrepetitive waveforms of digital signals, and it is usually capable of monitoring only two signals at a time, where there may be many more signals to be monitored at once, all of which may be involved in a single problem.

There will be occasions where the oscilloscope and a digital multimeter are the only instruments available, so we shall pursue further the use of the oscilloscope on digital signals.

Oscilloscope triggering — When using the oscilloscope on digital circuits, the instrument should be set for NORMAL or TRIGGERED triggering, whichever term is used on the triggering mode control. While most analog troubleshooting is done using the AUTO mode of triggering to give a baseline in the absence of signals, using AUTO on digital signals is not recommended. This is because AUTO will produce a baseline that may interfere with slow digital circuit troubleshooting. See Figure 3–44.

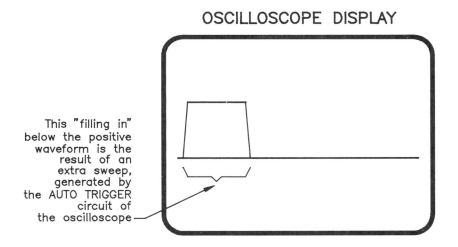

Figure 3–44 The effect of using the oscilloscope's AUTO triggering mode on digital signals that change at a rate of less than about 40 Hz

Advantages — There is one major advantage to using a dual-channel oscilloscope for digital troubleshooting over a simple logic probe. The oscilloscope can show the *timing* of two signals with respect to each other or to a common triggering event. Since timing is so important to the operation of digital circuits, this is a good capability to remember when and if it becomes necessary to establish the timing relationships of two signals.

An oscilloscope can be used as in Figure 3-45 to observe the relative timing of two different signals. The oscilloscope must be set to trigger on the input signal, which occurs first, in order to see both signals as they relate in time to one another.

The oscilloscope probe — The oscilloscope probe must be compensated if the instrument is to show the waveforms as they really exist. An improperly adjusted probe will show spikes or rounded edges on digital signals.

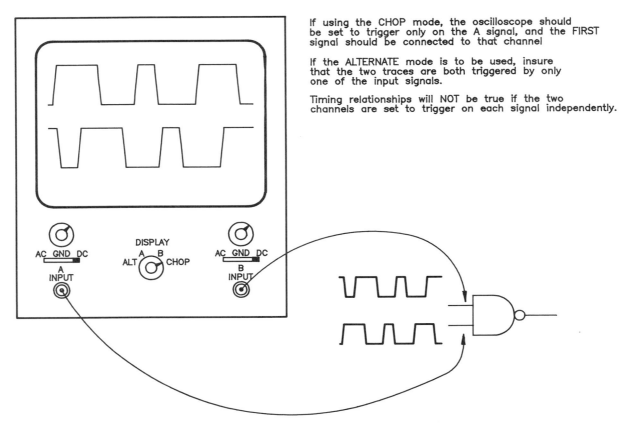

If using the CHOP mode, the oscilloscope should be set to trigger only on the A signal, and the FIRST signal should be connected to that channel

If the ALTERNATE mode is to be used, insure that the two traces are both triggered by only one of the input signals.

Timing relationships will NOT be true if the two channels are set to trigger on each signal independently.

Figure 3-45 Connecting an oscilloscope in this manner will allow viewing the true timing relationship between the two signals

The probe tip usually selected for troubleshooting is designed to open against spring pressure and clip to the test point under consideration.

It may be easier to use oscilloscope probes on ICs if an IC extender clip is used. See Figure 3-46.

Extra-small probe tips are sometimes available as a quick-change option for the basic oscilloscope probe. These are recommended when the use of the normal grip-tip of the probe carries a risk of shorting the circuitry on the PC board. See Figure 3-47.

Disadvantages — The major disadvantage to the oscilloscope is that a stable display depends upon a *periodic* waveform. If the input signal does not vary in frequency, waveshape, or amplitude, and if the waveshape occurs over and over, the oscilloscope will show a repeating pattern in the same identical position on the screen. This will appear to be a pattern that stands still.

Digital signals, however, are often *aperiodic,* meaning that they follow no repeating pattern. Figure 3-48 shows the different display that a signal of this type would produce.

After the initial triggering transition of the aperiodic signal, the time that the signal remains positive varies with each cycle of the waveform, accounting for the gradual dimming of the top horizontal portion of the trace. In a similar manner, the time that the negative portion begins varies, and so forth. The overall result is rendered almost useless.

Figure 3-46 Use the IC extender clip to get IC pin connections elevated to where you can attach instruments with less danger of shorting the pins (Photo courtesy of AP Products Inc.)

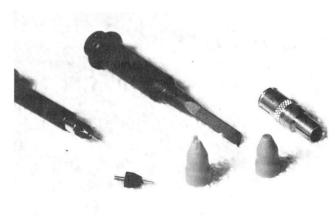

Figure 3-47 Optional tips and protection sheeves can be used instead of the standard probe tip

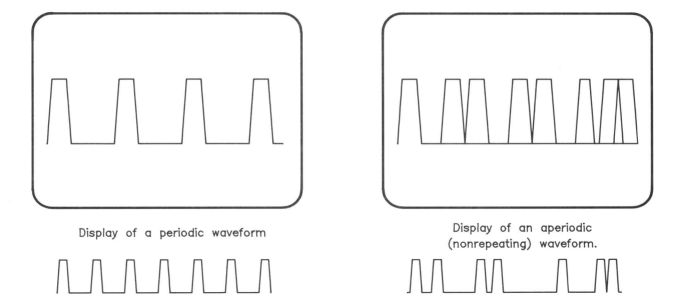

Figure 3-48 Oscilloscope displays for periodic and aperiodic digital waveforms

The oscilloscope and three-state ICs — A further disadvantage in using an oscilloscope for troubleshooting three-state chips is that the waveforms shown while monitoring them may appear to have bad logic levels part of the time. Three-state devices are usually turned on and off under the control of a microprocessor. There are times when none of the chips are transmitting data, and the bus lines may float to any voltage. The oscilloscope will show the logic levels as apparently bad, indeterminate levels. The only time the levels on a three-state bus must be at proper high or low logic levels is when data are actually being read by another chip. Levels at any other time are of no particular consequence. See Figure 3-49.

The oscilloscope hold-off control — Going a bit deeper into solving problems of using an oscilloscope on digital circuits, delaying the start of successive sweeps can help "line up" on a later portion of a waveform that *does* repeat. Magnified early portions can be seen using a special control on the oscilloscope. This control is the *hold-off* control. Varying

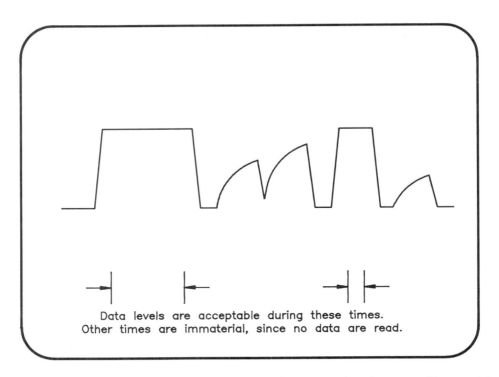

Figure 3-49 Logic levels on a three-state bus do not have to be proper unless data are read by a receiving chip

this control changes the time at which the sweep triggers to some time later than it normally would. This control will help only if you wish to see the early portions of a waveform that eventually repeats. A common hold-off time may vary from none to as much as perhaps three times the horizontal sweep time in use. See Figure 3-50.

Another problem with using an oscilloscope is that it can easily "miss" (make indistinguishable to the operator) signals of extremely short duration. Sometimes, the only clue to a technician that one of these extremely short signals is occurring is that the oscilloscope is triggering in the "normal" mode without an apparent vertical signal, yet removal of the triggering input causes loss of the trace. These very short signals could trigger the oscilloscope but not be visible because of their extremely short duration. If the input signal changes polarity and immediately back again, there will be no visible spot to show up on the CRT's phosphor to mark the signal.

Oscilloscope risetime — An oscilloscope with a lesser risetime specification will display digital signals more accurately than an oscilloscope of less capability. See Figure 3-51.

To get an idea of how much an oscilloscope might distort a waveform with a very fast change of state, it is necessary to know one of two specifications for the instrument. The risetime specification of the oscilloscope can be multiplied by five to give a figure which represents the *fastest* transition that can be shown with minimum (approximately 2 percent) distortion. Waveforms changing state at a faster rate (i.e., in less time) will appear increasingly distorted. In other words, if an oscilloscope is rated at 17.5 nanoseconds risetime, it would be a good instrument to view signal transitions of 90 nanoseconds or more. Faster risetime waveforms would not be shown accurately.

Oscilloscope bandwidth — The other specification, the bandwidth oscilloscope, can be manipulated to arrive at the risetime figure. Divide the oscilloscope bandwidth figure in MHz into the constant 350 to give an answer in nanoseconds. For an oscilloscope rated at 35 MHz, the risetime would be 10 nanoseconds. This instrument would be good for viewing waveforms of more than 50 nanoseconds.

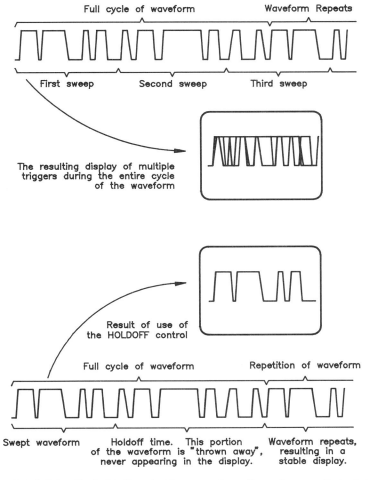

Figure 3-50 Using the hold-off control to see early portions of a repeating waveform

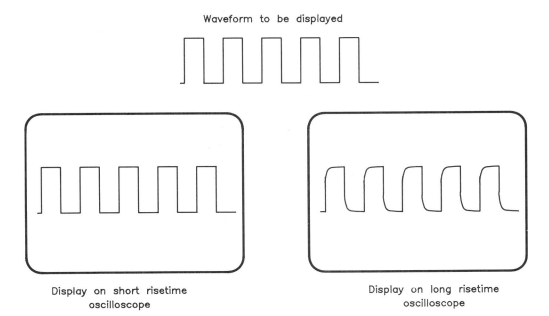

Figure 3-51 Display of a fast risetime digital waveform as seen on a short- and a long-risetime oscilloscope

3.10.10 Make Selection — Digital circuits are most often of two general types, TTL or CMOS. The way to recognize them is primarily by the marking on the chips, verified by the amount of voltage (Vcc) applied to them.

Derivations of these two basic types are also made, such as the 74H and 74F series, which are also covered in this text.

The following paragraphs give important details of all of these types and recount things to keep in mind while troubleshooting.

3.10.10.1 TTL NOTES — TTL (transistor-transistor logic) output logic levels are not 0 and 5 V. Contrary to what might be expected, the logic levels of a TTL circuit have considerable leeway before they become unacceptable. It is important to realize that a reading of 3.8 V, for instance, is a good high logic level for TTL. Even 2.5 V is acceptable as a high. See Figure 3-52.

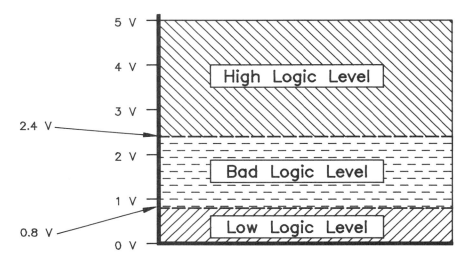

Figure 3-52 Graph of TTL logic levels

An important peculiarity of TTL circuits is that *an open TTL gate input will show a "bad" level of about 1.6 V* and will be interpreted internally to the chip as a high.

The basic family of TTL ICs is the 74XX or 74XXX series where the Xs are replaced by numbers on an actual chip. These are saturating, bipolar ICs. This family offers the greatest variety of functions. They are characterized by reasonable speed (about 35 MHz) and high power consumption (about 10 milliwatts per gate.) Special TTL chips with enhanced reliability are used by the military and are specially marked with 54XXX rather than 74XXX.

Another subfamily of TTL is the 74SXX or 74SXXX series. These are nonsaturating ICs, which gives them a higher frequency capability of about 125 MHz. The resistors used internally have about half the resistance of those in standard TTL; causing the power consumption to double to about 20 milliwatts per gate. These ICs use internal Shottky diodes.

The next subfamily of TTL is the 74LXX or 74LXXX series. These ICs sacrifice frequency capability, to typically only 3 MHz, for a decrease in power consumption to 1/10 that of standard TTL, a mere 1 milliwatt per gate. These are called low-power ICs.

A compromise was made in manufacturing yet another subfamily of TTL, the 74LSXX or 74LSXXX chips. These retain the nonsaturating feature of the Schottky series with internal resistors of five times standard TTL. These chips have frequency capabilities of about 45 MHz and power consumption of about 2 milliwatts per gate.

Later arrivals to the TTL family include the advanced low-power Schottky 74ALSXX series, having frequency capabilities to 5GHz and the power dissipation of the 74LXX series, or 1 milliwatt per gate.

The last TTL family is the 74FXX series. These are called "fast" TTL, and are pin-for-pin replacements for the 74SXX series ICs. They offer an improvement over them, however, of up to 1/4 the power consumption of that family, or about 5 milliwatts per gate.

3.10.10.2 CMOS NOTES — CMOS (Complementary Metal Oxide Semiconductor) digital chips have the considerable operating voltage range of anything from 3 V to 18 V. The acceptable logic levels for a CMOS digital signal are defined as a percentage of the supply voltage that is applied to the chip. See Figure 3–53.

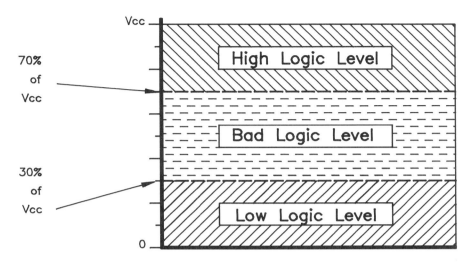

Figure 3-53 Graph of CMOS logic levels

CMOS chips are commonly labeled with a 4000 series identification, such as 4002. Motorola modifies this numbering scheme by adding a "1," labeling a standard 4005 chip a 14005.

CMOS chips must be handled with care. They are very sensitive to damage from electrostatic discharge. The fundamentals of ESD damage protection were covered in paragraph 2.0 and should be reviewed if they are new to the technician.

A peculiarity of CMOS circuits is that *an open CMOS gate will be erratic* in operation. If passing your hand within a few inches of a CMOS circuit or touching it with an instrument drastically affects the operation of the circuit, you can immediately suspect an open circuit to one of the inputs.

3.10.10.3 OTHER FAMILIES — The TTL ICs have an advantage of speed over CMOS chips. CMOS chips, on the other hand, require far less operating power than TTL. This situation has given rise to other families that have tried to marry the advantages of both and to provide chips that were directly compatible with the switching levels of both families.

The first attempt was the 74CXX series. These chips were available in the same pinouts as the 74XX series, but had the very low-power advantages of CMOS technology. Their inputs and outputs were only CMOS-compatible. As such, they were more similar to the 4XXX series, the pinouts being the major difference. Frequency capability extended to about 6 MHz.

The 74HCXX family offered higher speeds than the 74C series, up to 60 MHz. With this came a limitation of Vcc from 2 to 6 V maximum. The input levels were standard CMOS, and the outputs were Vcc minus 0.1 V for a high and 0.1 V for a logic low.

The next problem addressed was incompatible switching levels. CMOS can drive TTL when both families are operated from 5 V. But when TTL is connected to CMOS, the 2.4 V high logic level output of TTL is not compatible with the required 70 percent level (3.5 V required) for the CMOS input.

The 74HCTXX series has TTL compatible inputs and standard CMOS outputs, making them ideal interface chips between the two families. Vcc is limited to from 4.5 V to 5.5 V. Input levels are 2.0 and up for a high input, 0.8 and below for a low. The output levels are the same as given for the 74HCXX family. These chips make excellent replacements for LS TTL chips.

3.10.11 NOTES ON NONDIGITAL WAVEFORMS IN DIGITAL CIRCUITS — If the servicing technician is tracing digital signals with the oscilloscope, there are two signals common in digital circuits that will look as though they are defective when they are not. These signals come from the timing resistor/capacitor combination of a timing chip, such as a one-shot multivibrator, and any input to a Schmitt trigger gate or inverter. See Figure 3–54.

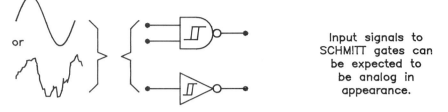

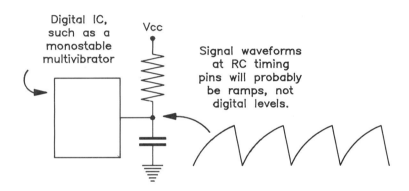

Figure 3–54 Two digital circuits that may not have digital on-off waveforms

3.10.12 NOTES ON USING THE LOGIC PULSER — The logic pulser is an instrument that looks very much like the logic probe. Refer to Figure 3–42.

The logic pulser can be used to inject a digital signal. Its principal use is to provide a digital pulsed signal for tracing signals from one stage to another. A good example is when the circuit normally has a string of pulses at the input, but the troubleshooting mockup does not provide this signal. Connect the pulser's power cord to a convenient source of power on the circuit under test. Then connect the logic pulser's short ground lead to the circuit ground and the pulser output to an appropriate point in the circuit. If a single pulse (of either polarity) is desired, press and release the button on the pulser. If a train of pulses is desired, hold down the button on the pulser. After a single pulse and a momentary delay, the pulser will then emit a continuous string of pulses. Other uses for the logic pulser are covered in paragraph 4.3.5.1.1.

3.10.13 Check for Vcc, etc. — Before getting deeply into the card, it is important to verify that the card is receiving adequate voltage from the power supply. Check that power is actually available on the card, not merely at the output of the power supply. Broken wires between them are thus easily noted.

Most digital circuits operate on a single supply voltage. Check for the presence of the proper voltage with a voltmeter. The digital voltmeter is becoming the standard instrument for measuring voltages and is recommended for this task. While at a power supply test point, switch your DVM or oscilloscope to AC and look for AC on the DC line. A reading of more than a few millivolts can suggest a problem in the power supply filter or regulator, or possibly a filtering problem on the card in question.

The Vcc line for TTL digital ICs must not vary more than 0.25 V from 5.0 V. Voltages lower than 4.75 V may cause erratic circuit operation because of the generation of "glitches," unwanted short signals caused by unmatched delay times within the chips. On the other hand, voltages above 5.25 V can not only cause glitches for the same reason, but they can endanger the chip by simple power overload. The chip can overheat and fail from this cause.

Don't forget to check the ground return for digital circuits, too. It is just as important as the Vcc line, but it is often overlooked. The ground lines for digital circuits *must* be solid and continuous. If there is any doubt about the integrity of the ground connection under a screwhead, loosen and then retighten the screw. This action will disturb any developing poor connection, and may make a change in the operation of the circuit, thereby indicating that this connection needs to be taken apart and thoroughly cleaned before reassembling.

Some circuits refer to the ground line as a "return" line. Thus, two wires labeled "+5 V" and "+5 V Ret" would actually be the plus 5 V supply line and the "ground" for that circuit. Using a separate line called "return" is a strong indication that the return line is *not* actually ground or the chassis connection. The technician must be aware of a very important fact at this point:

The ground return circuit for some equipment is not the same as the chassis ground.

Erratic or widely varying values of Vcc from that specified on the schematic is an indication that the negative lead of the voltmeter may not be connected to the proper "ground" for that voltage source. See Figure 3-55.

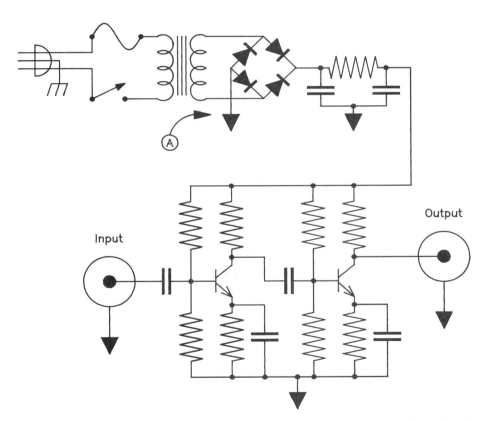

Figure 3-55 Example of a circuit where the chassis ground is not circuit common. A voltmeter for testing Vcc must be connected to point A, not to the chassis of the equipment

3.10.14 Check for Stuck or Missing Signals

— The logic probe will properly and automatically interpret voltage levels appropriate for the logic family for which it is set. This is the function of the TTL/CMOS switch on logic probes. This feature of the logic probe takes a considerable interpretive workload off the technician.

Interstage (inter-chip) signals can be rapidly sampled for activity from pin to pin on the ICs. Remember that the *IC pin numbers of the ICs goes counterclockwise on the top of the board* (the component side), where the activity reading should be taken. Readings taken at the pins of the IC on the top of the board will show poor solder joints and broken PC board traces to the pins affected. If a circuit begins working properly and suddenly when a chip pin is touched, this is a strong indication a poor solder joint is under the probe tip.

TTL inputs that are floating will measure as a "bad" level of about 1.6 V. The logic probe high, low, and pulse (or activity) LEDs will all be extinguished. Don't fall into the trap of interpreting such an indication as a voltage value of "nothing." (Nothing can mean either infinity or zero!) This pin will probably be interpreted within the chip as a logic "1," even though it is a "bad" level.

CMOS circuits that have floating inputs may be pulled high by the application of the logic probe to that gate. This is not a hard and fast rule, however, and will change with different logic probes. *Any CMOS circuit that changes state when touched by the logic probe should be suspected as having an open input on that line.*

3.10.14.1 Verify That Signal MUST Be Present
— Once an apparently missing signal is discovered, it is necessary to verify that the signal is really missing. The theory of operation, if available, is the best clue to the answer to this question. The mere presence of a constant low or a constant high signal in a digital circuit may only indicate a pin that has been deliberately tied high or low. An example would be either of the set or reset inputs of a set/reset flip-flop. If raising either of these pins to a logic high to force a set or reset is never desired in a specific circuit, these inputs must be connected to ground permanently. See Figure 3–56.

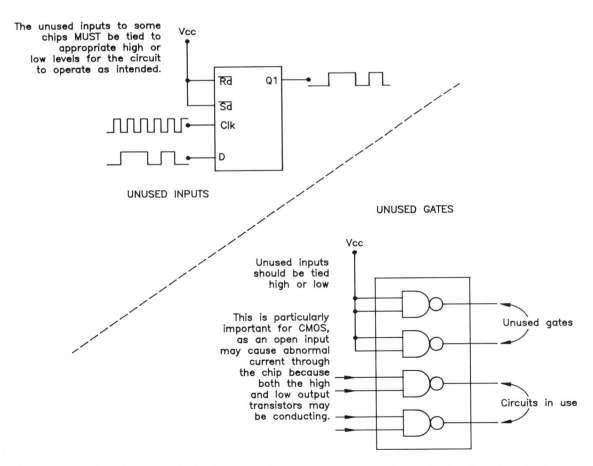

Figure 3-56 Unused logic inputs should be tied either high or low, depending on the function of the circuit

Remember that a floating TTL input will probably be interpreted internally as a logic high.

A floating CMOS input will be erratic in operation. If CMOS chip inputs are left to float to an indeterminate level, this may cause the output switching transistors to both conduct at the same time. This can cause unusually heavy current flow through the output stages. Unused CMOS inputs should be tied high or low as necessary.

Unused output lines, however, should be left open to assume either a logic high or logic low depending upon the state of the input logic levels. A good example is Figure 3–56.

3.10.14.2 IS THE STAGE ACTUALLY DEFECTIVE? — Just because an input signal is present does not necessarily mean it is the right one. The input signal could be arriving at the wrong time to be effective, causing no output from a gate in spite of good activity on all its input pins.

It is necessary to carefully apply *timing* criteria to the question of whether or not a missing output signal indicates a defective stage or not. See paragraph 3.10.4 for a review of digital timing requirements. Refer to Figure 3–35.

Overriding digital signals with the logic pulser — The circuit of Figure 3–35 could be verified as working without benefit of timing information if one input is overridden with the random pulses of a logic pulser. Force a string of pulses into one of the inputs of the gate. The unsynchronized pulses produced by the pulser will overpower the normal waveform produced by the previous stage without causing damage to that previous chip. Since the two signals at the inputs are now random with respect to each other, there will be times when the inputs to the gate are satisfied and an output will occur if the stage is good. If the defective stage is found, go to Phase 4, "Identifying the Defective Component."

3.10.15 Look for Inputs Being Ignored — If the digital circuit apparently has all necessary signals and is still not operating properly, there is a small possibility that one of the inputs to a logic chip is being ignored because of an internal failure in the IC. See Figure 3–57.

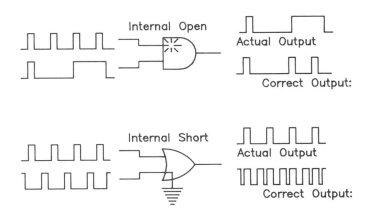

Figure 3–57 How an open AND input or a grounded OR input failure internal to a chip can cause an input to be ignored

Troubleshooting this kind of problem will require the use of an instrument that can analyze the timing relationships of all the active inputs to the chip in question. Most chips will have some inputs active, with other inputs tied high or low. The verification of the static high or low inputs can be done with the logic probe.

3.10.16 A Timing Problem is Indicated — When a timing problem is indicated, the best instrument to use is a logic analyzer. Since not all shops have one available it may be necessary to make do with an oscilloscope. These are the only practical instruments to

use in tracking down timing problems. The oscilloscope is acceptable for two input signals, and a logic analyzer for more, up to 16 or more signals at once.

Connect to the inputs and the outputs of the chip or chips in question and, with the aid of the chips truth table, determine which chip is not functioning properly.

3.10.17 ARE MORE THAN TWO INPUT SIGNALS INVOLVED?

If no more than two digital signals are involved, an oscilloscope can be used to find the timing problem. If more than two inputs are involved, however, it will be necessary to use a logic analyzer to see all of the timing relationships.

One way of looking at the relationship of three basic instruments — the digital multimeter, the oscilloscope, and the logic analyzer — is that the digital multimeter is the best instrument for static DC circuits. The oscilloscope is most useful when used on dynamic analog circuits. The logic analyzer is designed for use with the timing-critical multiple signals of digital circuits.

3.10.17.1 NOTES ON USING THE LOGIC ANALYZER —
The typical logic analyzer monitors logic levels at many points in a digital circuit simultaneously. The analyzer has many probes or pod connections that must be connected to the circuit in question at all of the circuit points of interest. See Figure 3–58.

The analyzer shows these logic signals over a relatively long period of time by capturing data in its memory much as a movie camera captures motion on film. The logic analyzer can be used only for digital troubleshooting because the input circuits respond only to the logic "1" or "0" levels, not analog values between. See Figure 3–59.

The connection of the logic analyzer starts with providing good ground connections for each of the probes or pods.

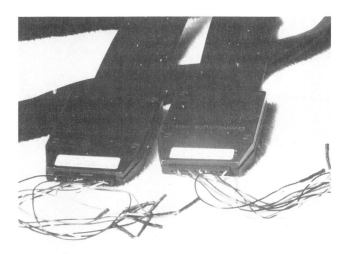

Figure 3–58 Logic analyzer pods are used to connect the instrument to the circuit under test

Figure 3–59 A logic analyzer (Photo courtesy of Hewlett-Packard Company)

Clocking the analyzer — The analyzer will take a separate "picture" of all of the logic levels at the probes at each pulse of a clock signal. The logic analyzer can use its own internal clock or the clock of the circuit under test. After the logic level "picture" is taken, the digital levels may change within the circuit under test, and the analyzer then takes another "picture" when the clock pulses again. The clock within the analyzer can be used, but if relatively slow digital circuits are being monitored, this results in recording many "frames" of the same information. This is rather like a movie camera taking more than one single-frame shot of that which is already frozen in motion. The result in the analyzer's case is that memory space is wasted in duplicate information. This situation would be encountered when troubleshooting slow circuits that do not have their own clock.

If the digital circuitry under test has a clock signal, that clock should be used to synchronize the analyzer. One of the input leads of the analyzer is designated "clock." Using the clock signal from the circuit under test will result in a single "frame" of digital logic levels for each beat of the test circuit's clock. Each time the logic levels change, a single "picture" is taken.

How the analyzer captures information — Let us assume here that the clock from the circuit under test has been selected and that all of the desired digital probes or pods are now connected between the circuit and the logic analyzer. It is time to push a button and begin gathering data.

The logic analyzer will record the logic levels at each clock pulse into its random access memory (RAM). When the last byte of memory is used, the first byte of memory is overwritten by the following byte. This makes an endless memory loop that is continuously recording the logic levels of the circuit under test.

Some way of automatically stopping this continuous recording process is needed because a human operator is far too slow to stop it at any specified time. The specified time, however, has yet to be defined.

Triggering the analyzer — An oscilloscope is triggered to begin a sweep, thus showing an event which occurs later in time from the trigger. An analyzer is triggered to stop recording data so that data can then be displayed and analyzed.

The logic analyzer can be set up to trigger a great many different ways. It can trigger on a single event, as does an oscilloscope. One of the leads to the analyzer is marked as a trigger lead. This lead may be connected to any point in the circuit that the technician wishes to use to *stop* the analyzer. This is the simplest way to stop the recording loop.

The analyzer can also be set to stop its recording loop by any combination of logic levels among the data lines it is recording. Suppose we want to know what happened at about the time the logic levels were "0" on the first probe, "1" on the second, "0" on the third, and "1" on the fourth. (This process can be extended to include all available data channels.) If we had only 4 of 16 channels connected, we would set up the logic analyzer to trigger on a "word" of xxxx xxxx xxxx 1010. Remember that binary values begin on the right, and the first probe is the least significant figure of the number. Remember also that an "x" means "don't care."

Once the trigger word is set into the logic analyzer and the button is pushed to begin gathering data, the data are constantly recorded into the memory loop. When the combination of logic levels at the probes matches the trigger word, the recording of the logic levels continues for a short time, typically an additional half of the total memory recording time, then it stops.

Using the analyzer data — Now that the digital information is recorded into the analyzer's memory, it can be displayed in many different ways. The principle way to display the information is in the timing display mode. See Figure 3–60.

The levels of logic are shown in the timing display as they relate to each other *in time*. The logic levels of each trace can now be matched with the circuit under test using the probe numbers as identification. Armed with this information, the technician can find the timing problems in very complex digital circuits. All that is needed besides this display is a thorough understanding of the timing and operation of the circuit.

Using the logic analyzer on microprocessor circuits — The logic analyzer is particularly well-suited to troubleshoot microprocessor circuits. Before it can be used with real effectiveness, however, the program operating a microprocessor must be available. The hardware will be merely following the dictates of the program, therefore the software information is a must.

Perhaps the greatest help in debugging a microprocessor problem would be the source code listing with the microprocessor codes included. This can be called the "print" or "list"

100 ☐ **Locating the Defective Stage**

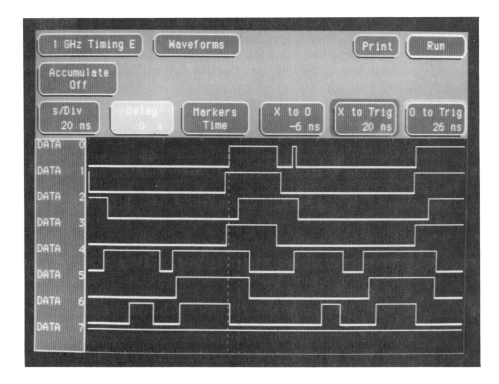

Figure 3-60 The timing display mode of the logic analyzer gives a graphical display of the logic levels encoutered by the analyzer probes (Photo courtesy of Hewlett-Packard Company)

file. One of these two files is generated by a computer when the program is assembled. See Figure 3-61.

The codes governing the operation of the microprocessor are listed down the left edge of our example in Figure 3-61. These codes can be seen on the logic analyzer as the microprocessor steps through them.

Logic analyzer summary — The logic analyzer is often used by manufacturers who wish to find problems while developing new digital circuits. While its use for finding problems in established circuits may not yet be common, it is undoubtedly the most powerful instrument available for finding complex digital timing problems. To fully utilize the instrument, it is necessary to have complete documentation on the suspected circuit, including a listing of the program.

3.10.17.2 Signal Trace with the Logic Analyzer, Find Ignored Input
— If the documentation for the bad digital circuit is available, using the logic analyzer is a good way to find the circuit that is ignoring an input. The timing display can even be printed out on a printer and studied at leisure to find which signal and, therefore, which chip is bad. The defective chip can then be deduced from a bad logic level.

3.10.18 NOTES ON USING THE OSCILLOSCOPE WITH MULTIPLE INPUTS —
If there are only two inputs to the stage under test, a dual-channel oscilloscope can be used to test the timing relationship of the input signals. Refer back to Figure 3-45.

With only two input channels, as the case of Figure 3-45, the output waveform cannot also be directly monitored on a dual-channel oscilloscope. The output could, however, be applied to the *external trigger* input of the oscilloscope, thereby using that as a sort of third input. By properly setting the polarity of the external trigger as appropriate for the circuit under test, the sweeps would be initiated. The scope will show the timing relationship of the input signals that produced the sweep. The polarity of the input signals must match the truth table for the gate in question. See Figure 3-62.

AVOCET SYSTEMS Z80 ASSEMBLER - VERSION 1.04M SERIAL #00413

SOURCE FILE NAME: DIAGS.ASM

```
                    ;
                    ; (DIAGS.ASM)
                    ;
                    ;This assembly language program will test all of the chips used
                    ; in the Microcontroller.  It is designed to rotate a "0" thru
                    ; all of the 24 output lines of the 8255.
                    ;
                    ;System Constants:
                    ;   ROM Addresses from 0000H to 07FFH.
                    ;   RAM Addresses from 0800H to 0FFFH.
                    ;   I/O Addresses - See System Equates, below.
                    ;
                    ;
1000                PORTA   EQU     1000H
1001                PORTB   EQU     1001H
1002                PORTC   EQU     1002H
1003                CONTL   EQU     1003H
0FFF                RAMTOP  EQU     0FFFH
                    ;
                    ; The main part of the program begins here:
                    ;
0000 31FF0F                 LD      SP,RAMTOP       ;Give the Micro a Scratchpad in RAM
                    ;
                    ; Initialize the 8255
                    ;
0003 210310                 LD      HL,CONTL
0006 3E80                   LD      A,80H           ;Code for all ports as outputs
0008 77                     LD      (HL),A
                    ;
                    ; Put all output lines high
                    ;
0009 3EFF                   LD      A,0FFH
000B 210010                 LD      HL,PORTA
000E 77                     LD      (HL),A
                    ;
000F 210110                 LD      HL,PORTB
0012 77                     LD      (HL),A
                    ;
0013 210210                 LD      HL,PORTC
0016 77                     LD      (HL),A
                    ;
                    ;
0017 CD3000                 CALL    WAIT
                    ;
001A 3EFF                   LD      A,11111111B
001C 210010      HERE:      LD      HL,PORTA
001F 77                     LD      (HL),A
                    ;
0020 210110                 LD      HL,PORTB
0023 77                     LD      (HL),A
                    ;
```

Figure 3–61 A "PRINT" file is the program listing and the hexadecimal codes that make up the microprocessor program

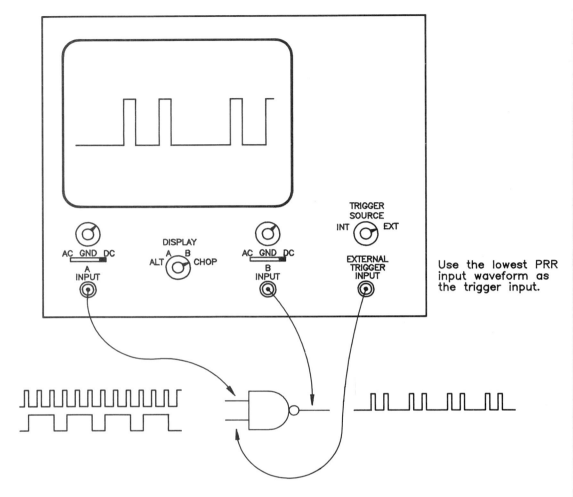

Figure 3-62 Using the external trigger input of the oscilloscope as a third signal input to the instrument

3.10.19 Signal Trace with the Oscilloscope, Find Improper Signals
— One of the input signals could be ignored because of a failure within the chip, a failure that would not show with a logic probe. Enter the dual-channel oscilloscope. By monitoring two points within the circuit under test it is possible to detect a difference between what the output *should* be, and what it *is*. See Figure 3–63. Which signal do you suppose is being ignored?

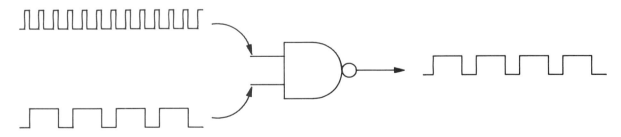

Figure 3-63 Which input signal is being ignored?

Although it may take considerable time, using the oscilloscope is a definite possibility for finding defective chips if a logic analyzer is not available.

3.11 NOTES ON PREPARING ANALOG SCHEMATICS FOR USE — Schematics in frequent and detailed use should be enhanced for more efficiency. Here are some tips that can be used to customize schematics for analog equipment.

☐ *Color code your schematics.* Color coding the lines of a schematic help to make troubleshooting faster because it is easier to follow colored lines across the schematic and there is much less chance of getting sidetracked onto the wrong line. It also enables the technician to more easily eliminate lines whose functions are fundamental. For instance, some schematics have a ground line that goes to many areas of the schematic. If this line is colored in a relatively wide, bold black, the technician can "feel" a ground whenever that color is encountered, even if it happens to be near the top of the schematic. In a similar manner, the main positive voltage for the circuits might be +12 VDC. This line should be a red line and all +12 VDC potentials colored in that color. We now have a schematic where we could expect all of the stages to eventually connect between a red and a black line.

Other lines should be colored in different colors. Highlighter pens are best, because they allow reading the original lines through the coloring without covering them. The black is, of course, an exception. Other secondary supply voltages, +5 VDC for instance, might be assigned a pink or orange color.

Signal flow is another item that is worthwhile to add to schematics. Signal flow is best done with the wide highlighter pens often used to mark over whole lines of text. The flow of signals need not be marked exactly through specific components. This type of marking is particularly helpful when several signals mix to produce a third, or when a signal splits into two others. Color combinations show mixing and splitting very well, helping to make the circuit functions that much clearer at a glance.

Be sure to include a key to the color coding somewhere on the schematic.

☐ *Mark the schematic with normal voltages.* Mark normal DC voltage measurements on the schematic. A small line and arrow can be used to refer the reader to the proper point on the schematic. Remember that DC readings are normally recorded *without* signals being present in the circuit. This is because signals frequently upset the DC readings by an amount somewhat related to the amplitude of the AC. It is better to eliminate this variable entirely. Put an appropriate note to this effect somewhere on the schematic.

The biasing, emitter, and collector voltages of transistor stages are particularly helpful when troubleshooting. The normal outputs of power supplies are also very important to know while troubleshooting.

☐ *Unusual waveforms.* Special waveforms, such as those often found in pulsed circuits, should be recorded at critical test points on the schematic. After drawing a likeness of the normal waveform, note the vertical amplitude at the peak and any other important points and the sweep time per division. Including these items is important to duplicating the original monitoring conditions for troubleshooting comparison. When recording these normal readings *be sure* that both the vertical and horizontal channels of the oscilloscope are always switched to the "calibrated" position.

☐ *Record normal resistance readings.* It may be a good idea to record normal resistance readings from various points to ground. For instance, the resistance from the Vcc line to ground might come in handy when working on a suspected shorted card. Note the polarity of the ohmmeter, the range, and the model of the ohmmeter on the schematic. An ohmmeter will often give two entirely different resistance readings depending upon the polarity of the test leads. This is caused by the many semiconductor junctions used in today's circuits. These junctions change resistance over a very wide range with the amount of ohmmeter current used. Since the range in use and even the model of the ohmmeter can affect the current used, these items should also be on the schematic along with the reading.

☐ *Solid state tester patterns.* These patterns are very helpful on both digital and analog schematics. Normal patterns may be shown for many points in the circuit. It would probably be best to use a ground for the common reference point unless the waveform is specifically needed otherwise, such as in a "floating" circuit. Phase 6 explains this use of the solid state tester in detail.

3.12 REVIEW NOTES ON USING THE OSCILLOSCOPE — The oscilloscope is the most valuable generic instrument for troubleshooting analog circuits. It can show many characteristics of a changing waveform: amplitude, frequency, period, wave shape, and even phase relationships. If the technician has not yet read "Notes on Using the Oscilloscope," paragraph 2.16, that material should be read before going further.

3.13 *Use Established Test Points* — Review the information in paragraph 2.15 in regard to using the test points provided by the manufacturer.

3.14 NOTES ON TYPICAL ANALOG STAGES

Stage Functions — There are several different general types of analog stages, each analyzed and operated differently.

The processing stage takes a single input signal and does something specific to it. Typically, the stage might amplify the input waveform in voltage or power. Audio amplifier stages and intermediate radio frequency stage amplifiers are examples. Input and output signals should be modified by normal action of the circuit, according to the purpose of the stage. See Figure 3-64.

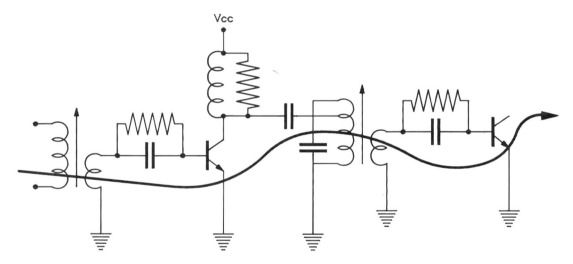

Figure 3-64 A typical processing (amplifying) stage

The mixing stage combines two or more inputs to produce a third output. The radio frequency mixer stage of a radio receiver is an example. Both inputs to a mixer stage must be present, one of which is sufficient to drive the mixing device (a diode or transistor, usually) into its nonlinear operating area, where the mixing action takes place. The output should be as specified for the stage. Refer back to Figure 3-20.

Oscillating stages take the pure DC supply voltage and originate a signal that may be analog or digital in nature. Testing an oscillator consists of checking its output for proper amplitude, frequency, and sometimes waveform. See Figure 3-65.

Signal splitting stages take a single input and generate two or more signals from it. A common example is the audio phase splitter used to take a single input signal and from it produce two similar signals at the outputs, each 180 degrees apart from one another. An example is the phase splitter circuit preceding a push-pull audio output stage. The push-pull stage requires these two signals for inputs. See Figure 3-66.

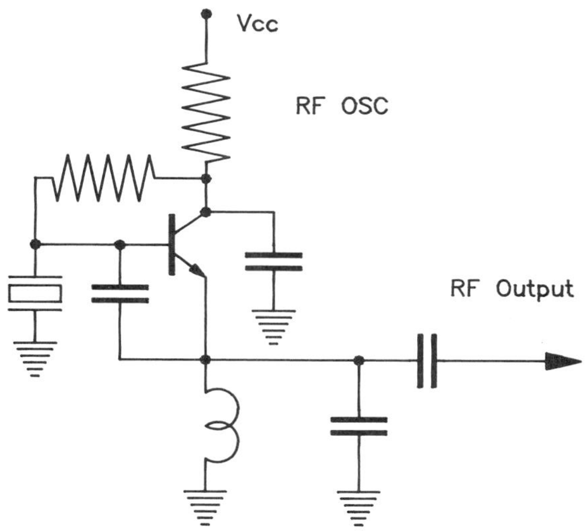

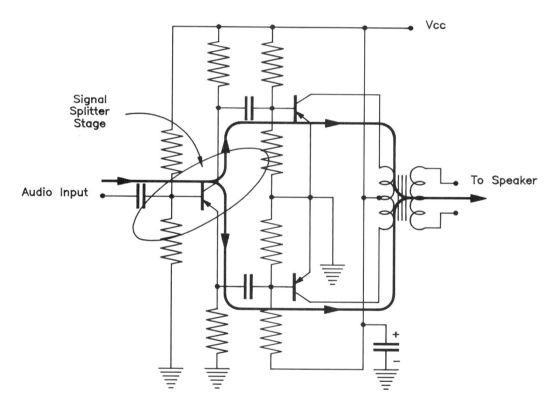

Figure 3-65 A typical oscillating stage

Figure 3-66 A typical signal splitting stage

Classes of operation — Amplifiers are classified according to how many degrees of a full 360-degree sine-wave input signal appears in the output signal. Different classes of operation have distinct characteristics and advantages. The choice of which to use is determined by the requirements of the circuit.

THE CLASS A AMPLIFIER — The Class A amplifier can be described as being half conducting without a signal being applied. Since it is in this halfway state, the incoming signal can shift the operation to a more or a less conducting state. This allows the amplification of all parts (360°) of the input signal and results in the best circuit to use for minimal distortion of the signal from input to output. Abnormal distortion in a Class A amplifier is due either to overdriving the input or by improper biasing of the transistor. The Class A amplifier is used in all small signal RF amplifiers, audio amplifiers, and for DC amplifier circuits. See Figure 3–67.

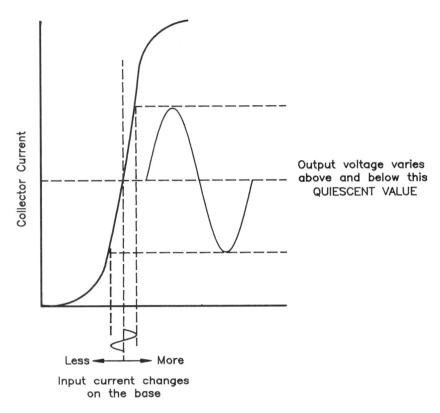

Figure 3–67 Graph of the Class A amplifier stage input and output voltages

THE CLASS B AMPLIFIER — The Class B amplifier operates at a point where it is just barely turned on without a signal being applied. Typically, the Class B stage amplifies about 180°, or half of the input signal. The proper polarity of input signal can then cause a large amplitude of output and great deal of power output from the stage. The power amplification of the Class B stage is its most important feature. When distortion of the input signal is of concern, two of these stages can be operated together with one of them amplifying one polarity of the incoming signal, and the other the remaining polarity. The two outputs must then be recombined to reproduce the original waveshape. The Class B amplifier is used mostly in power audio amplifiers. See Figure 3–68.

THE CLASS C AMPLIFIER — The Class C amplifier has an output only when the input signal reaches a sufficiently large amplitude to bring it into conduction. This stage amplifies less than 180°, half, of the input signal. This results in the most power efficient operation, but one which can greatly distort the waveshape of the input signal. The Class C amplifier is used when output waveshape distortion is not a concern, such as in non-critical pulse amplifiers and in radio frequency circuits where the required sinusoidal waveshape is restored by a flywheel circuit consisting of inductance and capacitance. See Figure 3–69.

Transistor amplifier configurations

There are three basic amplifier configurations. Each has advantages, and the technician must know what to expect when encountering them in working circuits.

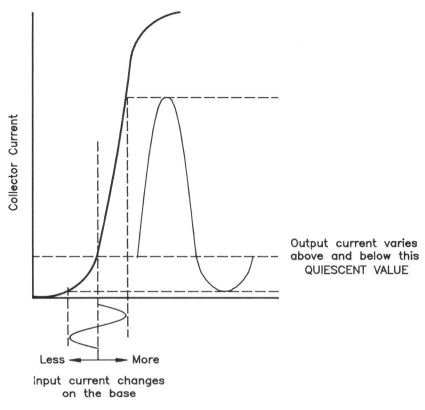

Figure 3-68 The Class B amplifier begins with very little output circuit current flow and amplifies only half the input signal

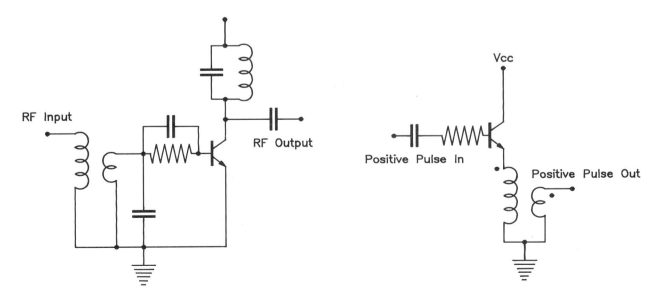

Figure 3-69 Two examples of Class C operation: RF and pulse amplifiers

The common emitter circuit is the most common circuit configuration for the bipolar transistor. It is characterized by high gain, both in voltage and in current. The output is reversed in phase from the input. In other words, a positive-going input results in a negative-going output voltage, and vice versa. It has a relatively high input impedance. See Figure 3-70.

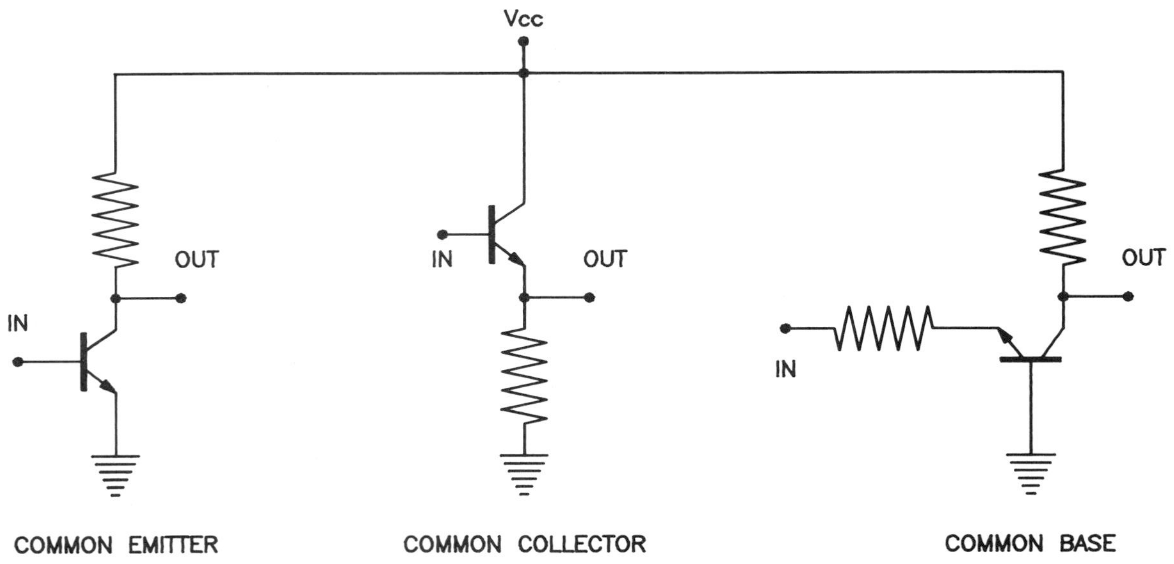

Figure 3-70 The three transistor amplifier configurations

The common collector circuit is used to obtain an increase of current from input to output. It does not provide any voltage gain at all. The circuit also does not produce a phase shift. It features relative high input impedance and very low output impedance. One common use is in voltage regulators as a constant-voltage source, since the voltage at the emitter is dependent upon the base voltage, not upon the voltage available at the collector. See Figure 3-70.

The common base configuration provides a large power gain. The input impedance is low, requiring considerable power to drive it. Voltage gain is very good but the input current is not amplified. This circuit is sometimes used as a constant current source, since the output current depends upon the bias current, not the supply voltage available at the collector. See Figure 3-70 again.

All these transistor configurations can be used with the biasing schemes of the various classes of operation, A, B, or C. To determine the configuration of a circuit, eliminate both the input and output elements of the transistor. For example, if the input is on the base and the output from the emitter, it is a common collector configuration.

3.15 Make Selection
— Analog circuit tracing can involve DC, AF, RF, or pulsed signals. Each of these stage-by-stage tracing techniques involves unique tips and tricks. These are covered in the paragraphs that follow.

3.15.1 NOTES ON TRACING SIGNALS THROUGH DC AMPLIFIERS —
Analysis of transistor amplifier circuits will require a thorough understanding of what happens when the voltage and current change through the circuit. Some simple rules will help the technician trace what happens in the three transistor configurations:

1. The **Common Emitter** and **Common Base** amplifiers:
   If the base of a transistor is pulled away from the emitter (more toward its collector), the transistor will reduce the collector-to-emitter resistance (conduct more heavily). If the emitter is pulled away from the base (and from the collector), the transistor will reduce the collector-to-emitter resistance (conduct more heavily).

2. The **Common Collector** amplifier:
   Whatever the base does, the emitter follows, volt for volt, with increased driving current.

DC amplifiers fall into three categories; the unbalanced amplifier, the balanced amplifier, and the push-pull balanced amplifier.

Unbalanced (single-ended) DC amplifiers — The input to an unbalanced amplifier is a voltage referenced to the circuit's ground and of a single polarity (usually positive). The output of the amplifier is always a voltage somewhere between the supply voltage and ground. An example of this kind of amplifier is the looped amplifier circuits of a discrete voltage regulator. Each stage of the amplifier is operating at somewhat fixed DC voltages, all of which are referenced to the common or ground point of the circuit. A change at the input of the amplifier string (in the case of the voltage regulator this would be a change of the output voltage sensed by the regulator) is reflected in shifts from previous DC values to new DC values throughout the stages of the regulator.

Troubleshooting unbalanced amplifiers can be done by supplying a suitable DC input signal by using a potentiometer voltage divider or a variable DC power supply. By varying the DC input voltage to the circuit, defects of the amplifier stages can be traced with a DC voltmeter. Since the DC amplifier can have a very high gain, operation of the stages under this condition of a dummy input can be expected to vary the final output voltage very quickly from full-on to full-off status. Note that the voltages within the unbalanced amplifier circuit do not attain zero or ground potential in normal operation. See Figure 3–71.

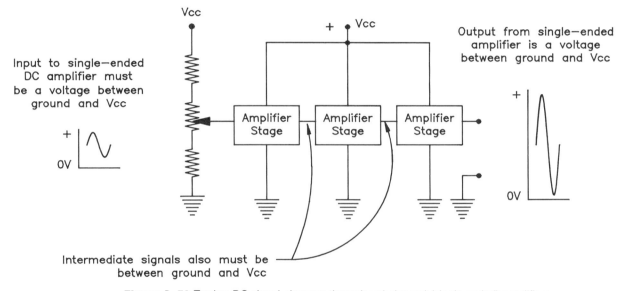

Figure 3–71 Tracing DC signal changes through unbalanced (single-ended) amplifiers

Balanced DC amplifiers — The balanced amplifier is given that name because it is "balanced" around zero volts (ground) as a neutral or quiescent condition. The input voltage to a balanced amplifier is referenced to the circuit ground and can vary above and below that point, positive or negative. The output of the balanced amplifier can also vary above and below ground, following the input signal with amplification of the input voltage level. There may or may not be a 180-degree phase shift from input to output. See Figure 3–71 (A).

Troubleshooting the balanced amplifier is relatively easy. Simply short the input to ground, and the output should measure very near zero volts. Some balanced amplifiers have an internal adjustment to insure that under grounded input conditions the output is exactly zero volts. This adjustment is often called a DC balance adjustment.

If the output voltage does not approach zero volts with the input shorted to ground, the defective stage can be isolated by opening the signal path between stages and applying a short to ground at the input side of the short. All of the following stages should now balance out to zero voltage if they are functioning normally. Move the shorting point until the defective stage is isolated. See 3–71 (A) for an example of this method of troubleshooting.

Push-pull balanced DC amplifiers — Push-pull balanced amplifiers are sometimes used in special high-precision amplifiers such as the vertical and horizontal deflection amplifiers of an oscilloscope.

110 □ **Locating the Defective Stage**

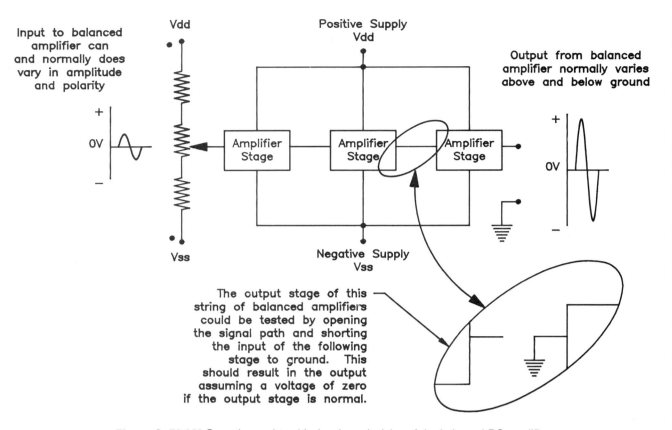

Figure 3-71 (A) Operating and troubleshooting principles of the balanced DC amplifier

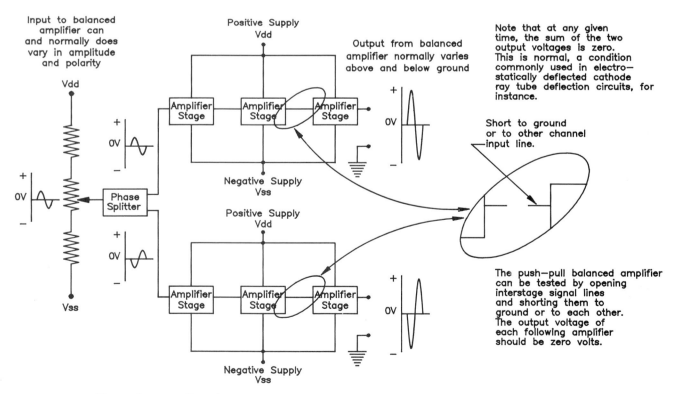

Figure 3-71 (B) Operating and troubleshooting principles of the push-pull balanced DC amplifier

These amplifiers have an input referenced to ground, and can handle input signals with both positive and negative polarities. This signal is immediately split into two signals, 180 degrees out of phase. One of these two signals is used to operate a standard balanced amplifier whose output goes above and below ground. An almost identical circuit is also driven with the phase-shifted signal. Thus, the outputs of the two circuits vary 180 degrees from one another. When one output goes positive, the other goes negative, and vice versa (see Figure 3–71 B).

Troubleshooting the push-pull balanced amplifier involves a technique similar to that given for the balanced amplifier. Zero volts into the input of the entire string of stages should result in both amplifier outputs to be very near zero volts, subject to any internal adjustments for balancing each of them. A defective stage can be isolated by opening interstage lines and shorting one input to the other, or by shorting each input to ground. See Figure 3–71 B.

3.15.1.1 NOTES ON TYPICAL DC AMPLIFIER STAGES
— DC amplifiers are almost always Class A amplifiers. The complementary-symmetry amplifier is used for low distortion power amplification.

The Class A common emitter amplifier — The Class A amplifier in the common emitter configuration was shown in Figure 3–70. The common emitter configuration is the most common configuration used in DC amplifiers, but common collector and common base are occasionally used too.

When troubleshooting these amplifiers with a DC voltmeter, it is important to know the common emitter amplifier circuit thoroughly. See Figure 3–72.

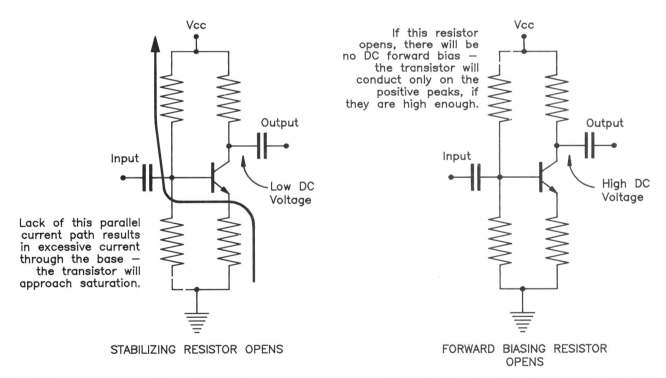

Figure 3–72 Effects that can be expected in a common emitter transistor amplifier circuit if either of the biasing resistors open

Biasing of a transistor is all important, and the sources for the bias should be thoroughly understood. The resistor in parallel with the base-to-emitter junction serves to stabilize and hold the voltage at the base constant should the transistor base current characteristics change slightly with variation in temperature. Since resistors commonly open rather than short, be sure to understand the effect on the bias if either of the biasing resistors should open. If the forward-biasing resistor should open, the transistor turns completely off, and the collector assumes the voltage of the supply. On the other hand, if the stabilizing resistor should open, there would be an excessive current flow through the base circuit and the transistor collector would conduct abnormally heavily, dropping the collector voltage to near zero.

The complementary-symmetry amplifier — The complementary-symmetry stage is commonly used in power DC amplifiers and in some high-power audio output stages. See Figure 3-73.

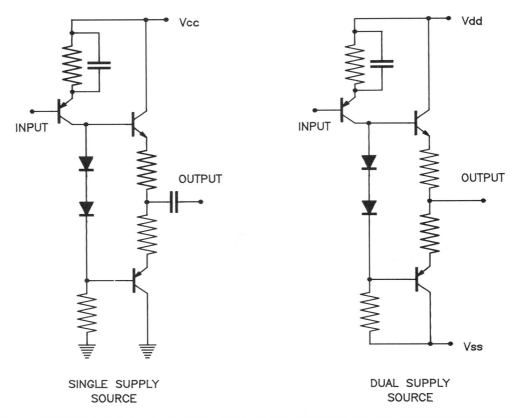

Figure 3-73 The complementary symmetry stage provides power gain and often uses two power supplies

The complementary-symmetry stage can be operated from two voltage supplies, one above and another an equal amount below ground. When supplied with power this way, the output voltage can go either positive or negative with respect to ground. This circuit is used in balanced amplifiers.

This circuit normally operates by having both of the transistors conducting Class B, both conducting just a little. An incoming signal will turn only one of them on, the opposite one off.

Failure patterns for the complementary-symmetry amplifier include the failure of both of the transistors in the output stage, either by opening or shorting. Since they are directly across the power supply with little resistance in series to limit current, if one transistor should short for any reason the next signal polarity to cause the remaining transistor to conduct may cause that transistor to fail due to the abnormally heavy current flow through it.

The Darlington amplifier stage — An amplifier stage might also use transistors in pairs with the collectors tied together. See Figure 3-74.

This circuit is ultra-simple and provides very high gain compared to a single transistor stage. The stage acts like a single transistor of very high gain.

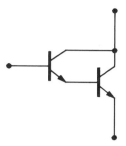

Figure 3-74 Transistors connected together in this manner are called Darlington Pairs

Multiple stage failures — One of the problems that the multiple stage DC amplifier may exhibit is the simultaneous failure of several of the semiconductors. This is because of the design of the amplifier—an overload at one point is amplified and constantly and relentlessly held as an excessive input at the next stage, very possibly causing a failure of that stage, too. The DC coupled complementary-symmetry used in audio amplifiers and the preceding stages seem very prone to this type of multiple failure.

3.15.2 NOTES ON TRACING SIGNALS THROUGH LOW-FREQUENCY AMPLIFIERS

— Test input signals for audio circuits can often be the signals normally used in the circuit. A stereo amplifier, for instance, can conveniently use the tuner or tape deck as an input for signal tracing purposes. If these inputs are not available, an audio signal generator can be used. The usual frequency for signal tracing in audio circuits is about 1000 Hz.

The ground side of the oscilloscope probe is connected to the chassis of the oscilloscope, which in turn should be connected to the neutral of the power line plug. Because of this, the oscilloscope should be used *only* on grounded circuits or circuits where the application of a ground will do no harm, such as battery-operated equipment and equipment that has a transformer in the power supply to provide power line isolation.

Audio and low-frequency signals can then be traced from stage to stage using an oscilloscope, the standard instrument for this job. As the stages amplify the signal from the input to the output, the overall amplitude can be expected to increase in voltage unless there is deliberate attenuation of the signal in the circuitry along the way. In addition to detecting missing and low-amplitude signals between stages, the oscilloscope is particularly good for showing moderate to severe cases of audio frequency distortion. Small amounts of distortion, however, can be difficult to detect using the oscilloscope. Refer back to Figure 2-35.

3.15.2.1 NOTES ON TYPICAL AUDIO & LOW-FREQUENCY STAGES

— The majority of interstage signal tracing on audio equipment involves only a few basic circuits. Variations on these circuits may account for 95 percent of the discrete circuits in common use today. See Figures 3-72 and 3-73 for examples. Note that the oscilloscope probe ground is understood to always be connected to the circuit common.

3.15.3 NOTES ON TRACING SIGNALS THROUGH RADIO FREQUENCY AMPLIFIERS

Tracing RF signals through small-signal amplifiers — The signal needed for tracing through small-signal RF amplifiers can be produced externally by a signal generator or by using available RF signals from an antenna and those signals normally within the circuit. The choice of which signal source to use will depend upon the sensitivity of the detecting instrument at the other end, usually an oscilloscope. If the sensitivity of the oscilloscope is marginal, remember that a signal generator can produce a much larger signal for troubleshooting than on-air signals for a receiver.

114 □ Locating the Defective Stage

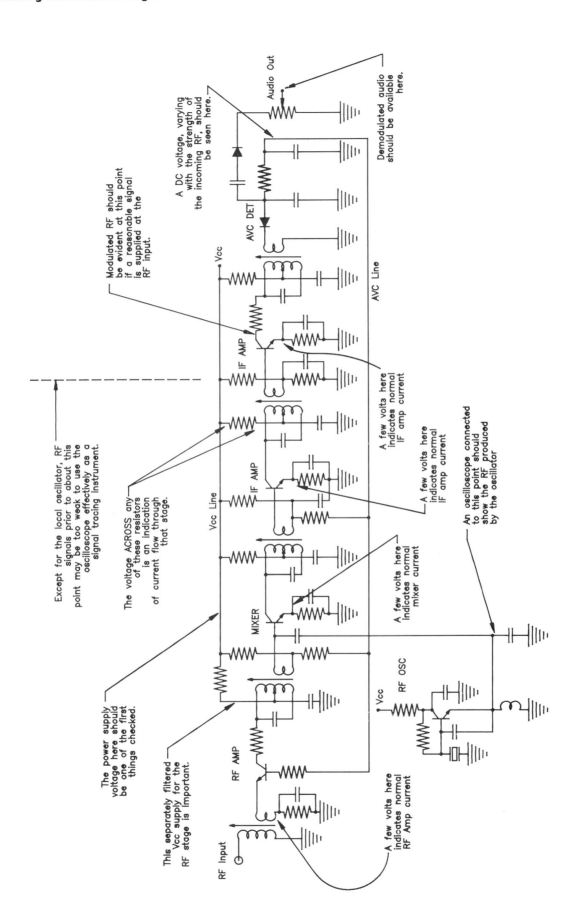

Figure 3-75 Typical test points for stage-by-stage signal tracing

The first few stages of a radio receiver must be operated Class A. They will generally have high gain and are intended to be used with very small signal inputs. Using an oscilloscope to try to trace signals in these early RF stages is not effective because of the extremely low voltage signal levels and the de-tuning effect of the oscilloscope probe when used in RF-tuned circuits. See Figures 3–75 and 3–76.

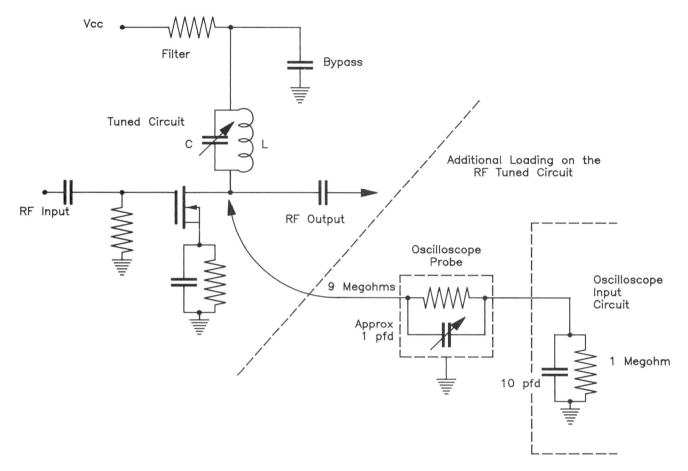

Figure 3–76 How the oscilloscope probe affects RF amplifier stage tuning

Even if the oscilloscope is connected across the output of an RF amplifier and a relatively large signal is injected into the circuit using a signal generator, the presence of an RF signal on the oscilloscope doesn't necessarily mean that the stage between them is operating. It doesn't take much leakage capacity from the input to the output circuit to couple high-frequency voltages right "through" a bad amplifier stage. Thus, being able to get a signal through an RF stage is not conclusive evidence that the stage is operating normally.

3.15.3.1 NOTES ON OSCILLOSCOPE RESPONSE — Radio frequency stages are not often directly tested for operation with the oscilloscope. The scope and/or probe can cause detuning of the circuit under test due to the internal capacity of the probe and the scope. Always use the proper oscilloscope probe. Connecting the circuit to the input of the instrument without a probe loads the circuit with about 10 pf of capacity and a megohm of resistive loading. This may not be much capacity, but it can severely detune high-frequency circuits. This loading effect can be reduced by using a times 10 (×10) probe. The probe will reduce the capacitive loading to about 1 pf and the resistance will increase to about 10 megohms.

The oscilloscope has definite limitations in regard to use with high-frequency signals. The scope's vertical amplifier is the major limiting factor on the highest frequency which the instrument will show accurately as far as *amplitude* is concerned.

If an oscilloscope has a bandwidth specification of 35 MHz, for instance, that instrument will accurately show sinusoidal waveforms of frequencies up to about half of this specification. The exact correction factor could be obtained from a graph, such as shown in Figure 3-77.

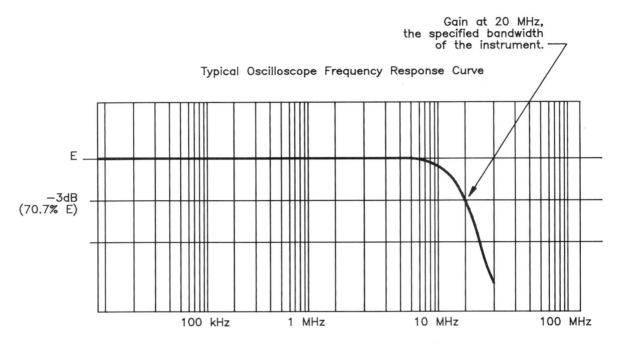

Figure 3-77 Graph of the frequency response of a typical oscilloscope

The important thing to note about this specification is that in the example of 35 MHz, the instrument does *not* show an accurate amplitude at this frequency. The bandwidth specification is that frequency at which the vertical amplifiers are severely affected enough to show the waveform at only 70.7 percent of its corrrect amplitude. As a constant amplitude input is increased in frequency above this value, the instrument shows increasingly "shorter" signals. See Figure 3-78.

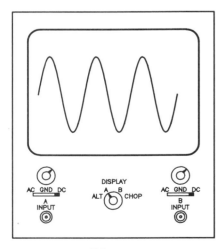 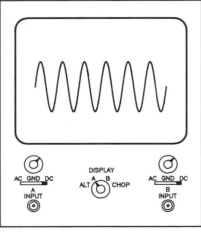 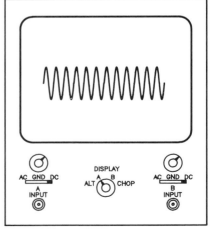

Below Bandwidth Frequency / At Bandwidth Frequency / Above Bandwidth Frequency

Figure 3-78 Effect of viewing equal amplitude signals of higher frequency than the oscilloscope bandwidth

If accurate voltage measurement is important, this frequency response matter must be taken into account as the bandwidth frequency limitations of the oscilloscope is approached.

For most troubleshooting, the oscilloscope is used as an instrument to indicate whether the expected signal is there or not. Using an oscilloscope for routine troubleshooting is not a critical operation and most adjustments of the instrument can be approximate.

3.15.3.2 NOTES ON TYPICAL RF STAGES — An oscilloscope cannot be expected to directly detect the very small signals present in the earliest stages of a receiver. These voltages are in the microvolt region, and the oscilloscope seldom shows signals less than a millivolt.

There is a partial solution, though. Both AM and FM receivers may provide a test point whose DC level will vary greatly with the "microscopic" RF signal input. The voltage present at this line is the *result* of the operation of all of the amplifiers ahead of it. Failure of any of the stages ahead of this monitoring point will cause a large decrease in the amount of DC output voltage with a given RF input level.

In the AM receiver, the automatic volume control (AVC) line voltage varies in approximate proportion to the incoming signal. Refer to Figure 3–75. This voltage may be either positive or negative with respect to ground, depending on the design of the receiver.

The FM receiver may have a test point or possibly two in the last IF amplifier stages. These are special amplifier stages called "limiters." The last stage or two of the IF strip can have so much RF signal fed into it that they are allowed to rectify, producing a DC voltage that varies with the strength of the incoming signal. Strong incoming signals may saturate the last stage completely. Further increases in signal begin to produce a limiter voltage at the previous stage. This gives two consecutive DC voltage indications of relative signal strength, one (the last stage) for weak signal inputs and the other for stronger inputs.

The AVC line or the limiter test points can be monitored to show the overall performance of all of the RF stages ahead of the monitoring point. This DC value will increase with the amount of RF at the input of the circuit. To use these test points, there should be a specified amount of AVC or limiter voltage for a specific RF input voltage stated in the specifications for that string of amplifiers. *If it takes an excessive amount of RF at the circuit input to give the specified DC voltage output, there is a problem in one or more of the stages between the test point and the antenna input. If any problem is indicated, the next challenge is to find which of the RF stages is causing the problem. Each of the stages must be checked as outlined beginning with Phase 4, "Identifying the Defective Component."*

Tracing RF signals through large-signal amplifiers — The large-signal RF amplifiers probably have enough RF to be easily seen on an oscilloscope, providing the oscilloscope has sufficient *bandwidth* to observe the waveform. RF signals with a frequency less than the bandwidth specification of the oscilloscope will have at least 70 percent of the proper amplitude, usually quite sufficient for troubleshooting.

The RF large-signal amplifier is operated Class C because of the increased efficiency and the availability of signals strong enough to bring the circuit into conduction. In so doing, the emitter resistor of the amplifier provides the DC (actually filtered pulses of RF) that is needed to evaluate the circuit operation. See Figure 3–79.

Connect a DC voltmeter to measure the voltage across the emitter resistor. As the amount of RF feeding the stage increases or decreases, the DC voltage reading will follow. Check the DC readings against those in the equipment specification to see if the stage is operating normally.

If the emitter of a Class C RF amplifier is connected directly to ground, the circuit can still be checked for normal operation by measuring the amount of voltage drop across a series filter resistor. The voltage drop across this resistor is a measure of the amount of current flowing through the stage just as the emitter resistor showed. The only difference is that the voltage measured is across the filter resistor, not referenced to ground. This is no problem at all for a digital voltmeter since it does not need to be referenced to chassis ground.

118 □ **Locating the Defective Stage**

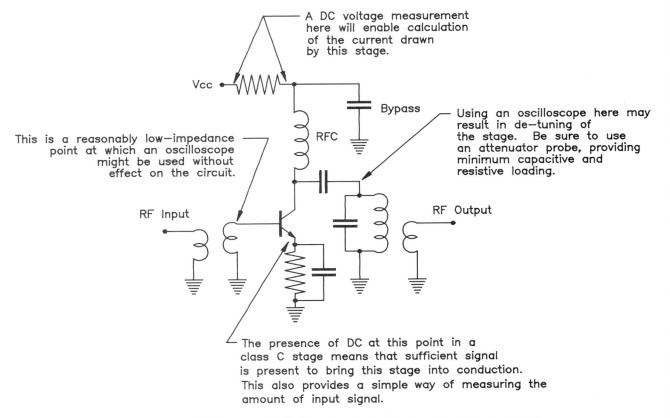

Figure 3-79 Typical large-signal RF Class C amplifier with test points for troubleshooting

RF Oscillators — RF Oscillators are often monitored for proper operation by measuring the amount of DC bias produced by their oscillations. The more powerful the oscillations, the more bias is produced. See Figure 3-80.

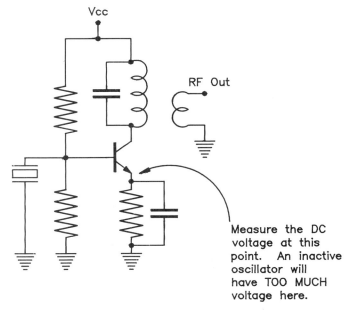

Figure 3-80 Measuring the bias of an oscillator is an indirect way of measuring its activity. Normal operation causes the oscillator to be biased in a direction to reduce current flow

The oscillator bias measurement is used for evaluating the performance of an oscillator because the DC meter does not affect the circuit operation. This is particularly beneficial at higher RF frequencies where only a few picofarads can seriously detune a reading taken across a tuned circuit.

3.15.4 NOTES ON TRACING SIGNALS THROUGH PULSE AMPLIFIERS —
The signals flowing from one pulsed stage to another contain three additional factors to consider over sinewave signals: waveshape, signal polarity, and timing. Where RF signals can tolerate distortion in many circuits and audio circuits generally regard distortion as bad, pulsed circuits often deliberately distort the waveshape.

Pulse waveshape considerations — The waveshape of a pulsed waveform may or may not be critical, depending upon the application. Television circuits use many non-sinusoidal waveforms whose waveshape is critical. On the other hand, the pulse used to initiate the firing of the horizontal or the vertical sweep circuits can be a simple, short pulse needing only a steep wavefront to trip the circuit. Schematics used frequently by a technician should include waveshapes in those applications where the waveshape is critical. If the manufacturer shows a waveshape, that waveshape is probably critical to the operation of the circuit.

Signal polarity considerations — Pulse circuitry most often uses Class C amplifiers for efficiency. If there is no signal, the amplifier draws no current. Only when a signal is present does the stage come into conduction.

Only one polarity of input signal will bring a Class C amplifier into conduction, thereby providing an output signal. Inputs of the opposite polarity serve only to confirm the cut-off condition of the stage, providing no signal output. See Figure 3–81.

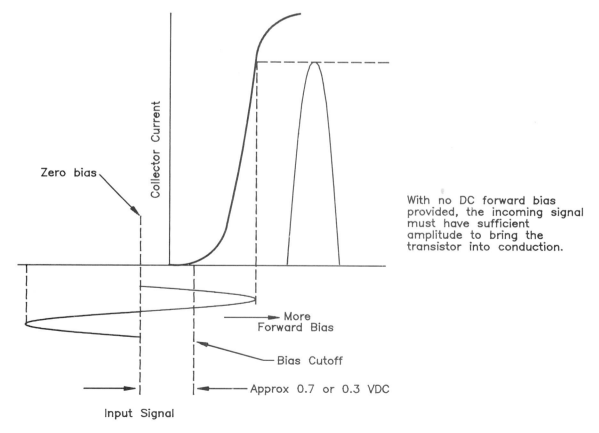

Figure 3–81 How the Class C stage responds to input signal polarity

Transformers also are used with Class C stages to couple the pulse from one stage to the next. A transformer can provide either a 0° or 180° shift of phase from input winding to output winding. The choice depends on how the windings are wired with respect to each other. See Figure 3-82.

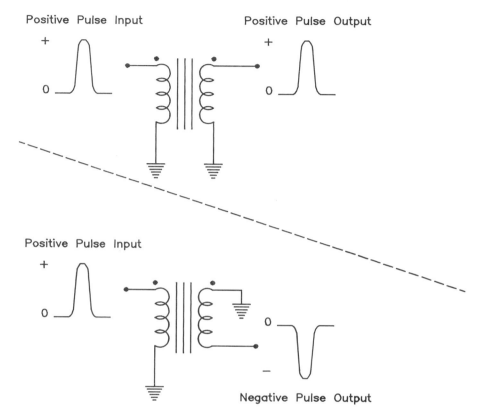

Figure 3-82 How changing the ground on the output winding of a transformer changes the output signal polarity

Since pulse stages are sensitive to signal polarity, the technician needs some means to identify the phasing of the transformers used with these circuits. This is done by marking like polarities of the windings with a phasing "dot." See Figure 3-83.

Phasing dots are important when replacing these transformers and when reading the schematics to understand the signal flows. It is necessary to note the position of the phasing dots when removing a defective pulse transformer and to replace the transformer with a new one phased in a similar manner.

The phasing dots are also important when tracing signals through pulsed circuits. When the primary winding is subject to a voltage of a specified polarity, the secondary windings of like marking will assume the same polarity. See Figure 3-83.

Signal timing considerations — Some pulsed circuits are still used where timing of two or more signals with respect to one another determines the operation of the stage. In such cases, the instruction manual should explain the timing relationships of the signals that are necessary for the circuit to operate normally. Such circuits are best analyzed using an oscilloscope, which will show these timing relationships.

3.15.4.1 NOTES ON PULSE CIRCUIT TERMINOLOGY — A review of the unique terms used in pulsed circuits is important to the technician's understanding. See Figure 3-84 for an explanation of some of the more important terms.

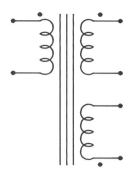

If this pulse transformer is excited by a pulse the following outputs could be expected:

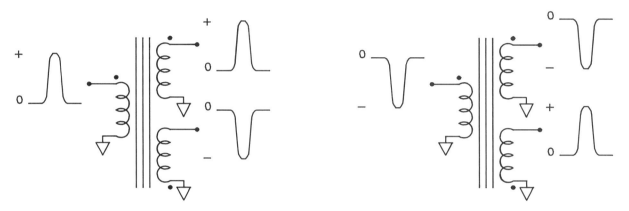

Figure 3-83 Pulse transformer polarity markings and instantaneous polarities

A term often used with pulsed waveforms is *duty cycle.* Duty cycle, in mathematical terms, is the amount of on-time divided by the total time of the cycle. When the result is multiplied by 100, the result is a percentage.

$$DC = \frac{\text{On-Time}}{\text{Total Time}} \times 100$$

For example, if you worked a total of 8 hours out of every day, the on-job time is 8 and the *total cycle time* before repeating is 24 hours. Thus, 8 ÷ 24 = 0.3333. Multiply this by 100 to obtain a working duty cycle percentage of 33.33%. Be careful not to make the common mistake of dividing the off-time into the on-time. This would result in an erroneous calculation of the above problem as 8 ÷ 16, giving the result that you worked 50% of your time!

Cautions — The oscilloscope is truly valuable for troubleshooting pulsed circuits. Only the scope can show the fine detail of waveforms and the differences between what really is there as opposed to what should be there. However, there are some special cautions the technician should observe.

There are voltage limitations to both the oscilloscope probe and at the oscilloscope input connector. It is appropriate to mention this at this point in the troubleshooting sequence because pulsed circuits can generate very high voltages in spite of a relative low supply voltage. These high voltages are produced by the collapsing field of an inductor whose exciting current has been suddenly interrupted. Although these voltages last for a very short time,

122 □ **Locating the Defective Stage**

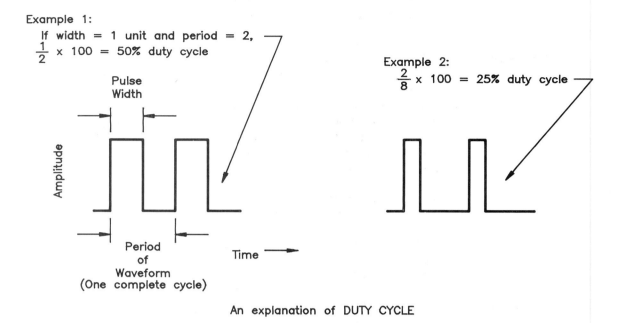

An explanation of DUTY CYCLE

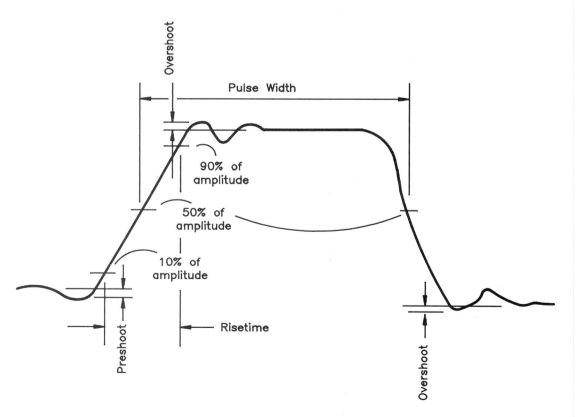

Figure 3-84 Terms commonly used with pulsed waveforms

they may be of sufficient amplitude to damage test equipment. The common 10× oscilloscope probe usually has a voltage limitation of no more than 500 V. A spike of voltage of 1000 V can easily puncture the insulation of the tiny capacitor in this probe, rendering it useless until repaired or replaced. The input of the oscilloscope itself, without the protection of a probe, may be as little as 150 V. Before connecting the oscilloscope or probe to a pulsed circuit be sure the voltages encountered cannot damage the instrument.

3.15.4.2 NOTES ON TRIGGERING THE OSCILLOSCOPE IN PULSED CIRCUITS

Where AF and RF circuits have recurring, periodic signals (each cycle repeated immediately without change from succeeding cycles), pulsed circuits often have a pulse of activity followed by a relatively long time of inactivity. Using the usual AUTO triggering function of the oscilloscope may result in a double pattern. Refer back to Figure 3–44.

This double pattern can be avoided by using the NORM or TRIGGERED mode instead of the AUTO mode. The AUTO mode will always give a baseline even when there is no signal. If the input pulses are less than a given amount (typically 40 pulses per second) an extra sweep is automatically initiated by the AUTO circuitry. Observing a pulse occurring only 20 times a second with the auto function on will make an extra baseline in the display. Switching to NORM and setting the trigger amplitude and polarity at the proper values will allow a sweep to be initiated only when triggered by the incoming pulse.

Pulsed waveforms can have very fast risetimes. Be sure to review *risetime* and *bandwidth* in paragraph 3.10.9 before using an oscilloscope for precision readings on pulsed waveforms that have very fast transitions.

3.15.4.3 NOTES ON THE ULTRA-HIGH VOLTAGE PROBE

Televisions and computer monitors use a flyback transformer circuit to produce the high voltage necessary to operate the cathode ray tube. See Figure 2–24.

This voltage is uniquely high, and there is a unique instrument to measure it. See Figure 3–85.

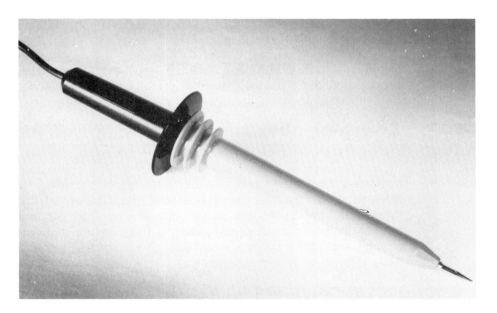

Figure 3–85 The ultra-high voltage probe used to measure cathode-ray tube high voltages (Reproduced with permission from the John Fluke Mfg. Co., Inc.)

The only safe way to measure these high voltages on CRTs is with the Ultra-High Voltage Probe. Be sure to connect the negative lead of the voltmeter to a solid chassis ground *before* touching the probe to the high voltage. Failure to do so will raise the entire voltmeter case to thousands of volts above ground.

There are two principal dangers when working with cathode ray tubes (CRTs): One is implosion, with the resulting danger of severe cuts if the CRT is broken; the second danger is posed by the extremely high voltages used with CRTs.

Implosion is a danger because the CRT has been evacuated of almost all air. This creates an inward pressure of about 14 pounds per square inch over the entire area of the CRT. While under this stress, any fracture of the glass can cause the entire glass area to break up and smash inward, disintegrating even more as the pieces strike one another. The pieces can then be thrown many feet away from the CRT, constituting a severe hazard to a person. The inner coatings on the glass of a CRT can also be dangerous if they enter the bloodstream through a cut. The second danger is high-voltage shock. Accelerating voltages of a CRT can be as high as 20,000 V or more, depending partially upon the size of the CRT. While the high voltages are not especially life-threatening because the current capability is usually quite low, the reaction to getting "bitten" can cause injury. The reaction to a shock can involve a cut on a sharp chassis edge when a hand is jerked back.

3.15.4.4 NOTES ON TYPICAL PULSED STAGES — Pulsed stages operate Class C, similar to RF large-signal amplifiers. However, they do not usually have LC output circuits to make the output sinusoidal.

Remember that pulsed circuits are sensitive to the polarity of signals. For example, if it takes a positive pulse to turn on a transistor, a negative pulse will not affect the circuit at all. *Be conscious of signal polarity as you troubleshoot in pulsed circuits.*

Pulsed circuits often take advantage of the charge and discharge of capacitors through resistors to determine timing and waveshape generation. See paragraph 4.12 and, in particular, the section that reviews using capacitors and resistors, and inductors and resistors, in pulsed circuits.

Finding the bad stage in a string of pulse amplifiers should be no problem. The interstage voltage levels are quite high enough to use an oscilloscope. Just look for any deviation from the normal waveforms in the circuit. Watch for *sufficient pulse amplitude and waveshape* very close to that in the documentation for that point in the circuit. When a pulse amplifier gives a problem, it usually just stops working: the input pulse may be correct but the output is missing entirely.

3.16 FIND THE BAD STAGE?
— If the defective stage has been positively identified, proceed to Phase 4, "Identifying the Defective Component." If it has not been identified, there is another possibility that may help to isolate the defective stage.

3.17 Consider the Degree of Original Problem; Perhaps Alignment May Correct
— Here is a case where the judgment of the technician comes very much into play. The degree of the original problem should be considered: Was the problem a complete failure, or is the equipment *almost* working properly? The closer the equipment comes to normal operation, the more likely it can be corrected by a simple internal adjustment or two. Radio frequency and high gain DC amplifier circuits are particularly likely to degrade in performance when slightly out of proper adjustment. Pulsed circuits, digital circuits, and audio circuits are less likely to need such adjustments and therefore are not as likely to have alignment adjustments.

3.18 NOTES ON USING A FREQUENCY COUNTER
— The frequency counter is a simple instrument to use, but its limitations must be kept in mind to prevent damage to the counter or inconclusive readings.

The maximum allowable input signal to a counter can be as low as 2 V. The input to some counters goes directly into an amplifier, which can be easily damaged by overvoltage. Other, more forgiving counters may be fitted with a signal attenuator that will give some degree of protection against input overvoltage. The general rule when using frequency counters is to use the minimum amount of signal input that will produce a stable count. A stable count is one that is consistent from one sample to the next, varying only in the last one or two digits.

Be sure to check the specifications of the counter that you will be using for its maximum input voltage limitation.

Frequency counters must be used on signals that do not vary in average amplitude. An amplitude-modulated radio signal is an example of a signal with varying amplitude. Only the RF cycles above a certain level would be counted.

3.19 Check for Proper Frequencies Involved — Radio frequency amplifiers often depend on L and C resonance for proper operation. Anything that can disturb the resonance of such a circuit can easily cause that stage to become inoperative. Yet, since resonance is a property evident only at a specific frequency, detecting these problems is difficult with most instruments.

A tuned radio frequency amplifier stage, for instance, may work quite well with the proper frequency. If the wrong input frequency is applied because of a malfunction somewhere else ahead of the stage, it may result in a very weak or nonexistent output from an otherwise good stage. The same symptoms would result, too, if the stage was out of resonance with the input signal because of a defect in the tuned circuit. Unless the frequency input to tuned RF stages is checked, the symptoms could be very frustrating. If the input frequencies are proper, then something causing de-tuning of the circuits is a likely possibility.

If the stage causing the problem can be identified, proceed to Phase 4, "Identifying the Defective Component." If not, go back to the start of Phase 3 and begin again, more carefully this time.

3.20 WAS IT A FREQUENCY PROBLEM? — If the frequencies coming into the circuit are not correct according to the frequency counter, the frequency problem must be resolved. Because of the range of possible problems that might meet this criteria, the cure can only be summarized here. The source of the frequency feeding the circuit in question should be examined to find out why the wrong frequency is being applied. This could be an LC or RC problem in the frequency-determining elements of the frequency source.

Obviously, the operating frequency must be correct for an RF stage to operate properly. If the incoming frequency can be changed to see if the circuit will operate normally at an *improper* frequency, and it does, then off-resonant operation of this stage is confirmed. Again, this is often caused by a bad circuit board connection, but capacitors, resistors, and inductors also fail. Replacing the frequency-sensitive elements of the circuit under test—L, C, or R—should correct the problem in this case.

3.21 Verify that Circuit is Still Bad — If the problem seems to come and go while trying to track it down, it may be an intermittent situation. Intermittents are categorized and covered beginning with paragraph 1.2.

☐ Phase 3 Summary

At this point the technician is sure of the defective stage on the circuit board. The tests and notes given in this phase can be reviewed and the tests made a second time to be absolutely sure that the suspected stage is indeed the bad one. Unless the technician is *sure* that the bad stage has been *definitely* identified, these tests and checks should be made again. A great deal of time can be wasted by taking further troubleshooting action based only upon a "feeling" that a particular stage is "probably" the cause, only to find out later that this was not the case at all. The next step is to look into the techniques available to find the particular component in this stage that is actually causing the malfunction. Phase 4 continues our progress in troubleshooting to the component level.

Review Questions For Phase 3

1. Discuss the types, advantages, and disadvantages of power supplies that may be available to the technician.
2. What would the terminals of a power supply marked "+ Sense" and "− Sense" be used for?
3. Discuss the difference between "+", "−", common, ground, and chassis as they relate to bench power supplies.
4. When are extender boards necessary?
5. What is the difficulty in troubleshooting looped circuits?
6. What is the key to troubleshooting looped circuits?
7. What is the purpose of, and a typical value for, a DC blocking capacitor for use at audio frequencies?
8. Name the principal parts of equipment documentation, such as a maintenance manual, and the uses of each. Bring samples of instruction books and evaluate the thoroughness and ease of use of each.
9. How is the parts layout diagram used during troubleshooting?
10. When comparing interstage signals, you find that the good unit has no signal at a specified test point. The bad unit does. Does this indicate the good unit is defective?
11. You must repair a given circuit within two weeks. No documentation is on hand at all. You do not know how the circuit works or what it does, therefore you cannot apply power. What alternatives do you have to troubleshoot it?
12. Lab exercise: Obtain several different PC circuits of preferable 12 components or less. Draw the physical layout diagram and schematic for each.
13. Collect and maintain a library of databooks for digital and analog integrated circuits, transistors, and related components.
14. List the improvements you could make to digital schematics to make them easier and more efficient to use.
15. Draw a schematic of how a NAND gate could be used as an inverter.
16. Name two reasons why two inverters might be cascaded.
17. The truth table of a NAND gate with inverting inputs is the same as the truth table for which other gate? Why would one gate be drawn over another if the truth tables are the same?
18. Name the three types of digital IC output stages.
19. What very important factor of circuit operation is not obvious by inspection of a digital schematic?
20. The school for a specific piece of equipment will probably center around what portion of a maintenance manual?
21. Figuring out a schematic and reasoning out how it works without having the theory of operation for the circuit is called _____ the circuit.
22. Look up "correlation" in the dictionary. How might the paragraph 3.10.4.1 be reworded to make the concept clearer for you?
23. Discuss the differences between circuits using software and those using only logic gates.
24. On what kind of circuits is the logic clip most useful?
25. Why cannot a simple mechanical switch be used to provide the 1's and 0's of a digital input signal?
26. What commercially available instrument can provide a single digital pulse needed to troubleshoot a specified stage?
27. How many different indications can the logic probe show, and what are their significance?

28. What are the most effective instruments in troubleshooting open-circuit type failures? Short-circuit type failures?
29. Explain why the oscilloscope is not effective in showing aperiodic waveforms.
30. What would be the fastest rise time signal that could be viewed with acceptable (<2%) distortion on an oscilloscope with a 50 MHz specification?
31. An oscilloscope with a poor rise time specification will distort digital signals. How will these signals appear?
32. What is the minimum voltage that a TTL chip will interpret reliably as a logic high?
33. What is the maximum voltage that a TTL chip will reliably interpret as a logic low?
34. What is the significance of a voltage level of 1.6 V as read on the input of a TTL logic gate?
35. How will a disconnected input pin to a TTL gate be interpreted?
36. CMOS logic levels are defined as a percentage of what voltage?
37. What is the principal characteristic of an open circuit to a CMOS input?
38. Discuss IC gate families of the TTL and CMOS families, including those meant for interfacing the two families.
39. You see a sawtooth waveform on a digital chip, a multivibrator. Is the oscilloscope showing what is really there?
40. A schematic indicates two lines into a circuit, one labeled "+12 V" and the other "12 V Return". What does this mean?
41. At a specified point, you have a digital signal. Does this mean that the signal is normal?
42. You have determined that a particular signal is not timed properly for the circuit to operate. Where would you look in the circuit for the problem, in relationship to this point?
43. What chip defect could cause an input to a chip to be ignored?
44. What are the two principal setup items for a logic analyzer?
45. List the improvements you could make to analog schematics to make them easier and more efficient to use.
46. Name the common types of analog stages.
47. Discuss the differences between an input sinewave and the output for classes of operation A, B, and C.
48. Class A or B bipolar transistor operation is characterized by a resistor connected from the _____ circuit to the source voltage for the _____ circuit.
49. What class of operation uses no forward-biasing resistor?
50. What class of operation is used for large-signal amplification of RF signals?
51. What class of operation is used for small-signal amplification of RF signals?
52. What class of operation is characterized by minimum distortion?
53. What class of operation is most used in pulse-amplifier circuits?
54. One class of operation is efficient because it draws no current in the absence of a signal. What is this class?
55. What amplifier circuit is common for high-powered audio amplification?
56. How is the class of operation for a given transistor determined?
57. What is the peak amplitude in volts of an RF signal if the oscilloscope has two divisions from top to bottom of the sinewave, the frequency of the signal is 30 MHz, and the oscilloscope bandwidth specification is 30 MHz? A 10× probe is being used and the vertical amplifier is set for 0.2 mv/div.
58. Can you expect to see incoming RF signals in the early stages of a radio receiver on an oscilloscope?
59. The common signal strength test point for an AM receiver is called the _____.
60. The common signal strength test point for an FM receiver is called the _____.

61. Where is the problem located if you are testing an AM receiver and find that it takes 50 microvolts to produce an AVC reading of 2 V, and the specification says that it should take 5 microvolts for this output?
62. Where is the best place to measure the performance of a large-signal bipolar RF amplifier stage?
63. How may the amount of activity of an oscillator be measured?
64. What class of operation is most common for pulsed stages?
65. The output winding of a transformer can provide a ____° or a ____° phase relative to the input signal.
66. If a soldering gun is rated at 20 percent duty cycle and is turned on for 20 seconds, what time should be allowed before using it again?
67. What is the definition of risetime?
68. What is the definition of pulse width?
69. What test instrument shows the true shape of waveform?
70. The AUTO triggering should not be used on signals of less than _____ because of the double baseline that will be produced.
71. Name the two dangers of the cathode-ray tube.
72. What signal characteristics should the troubleshooting technician pay particular attention to as a signal is traced through pulsed circuitry?
73. If the equipment for repair is working almost as good as it should, what should the technician consider first as a possible cure?
74. The input voltage limitation of a frequency counter can be as low as _____ volts in some cases, thereby warranting care in applying unknown voltages.
75. What might the technician suspect if the circuit works well after doing nothing to correct a problem?

PHASE 4
IDENTIFYING THE DEFECTIVE COMPONENT

☐ **PHASE 4 OVERVIEW**

Once the technician identifies the stage causing the problem, the next step is to find the specific component within that stage that has failed. Many components interconnect to perform the functions of each stage, so isolating the defective components requires more detailed work before the job is done. Special techniques and specific generic instruments are used for component-level troubleshooting.

4.0 Look for a Visible Fault — The defective stage that has been identified is only a small area to inspect for visual problems. Look over the card very carefully, on both sides. Look for small solder splashes, ICs with a slightly raised dimple in the center, obviously overheated parts, or bad solder joints. Components that have physical stresses on them in normal operation, such as connectors, will often develop bad solder joints. See the example of Figure 4–1.

Another problem that is common is the broken PC board trace. This, too, is often caused by mechanical stressing of the board. If such a fault is visible, the cure is to solder a bridge over the break. A "splint" can be made of a short piece of wire, soldering it firmly to the board on both sides of the break.

4.1 Review Block Diagram of Stages — Reviewing the block diagram for the PC board in question will help the technician understand the function of the defective stage. *The operation of the stage involved must be understood before troubleshooting to the component level can be efficient.* The timing of the various inputs and the resulting outputs of the stage are of particular importance.

If there is no stage-by-stage block diagram, it may be possible to deduce the stage-by-stage operations by carefully examining and interpreting the schematic drawing.

4.2 IS A SCHEMATIC OF THE BAD STAGE AVAILABLE? —
At this point, it is necessary to have a schematic, at least of the stage which is suspect. This diagram will help the technician understand the various functions of the components in the circuit and estimate normal DC voltages and signals to compare with the voltages and signals actually measured.

If no schematic is available, refer to paragraphs 3.9.2 and 3.9.3 to review what is required to draw a schematic from the actual circuitry.

4.3 IS THIS A DIGITAL CIRCUIT? - Analog and digital circuits are analyzed differently when troubleshooting to the component level. Integrated circuits are themselves stages. ICs can contain a great many transistors and resistors, but we cannot go inside them to repair the problem. We replace the whole "stage" as one unit.

Digital circuits most often fail because of the lack of a signal— a "stuck" signal line somewhere on the PC board. The problem is usually found by looking for a missing signal

130 ☐ **Identifying the Defective Component**

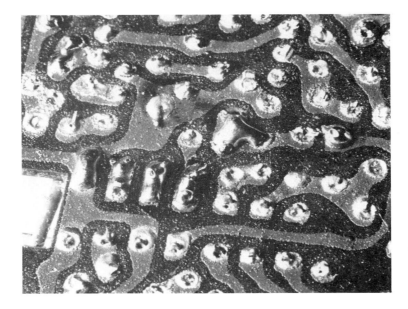

Figure 4-1 Magnified example of poor solder joints, the probable result of mechanical stress on the connection and/or poor workmanship

and curing it. Analog failures, on the other hand, are most often found by the shifts they produce from normal DC or signal amplitude readings.

If the bad stage is an analog circuit, go to paragraph 4.4. Otherwise, continue here with the digital circuit troubleshooting alternatives.

4.3.1 Check for Proper Grounds and Voltages to the Chips —
A circuit cannot be expected to operate without proper supply voltages. It is simple, quick, and easy to check for proper supply voltage to the chips. Don't rely on a digital probe to give a good reading of Vcc, however. It would indicate a healthy high if the supply voltage to a TTL chip was only 2.5 V. Use an accurate voltmeter and check for voltage on the component side of the board. Checking for supply voltage and a good ground connection from the top of the board also checks the condition of the board-to-IC solder connection. The supply line for a TTL chip should be 5.0 VDC, + or − 0.25 V. A reading of more than a few millivolts from the power supply ground to the ground pin of an IC is an indication of a poor or open connection from the pin to ground.

A digital signal line that is found to be stuck low—at a 0 voltage—is another good reason to look for the proper supply voltages to the chip that drives that particular line.

Positive power supply voltages are commonly labeled Vcc. The negative pin is sometimes called ground and sometimes Vss. Other circuits besides IC chips may use Vdd for the positive supply voltage and Vss or Vee for the ground.

4.3.2 ICs OR COMPONENTS IN SOCKETS? —
It is less expensive for a manufacturer to produce circuits without sockets. Because of this, the technician may not have the option of simply unplugging an IC and replacing it. To reduce the risk of potential board damage due to soldering operations, it is advisable to track down exactly where the problem lies. Do not desolder a component unless reasonably sure that it is indeed the bad one.

If the technician is fortunate enough to be working on circuitry that has provided sockets for all of the ICs involved in the failure, it is a quick and easy matter to replace either or both of the suspected chips, source and load, to see if that cures the problem. See Figure 4-2.

4.3.2.1 Replace Source and Load Components —
Use care when replacing components in determining which is defective. The substitute parts must be good ones. Replacing doubtful components with other doubtful components is a waste of time.

Figure 4–2 If the ICs of the suspect stage are in sockets, they can be easily replaced without a lot of measurements

Be sure the original pin #1 or other locating indicator, if applicable, is oriented the same when replacing the part. Polarized components, such as transistors, electrolytic capacitors, and ICs, are unforgiving if inserted into a circuit backward.

There is always a risk that the replacement part might be damaged if a second component failure has caused the original part to fail. Additional checks should be made before applying power to eliminate this possibility. This is particularly important when the components are expensive or difficult to obtain.

4.3.2.2 Retest Circuit — The card will have to be put back into operation to prove the repair, at least enough to check the stage in question. If only one IC has been replaced with no success, the other of the two ICs should be replaced. This should cure the problem if your troubleshooting has been correct, and the card will be ready for normal operation.

4.3.2.3 Timing Problem Indicated. Look for Problem in Earlier Stages — If replacing both of the suspect ICs does not cure the problem, then the technician should redefine the bad stage again, starting at the beginning of Phase 3. If the stage in question still does not operate properly, this indicates that the *timing* of the circuit signals may be at fault. The problem must then be traced to its source in earlier stages.

4.3.3 Determine If a Source or Load Problem — The terms source and load refer to the functions of the chips or discrete devices involved. One of the ICs or discrete devices *puts out* a signal to others. This would be a "source" component. The ICs or other components that take this signal *in* for further processing are "load" devices. The paragraphs to follow discuss variations on this basic source-load concept and the possible variations of problems that can occur.

4.3.4 REVIEW NOTES ON USING THE LOGIC PROBE — Review paragraph 3.10.8, returning here when completed.

There is a simple procedure to analyze digital problems. *A great deal can be told about a stuck line between ICs by first using the logic probe.* Take a reading at the source *and* at the load IC, being sure to take the readings on the pins of the IC, not on the circuit traces. See Figure 4–3.

Each TTL totem-pole output stage has a transistor for pulling the output high and another transistor for pulling it low. Only one is "on" at a time, except during the very brief

Figure 4-3 Preferred method of taking digital signal activity reading. Reading the pin rather than the circuit will reveal circuit board trace problems

time when switching from one logic state to another. Refer back to Figure 3-32 for an example of the internal TTL output stage.

4.3.5   Select Circuit Category — Troubleshooting techniques vary with the kind of digital circuit in use. ICs are most often directly connected to each other. In other circuits, ICs are interconnected to transistors, diodes, and optical isolators. Circuits that involve ICs and components other than ICs may not use standard logic levels. Troubleshooting interface circuits between ICs and other components will require instruments other than the logic probe. See the individual category of circuit for the fine details of troubleshooting, paragraphs 4.3.5.1, 4.3.5.2, and 4.3.5.3.

4.3.5.1   NOTES ON IC-TO-IC STAGES — Most digital signals are passed from one digital IC to another digital IC. These signals are considered here first. A typical output and input stage is shown in Figure 3-32.

A review of logic probe indications, paragraph 3.10.8, may be beneficial at this time.

Notice that using two transistors in the output stage of a digital IC allows a forced high logic level output or a forced low. See Figure 4-4.

There are a multitude of possible failures that can occur with this arrangement. See Figure 4-5.

Probing a digital circuit with a logic probe will likely produce one of the following situations.

1. If the source IC output pin indicates *the signal is being generated by the source but is not reaching the load or input pin* of the next IC, then there is an open in the PC board trace between the ICs. See Figure 4-6.

    In TTL circuits with an open trace, the input pin will show no activity on the "pulse" LED and a "bad" digital level where neither the high nor low LEDs on the digital probe will light.

    In CMOS circuits, the input pin will probably be very sensitive and will act erratically, and it may switch logic states when the probe is touched to the circuit.

    Tracing the open is easy. Simply progress along the trace from the source IC until activity is lost.

2. Failure of either of the transmitting transistors may pull the output line high or low. For instance, activity normal, high LED normal, low LED out = a) A *normal* signal,

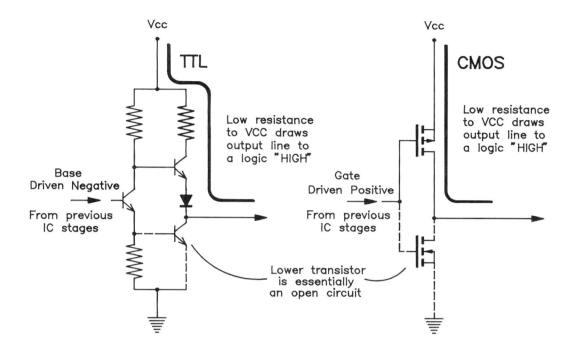

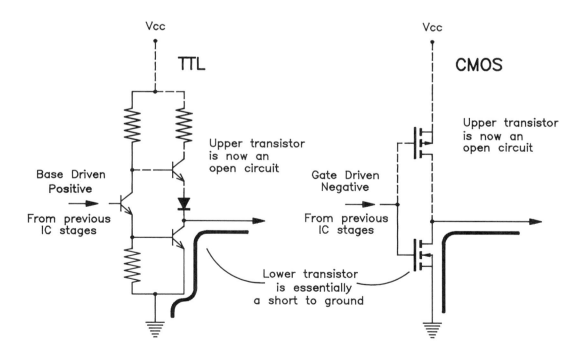

Figure 4-4 How the transistors in a digital IC make the logic high or low levels at the output line

134 ☐ Identifying the Defective Component

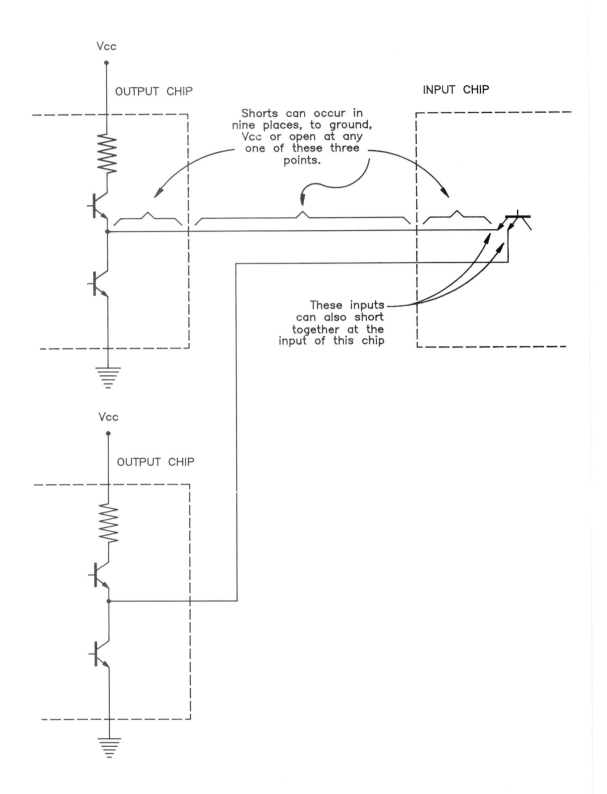

Figure 4-5 Some of the possible faults that may occur when one IC transmits to another IC

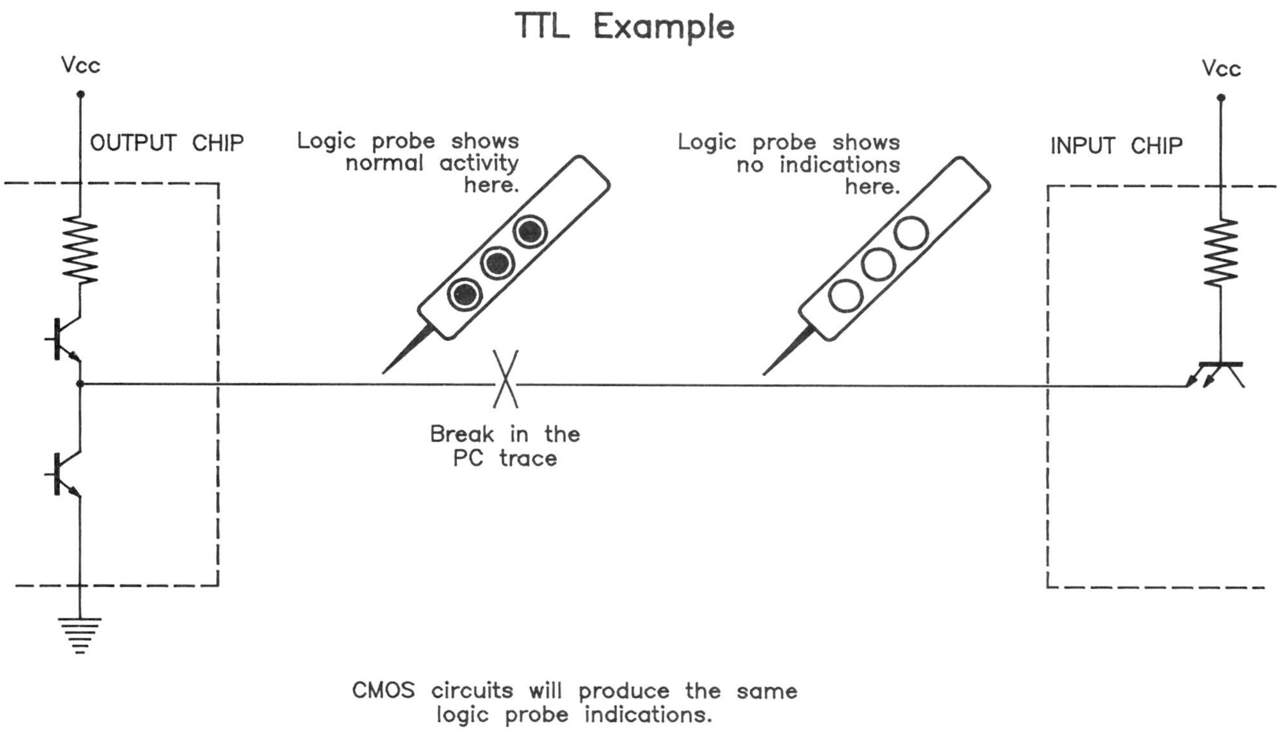

TTL Example

Figure 4-6 Tracing along an open PC board trace to find the open point

a high-duty cycle waveform showing low LED too dim to see, or b) open low source transistor. This results in the logic level driving from high to intermediate logic levels. If TTL, the receiving IC will interpret this as a constant high logic input. See Figure 4-7.

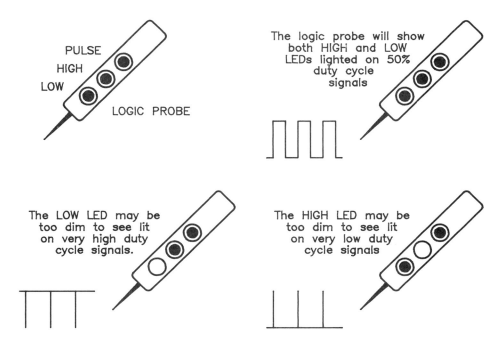

Figure 4-7 How the duty cycle of a digital signal may present confusing indications

In another instance, if the technician finds activity normal, low LED normal, high LED out = a) A *normal* signal, a low-duty cycle waveform showing high LED too dim to see, or b) open source high transistor. This results in the logic level driving from logic low to intermediate logic level. The receiving IC may interpret this as a normal signal since the receiving IC "floats" its input high. See Figure 4–7.

3. Failure of an open of both output transistors or an internal open lead to the output pin of the source. This will, in the case of TTL, result in a constant intermediate logic level at the source IC. See Figure 4–8.

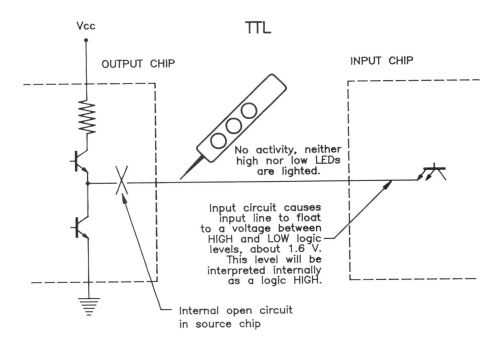

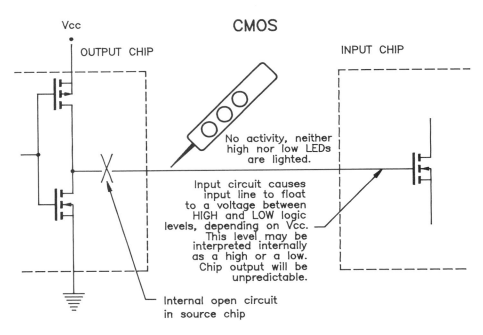

Figure 4–8 Lack of activity at the source IC with a constant intermediate logic level is an indication of an internal open in the source IC

4. If there is no digital activity and a constant high or low logic level at the source pin, this is called a "stuck bus." See Figure 4-9.

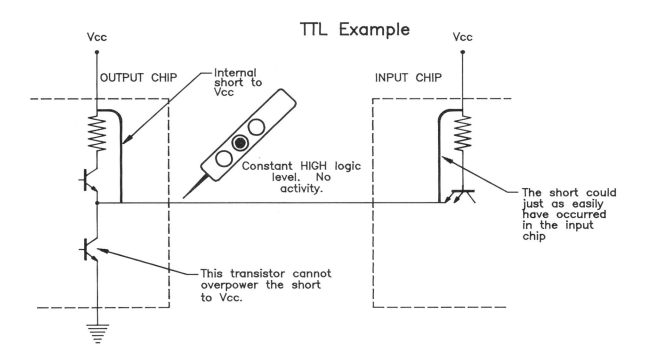

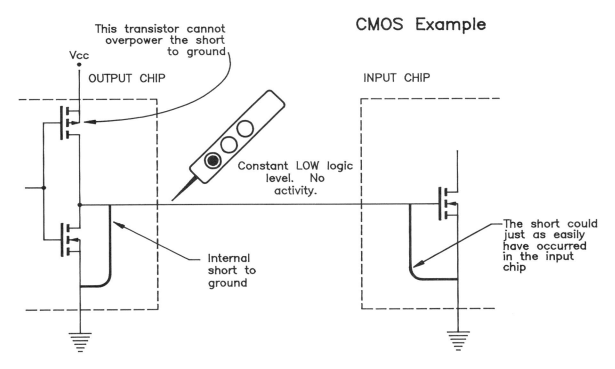

Figure 4-9 How an internal short in either the source or the load IC can cause a "stuck bus"

An internal short to either ground (the digital signal is stuck in a low state) or to Vcc (stuck in a high state) can cause the line between ICs to be held at these levels. The problem is, the actual fault could be caused by either the source or the load IC.

Before proceeding, *be sure that the right combination of correct digital signals is present at the input to the source IC* and that the output should indeed have activity. Remember that the *timing* of the input signals in relation to one another is of utmost importance in determining the output of a digital IC.

A stuck bus due to an internal short in either one of the ICs is part of the 25 percent of digital IC failures that cannot be directly detected by using a logic probe. See paragraph 4.3.5.1.1 for more help in troubleshooting this failure.

5. *If the signal between ICs appears normal* yet is possibly being *ignored* inside the load IC because of an internal failure, the signals may appear as back in Figure 3–57.

    This is a more difficult situation. If the chip truth table allows an output from the chip with all other outputs tied low, the other inputs to the chip can be deliberately tied low (to ground) temporarily to eliminate possible confusion with these signals. Then apply a signal to the suspected input with a logic pulser, looking for activity at the output. See Figure 3–57.

    If the IC is suspected as ignoring one of its inputs, this possibility can often be verified with an oscilloscope, as explained in paragraph 3.10.15. Refer to that paragraph for further troubleshooting of this ignored input problem.

6. A less common fault can occur when two inputs to a chip are shorted together inside the load chip. See Figure 4–5. Verification of this problem is done by applying a pulser signal at one input, then detecting the same signal coming out the other input by using a logic probe.

4.3.5.1.1 NOTES ON USING THE LOGIC PULSER
— A basic description of the logic pulser was given in paragraph 3.10.12. We shall consider here some additional uses of the logic pulser.

Is this a semi-short or hard-wire short? — The logic pulser can be a big help in troubleshooting a stuck bus. If pulsing the line results in a signal on the line, a semi-short is indicated. See Figure 4–10.

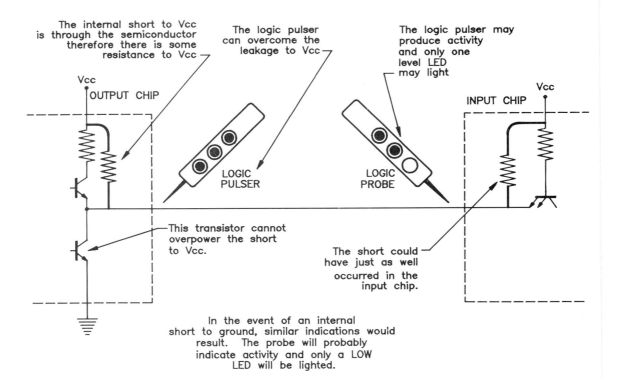

Figure 4–10 The logic pulser can overcome an internal IC short because it is not a zero resistance path

If a stuck bus shows no activity even when pulsed, the trace is a hard-wire short, directly shorting to Vcc or to ground.

The current tracer can be used with the logic pulser to locate either kind of short that is causing the stuck bus.

4.3.5.1.2 NOTES ON USING THE CURRENT TRACER — If a current tracer, such as pictured in Figure 4–11, is available, the technician can find which of two chips involved is causing a stuck bus problem.

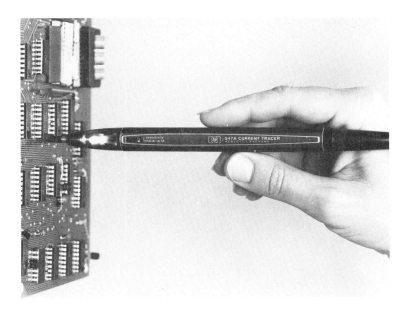

Figure 4–11 The current tracer

The current tracer has a tiny magnetic pickup coil in its tip. This coil is very sensitive, but only to *pulsed* currents. The current tracer will not indicate constant DC current flow.

The current tracer must be adjusted so that *when physically aligned* with the current flowing from a *normal* source IC, the indicator light shows a medium-intensity glow. Holding the tracer over the trace at the source IC with the proper alignment to the current, the light is again observed. *If the short is in the source IC, there will be no glow; if the short is directly on the trace or in the load IC, a very definite glow* will show. The indication of excessive current holds true if the internal short is to Vcc or to ground. See Figure 4–12.

If, for some reason, normal signals are not available at the source IC output pin, the logic pulser can be used as a source of digital pulses. A short anywhere outside of the source IC will cause a glow of the current probe. Of course, the probe must be adjusted for a "normal" current before this test is valid. See Figure 4–13.

4.3.5.2 NOTES ON IC-TO-DISCRETE STAGES — Digital circuits communicating with the "outside world" usually do so with the help of components that are not ICs. Transistors are common components used in this application. These circuits are often very simple, since they operate Class C, either full-on or full-off. In this way, transistors are used as a switch. Many of the transistors used with digital output circuits are therefore called *switching transistors* in the data manuals. Figure 4–14 shows some representative circuits of how a digital IC might be connected to a transistor.

In these circuits, the digital IC is the source component. Note that in some cases, the digital IC is shown connected to a transistor. The transistor provides the current needed to operate the output device. *The series resistor in the transistor base circuit provides the technician with an excellent monitoring point* because it effectively isolates the digital IC from problems that the transistor might develop. For instance, if the transistor shorts from base

140 ☐ **Identifying the Defective Component**

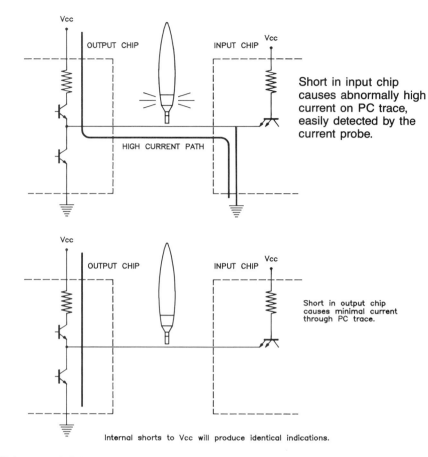

Figure 4-12 Using normal digital signals to determine which of two ICs is shorted with the help of the current tracer

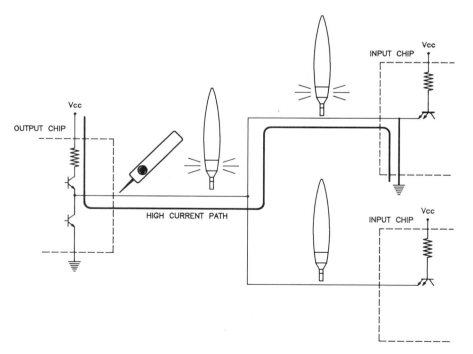

Figure 4-13 Using the logic pulser and the current tracer together to find a short in the load IC

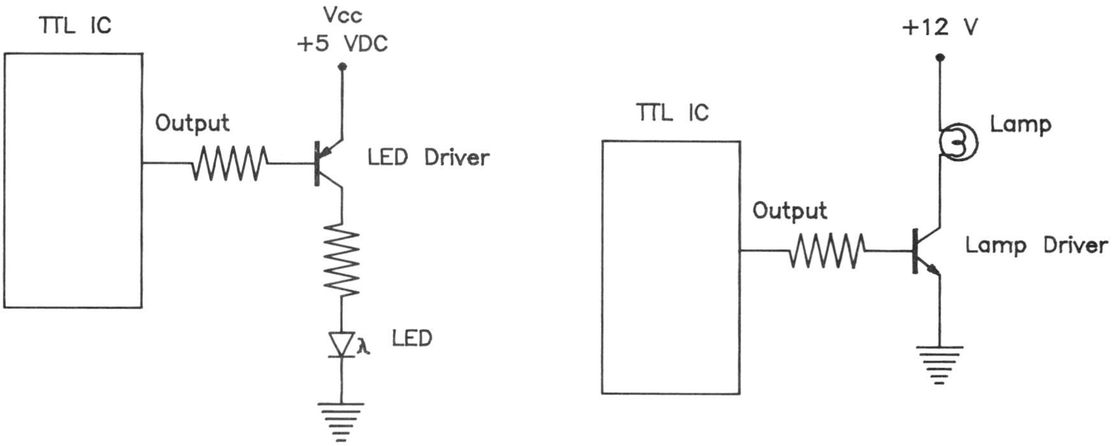

Figure 4-14 Sample circuits for connecting digital ICs to transistors for interfacing purposes

to emitter, the signal fed to the input end of the resistor from the digital IC would not appreciably change from what it would be in a normal circuit. The voltages on the base circuit end of the resistor, however, would change considerably.

Check the logic levels at the output of the digital IC with a logic probe. If the logic levels are good, the problem must lie with the transistor or components later on in the signal path. If the logic levels are not good, then the problem is almost certainly within the digital IC or prior to it.

Checking the voltage levels within the transistor amplifiers should be done using an oscilloscope. If the transistor is converting the logic levels to another voltage, or worse yet, to a negative voltage, the logic probe will not give valid indications. Remember that the logic probe is intended for use in circuits of from 0 to 5 VDC in the case of TTL, or from 3 to 18 V with CMOS circuits. The high- and low-voltage levels reported by the probe depend on two definite thresholds. These thresholds will not apply to the transistor circuit under test. See Figure 4-15.

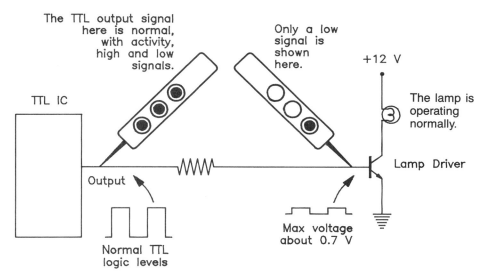

Figure 4-15 Why a logic probe will not indicate properly in a transistor circuit

In summary, when dealing with interfacing circuits that are not IC chips at both source and load ends, the logic probe must be used with caution and the technician must be able to interpret indications correctly.

Open collector output stages — Another circuit that the technician may encounter in interfacing circuits is the IC with an open collector output stage within the chip.

Compare the open collector IC of Figure 3-33 with the totem-pole output of Figure 3-32. Notice that there is no transistor for pulling the output pin up to a high logic level.

The open collector IC can usually be recognized in a circuit, even if not specifically labeled as such on the schematic, by the presence of a resistor tied from the output of the IC to a source of voltage. This voltage supply can be a different voltage source than that powering the remainder of the chip. This resistor is called a *pullup* resistor. Its function is to provide a logic high on the output line when the internal transistor open-circuits. When the internal output transistor conducts, the output line is forced low.

If the output transistor of the open collector chip should develop an open circuit, the chip will not produce a logic low, and the output will always be a high. On the other hand, if the internal transistor should short, the symptoms would reverse, with the output always being a low. Failure of a different power supply than that powering the chip can be easily overlooked, making a good open-collector chip seem defective. Failure of the +12 V supply of Figure 4-16 would cause an apparent failure of the IC.

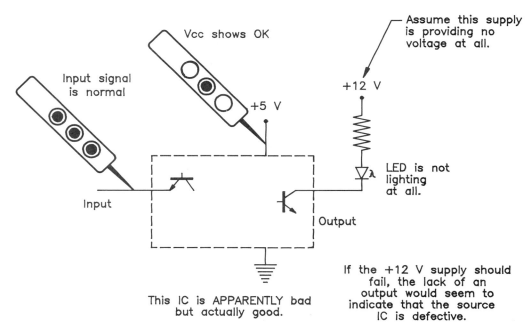

Figure 4-16 Circuit showing how failure of a different supply can cause the output of an open collector IC to appear bad

4.3.5.3 NOTES ON DISCRETE-TO-IC STAGES — Signals applied from an analog world must be converted to definite on or off states when connecting them to digital circuits.

One would think that something as simple as a switch, for instance, would provide a good on or off condition that could be properly interpreted by the digital circuitry. See Figure 4-17.

If the switch itself were perfect, then this circuit would work fine. Switches, though, are not perfect. They consist of metal contacts that are made to come into physical contact with each other. This coming together results in bounce. See the example waveforms of Figure 4-18.

The switch is only one example of the many different inputs that must be *conditioned* to provide the sharp on/off transition needed by digital circuits. A failure of conditioning circuitry will cause the digital circuits to malfunction.

Another example of signal conditioning is the Schmitt trigger circuit. See Figure 4-18. These trigger gates are available in inverters, buffers, NAND, AND, NOR, and OR configurations.

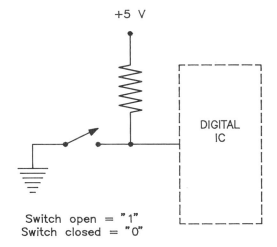

Figure 4-17 A switch and resistor combination that supplies either a high or a low logic level

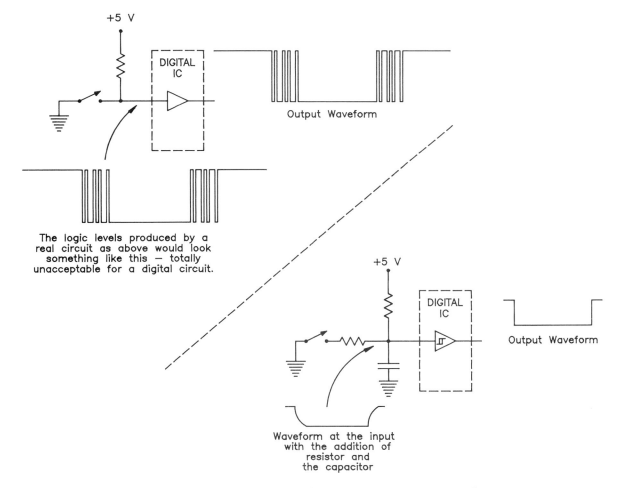

Figure 4-18 Waveforms of switch bounce and the Schmitt trigger used to "clean up" the waveform

The logic probe is inconclusive in interpreting analog signals into Schmitt triggers because the probe is digital in its interpretations. The oscilloscope must be used to show if the incoming signals to a Schmitt trigger gate are high enough in amplitude for the circuit to operate. The Schmitt trigger requires definite levels of input signals before the output will toggle. The exact levels at which the circuit toggles depend upon the Vcc supplied for CMOS circuits. For TTL, the levels are typically 1.6 V for a positive-going signal and 0.8 V for a negative-going signal. Without *at least* this much input voltage swing, the circuit will not toggle. If the input signal is of sufficient amplitude and the IC still does not have the proper output, the IC is suspect and should be replaced.

The dual NAND gate signal conditioning circuit is sometimes used to eliminate the effects of switch contact bounce. See Figure 3–41 for an example. The dual NAND gate circuit will toggle on the very first pulse of a switch bounce and hold that logic state whether or not the switch then opens the circuit. It will toggle the other way only if the switch actually contacts the opposite side, when the switch is changed to the opposite state.

Troubleshooting this circuit is easy, using a logic probe on the input and output sides of the gate.

4.4 NOTES ON SAFETY MEASURES FOR VOLTAGE TESTS — The safety discussion of paragraph 2.0 should be reviewed, particularly if voltages in excess of 60 VDC or 30 VAC could possibly be encountered.

In addition to these safety tips, the technician is cautioned to use test leads with sharp, narrow points. A blunt probe can accidentally short the circuit and cause a great deal of damage. See Figure 4–19.

Figure 4–19 Using narrow and sharp probes such as these Huntron probes will reduce the possibility of circuit damage caused by a slipping probe

Although Huntron probes are good for routine resistance and voltage measurements, they should not be used for current measurements over about an ampere. Forcing high current through the fine tips will overheat and burn them.

4.5 Remove All Signal Inputs

— It is necessary to remove all incoming signals to avoid a likely shift in the DC voltage readings within the stage. Taking DC readings is the next step in troubleshooting from the stage level to the component level. This is sometimes called *static troubleshooting,* since the voltages are not changing.

If the incoming signals were left connected and turned on, the incoming signals might produce effects such as shown in Figure 4–20.

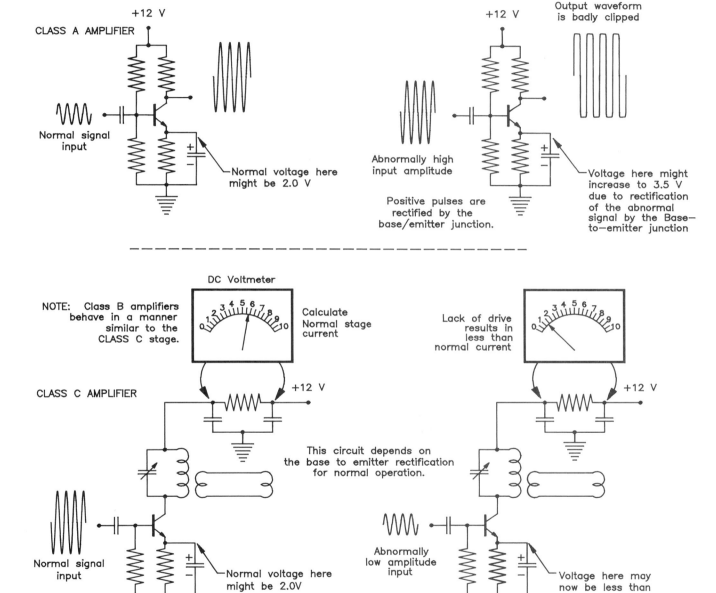

Figure 4–20 Abnormal signals can shift the circuit DC voltages from normal values

4.6 NOTES ON USING THE DC SCALES OF THE DIGITAL MULTIMETER

Only a few years ago, the VOM (Volt-Ohm-Milliammeter) was the instrument of choice for electronic troubleshooters. For a number of reasons, the VOM is now falling into disuse. Among the reasons are:

☐ Fragility, both mechanical and electrical
☐ Inaccuracy, compared to digital instruments
☐ Hazardous operation, uses too high a current and voltage on the ohms scales for many modern components
☐ Loads circuit voltages, particularly on low ranges

The modern troubleshooter will favor the use of a digital multimeter (DMM). See Figure 4-21.

Figure 4-21 The modern digital multimeter is the best instrument to use with modern circuitry (Reproduced with permission from the John Fluke Mfg. Co., Inc.)

The DMM is rugged mechanically and more forgiving of an electrical overload than a VOM, such as applying voltage to the ohmmeter function; is extremely accurate; uses low, safe currents for the ohmmeter functions; and has no loading effect on 99+ percent of the circuitry the technician may encounter.

One of the most important specifications of a voltmeter is the amount of resistance that it must place across the circuitry under test. See Figure 4-22. The higher this resistance value the less loading effect the meter will have on circuit voltages. Where a good but older analog meter may have had as low as 60 K of loading effect using a 3 V range, the digital multimeter typically has 10 megohms of resistance regardless of the range in use.

4.7 NOTES ON HOW COMPONENTS AFFECT DC STATIC READINGS —

Schematics with numerous published voltage readings are rare. Usually, only principle voltage readings are given, such as the power supply output voltages. Occasionally, there may be specified DC voltage test points, complete with the normal DC voltage readings, but this is the exception. It generally falls to the technician to "dope out" what the voltages in the circuit should be, then see if what is measured matches those estimates. Note the desired order: *make the estimate, and then make the measurement.* Once estimates are confirmed to be correct, record them on the schematic for future reference. Recorded readings must be taken from a circuit that is working properly. See Figure 4-23.

Without benefit of prior experience with a circuit and without the normal voltage readings provided on the schematic, *the technician must estimate the DC voltage readings of importance*

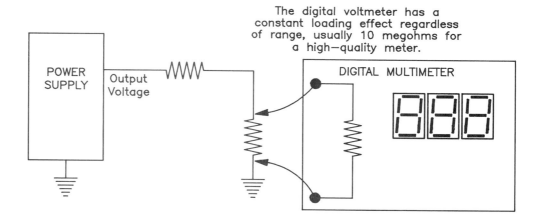

Figure 4-22 The voltmeter places a specified resistance across the circuit under test

according to circuit values. This takes some practice. A good feeling for ohms law and ratio-and-proportion will be necessary.

It is almost never necessary to calculate the normal voltages that should appear in a circuit. Most failures are complete failures or nearly complete. Such failures most often cause a large departure of the circuit DC voltages from normal readings. Because of this, it is usually sufficient for the technician to merely *estimate, not calculate,* the voltages that should be in the circuit. See Figure 4-24.

Once the technician has estimated voltages, the DC voltage is then measured with a voltmeter. *If there is a large difference* (a matter of judgment) *between the estimated and the actual voltages, concentrate on this discrepancy only* until the problem is resolved or until another, more seriously variant voltage is discovered.

When there is a large difference between the estimated and actual voltages, reestimate the voltage. Look carefully for paralleling paths of current flow that will affect the voltage measurement. See Figure 4-25.

Don't fall into the trap of finding a DC voltage reading that differs substantially from what you think it should, and then ignore that reading. This is a sure way to confuse the issue. *Concentrate on one suspicious reading until a) its value is verified as correct, b) another reading is discovered to be a higher percentage from normal, or c) the problem is solved.* If necessary, put a large star or other mark on the schematic where the reading doesn't seem right and focus your attention and troubleshooting efforts on this one point.

148 □ Identifying the Defective Component

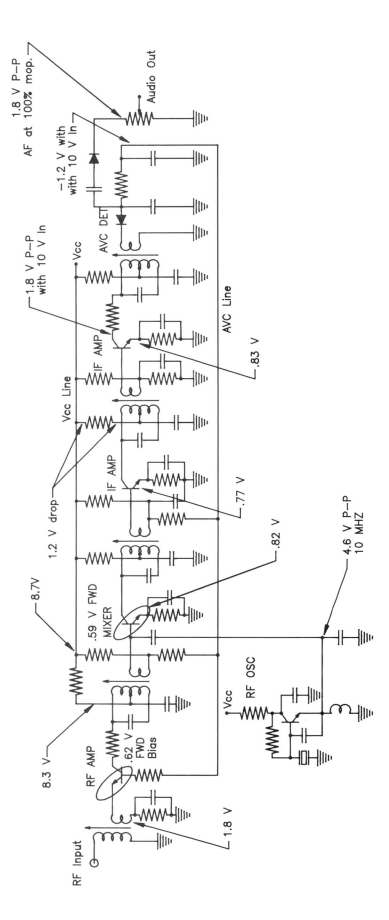

Figure 4-23 Examples of normal voltages are noted on the schematic for future use. Note that most transistor voltage readings should be taken in reference to the emitter, not to ground

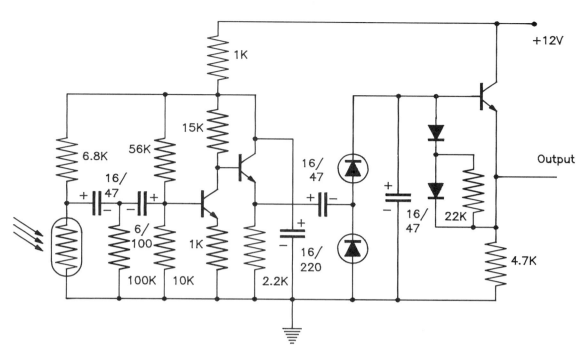

Figure 4-24 Example of a circuit for which little information is available on normal DC voltages. The technician must supply the approximations of these voltages

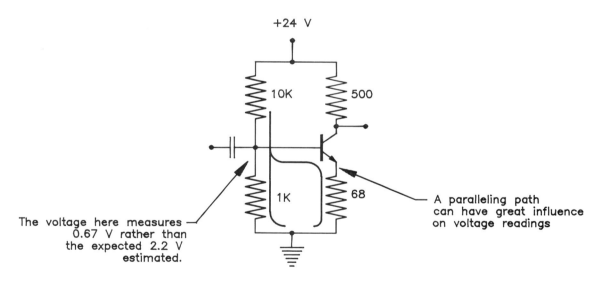

Figure 4-25 An example of a "hidden" paralleling path for current flow

Estimating DC voltages

EFFECTS OF RESISTORS IN DC CIRCUITS — Estimating DC voltages in a stage is primarily a matter of having a thorough understanding of series and parallel resistor networks. See Figure 4-26.

Remember that most voltage readings are taken with respect to the circuit common, or ground point. *The chassis is not always the circuit common or ground for voltage readings.* If your voltmeter shows very low DC readings or if your oscilloscope shows a large sinewave component, this is a strong indication that the ground side of the test instrument is not connected to the proper circuit common.

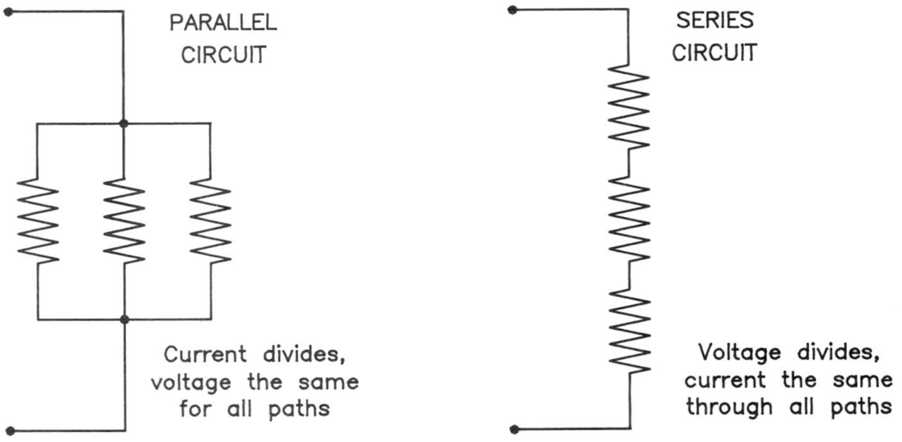

150 □ Identifying the Defective Component

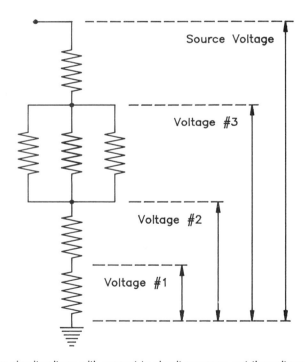

Figure 4-26 Parallel and series resistive circuits must be thoroughly understood to be able to estir te voltages not provided on schematics

Knowing how the voltage changes across resistors in series or parallel combinations is only a partial solution. Add voltage drops as necessary to get the voltage between resistors *in reference to ground*. See Figure 4-27.

Figure 4-27 Estimate the circuit voltage with respect to circuit common, not the voltage across each component

Measuring voltages with reference to the circuit common is particularly important if the oscilloscope is used as a measuring instrument. The oscilloscope ground is connected to the power line ground, which could damage some circuits if improperly connected to circuit points other than a power-line ground.

Taking a voltage measurement across a resistor can prove, under the proper conditions, that there is current flow. See Figure 4-28.

The presence of a voltage drop across a resistor proves there is current flow through the resistor *only* if the voltage is less than the source voltage and more than zero.

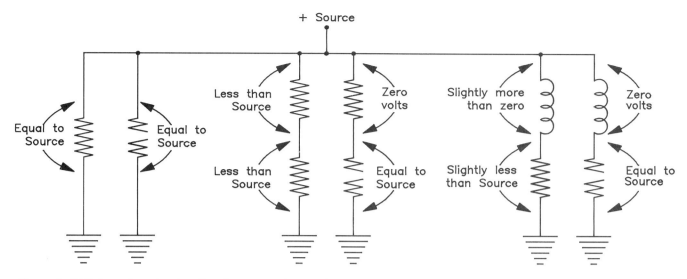

Figure 4–28 Only a voltage drop less than the source voltage and greater than zero proves that there is current flow through that component

A caution about voltage readings across resistors: Resistors in parallel with other components can be open yet show an almost normal voltage. See Figure 4–29.

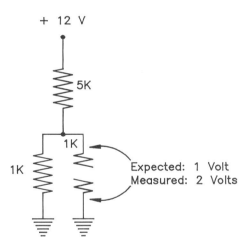

Figure 4–29 Paralleling resistors with other components can produce near-normal voltage readings when the resistor is actually open

There are two situations where resistors are commonly used that might confuse the technician. One application uses resistors in parallel with series-connected capacitors to force the DC voltage to divide more equally between them. See Figure 4–30.

Since the leakage of a good electrolytic capacitor at normal operating voltage may be 10 megohms or more, the 100 K resistors in parallel are the dominant factor in the division of DC voltage. Without them, two electrolytics of 10-megohm and 25-megohm leakage, for instance, would put the voltage across one of the capacitors far over its proper voltage rating, likely destroying it by causing it to short.

The second application is the inverse of the above: Two resistors of very low value are used to equalize currents that might otherwise be very different. See Figure 4–31.

When one of two paralleled transistors normally has a lower resistance than another, it is less significant in determining the current through that branch if a resistor several times larger is in series with it.

152 □ Identifying the Defective Component

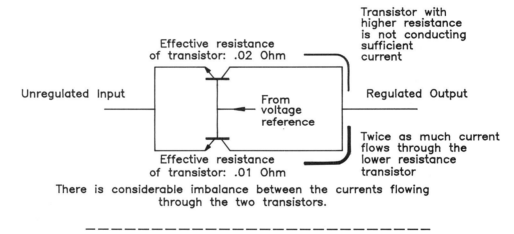

Figure 4-30 High-value resistors across series connected electrolytics force the DC voltage to divide more equally

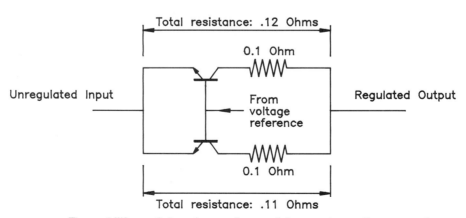

Figure 4-31 The use of low-value resistors is common when paralleling transistors to help in dividing current equally between the two paths

EFFECTS OF CAPACITORS IN DC CIRCUITS — Capacitors should be "invisible" to DC measurements. The DC voltages, in other words, should be the same whether or not the capacitors are in the circuit. This can only be true, however, if the capacitors are really open circuits and are not leaky. Some large electrolytic capacitors could conceivably disturb DC resistances because of their internal leakage, but this would be very rare in a practical circuit. Modern techniques of producing electrolytics make them *almost* perfect in any practical circuit. Old electrolytic capacitors that have been in service for many years may have developed leakage or they may have dried out inside and lost most of their capacity. This makes them ineffective as a large-value capacitor.

Once the DC values of an AC power supply have had sufficient time to settle down to their normal operating values (typically within a few 60 Hz cycles), the DC circuit voltages should remain constant. It is these values that the technician attempts to estimate.

A faulty electrolytic capacitor can upset DC circuit values when it develops internal leakage. Most capacitors short when they fail. The shorted capacitor greatly disturbs the circuit DC values, and is easily identified. See Figure 4–32.

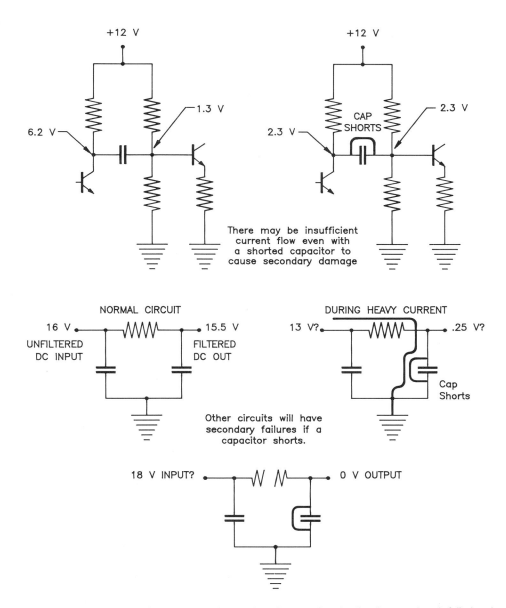

Figure 4–32 Two examples of how a shorted capacitor changes the circuit voltages when it fails by shorting

In summary, the capacitor "shouldn't be visible" to DC voltage readings. See Figure 4-33.

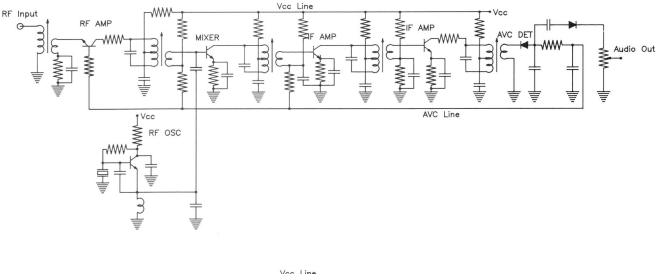

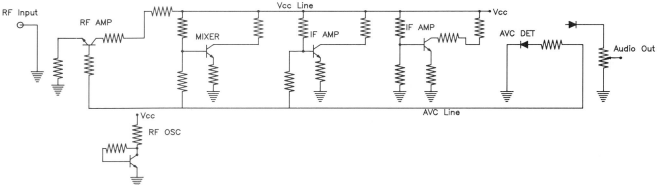

Figure 4-33 Example of a schematic and how the technician should "see" the circuit when estimating the DC circuit voltages within it

EFFECTS OF INDUCTORS IN DC CIRCUITS — There are three general sizes of inductors. These are the power supply or filter inductor, the audio inductor, and the RF inductor. Their names imply the frequencies at which they will develop substantial reactance.

Inductors fail by three means: to short out one or more turns, to short from any turn to case ground, or to open circuit internally. The open circuit is the easiest to locate because the component no longer passes the DC current that it should pass. See Figure 4-34.

Finding an inductor that has shorted from a winding to case ground is also easy. Dismounting the inductor from the chassis clears the short. Along with this failure there is often other internal damage, including shorted turns. Replacement of the inductor is the proper cure. In an emergency, an inductor can sometimes be operated in spite of a short to the case. This can also create a safety hazard if the inductor is shorted to its case, but insulating it from the chassis may still be an option.

Detecting shorted turns in an inductor, on the other hand, cannot be done effectively using static troubleshooting methods. See inductors under paragraph 4.12, "Notes on How Components Affect AC Dynamic Readings" for help in this case.

Inductors have the opposite effect of capacitors in DC circuits—they approach a short. Large inductors, such as those used to filter low frequencies in power supply filters, have a low resistance in the order of a few ohms or less. A very large inductor could have a resistance of much less than one ohm. See Figure 4-35.

On the other hand, a small inductor using many turns of very fine wire will have a much larger resistance.

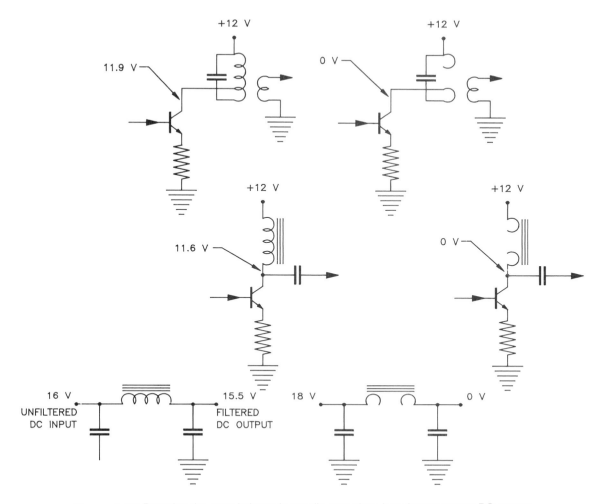

Figure 4-34 Detecting the open inductor is usually easy since it no longer passes DC current

Figure 4-35 This filter inductor has a DC resistance of only a few ohms

Most practical inductors will have resistances that fall between these two extremes. It is up to the technician to determine in any particular case the normal resistance of the inductors in the circuit and whether or not they will significantly upset the DC readings. This will also depend upon the resistance of other components in the circuit. If the circuit resistance is in the order of thousands of ohms, an inductor of less than a hundred ohms can be considered to be a short.

On the other hand, even a large inductor having a few ohms of resistance can upset DC readings if there are sufficiently low circuit resistances and heavy currents are involved. A high-current battery charger would be such a circuit.

In summary, when estimating voltages in circuits including inductors, view the inductors as a short circuit or a resistor of a few ohms as appropriate.

EFFECTS OF TRANSFORMERS IN DC CIRCUITS — The windings of transformers are just like inductors, discussed in the previous section, for the purposes of DC voltage estimation. As a general rule, the higher the voltage of a winding (either primary or secondary) and the physically smaller the transformer, the higher the relative DC resistance.

As an example, consider a transformer of 10:1 voltage reduction, intended for 120 VAC input and 12 VAC output. This transformer would have much more resistance across its primary winding than it would have across its secondary winding. There are two reasons for this: The number of windings in the primary is 10 times that of the secondary, which alone would account for a primary reading of 10 times the resistance of the secondary winding. In addition to this, a smaller wire is used on the primary because of the much lower current in the primary winding. This wire size factor accounts for a further increase of resistance of the primary. These two factors taken together might amount to a total resistance ratio of perhaps 100:1 for this example.

Transformers are used only in pulsed or alternating current applications. Their interpretation for estimating DC circuit current is similar to that which would be expected for two or more separate inductors in the circuit.

EFFECTS OF THE FUSE IN DC CIRCUITS — The fuse is really a very low resistance that is supposed to melt when too much current flows through it, thus opening the circuit and preventing further current flow. See Figure 4–36.

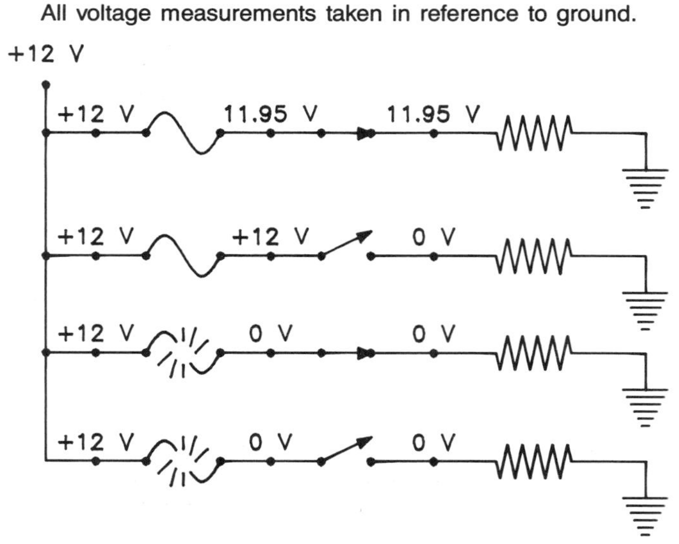

Figure 4–36 How the fuse and the switch appear to DC circuits

Remember that a "blown" or open fuse has a voltage across it, typically the full source voltage (if all switches are closed).

EFFECTS OF SWITCHES IN DC CIRCUITS — The switch should be either an open circuit or a short, depending upon the position of the control. When the switch is "off," there should be full circuit voltage across that switch. See Figure 4-36.

Note that the circuit voltage changes position from appearing across the switch to across the resistor when the switch position changes.

Defective switches usually refuse to change state internally, and thus are easily detected in a "dead" circuit with an ohmmeter or in a live circuit with a voltmeter as shown above. Switches in high current applications can overheat and develop an internal resistance. This kind of failure is readily identified by the presence of heat. A good switch should not become hot.

EFFECTS OF THE SEMICONDUCTOR DIODE IN DC CIRCUITS — The influence of the semiconductor diode must also be included in the analysis of DC circuits. The thing to remember about a semiconductor junction is that *if there is a varying forward current flow through a diode junction, the voltage across that junction is almost constant.* See Figure 4-37.

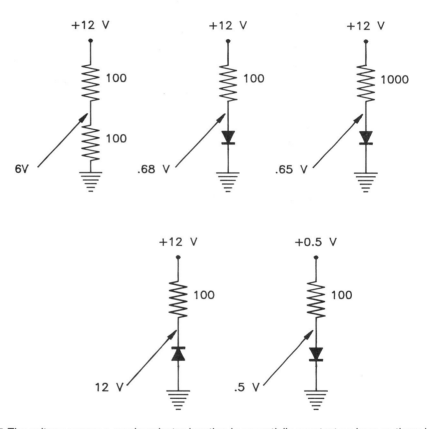

Figure 4-37 The voltage across a semiconductor junction is *essentially constant* as long as there is sufficient voltage and proper polarity for forward conduction

Diode failures can cause the voltage across the diode to be very low, indicating that a short has occurred within the junction, or the junction voltage may be far above the normal 0.3/0.7 V values for germanium or silicon junctions, respectively. A substantially higher-than-normal voltage across a junction with a positive polarity on the anode is a sure indication that the junction is open. See Figure 4-38.

If the forward voltage (positive on the anode) is insufficient to make the diode begin conducting, the diode remains an open circuit.

Circuit polarity will indicate if a diode is forward- or reverse-biased. If reverse-biased, the positive will be applied to the cathode (bar) end of the diode symbol. The diode is now

158 □ Identifying the Defective Component

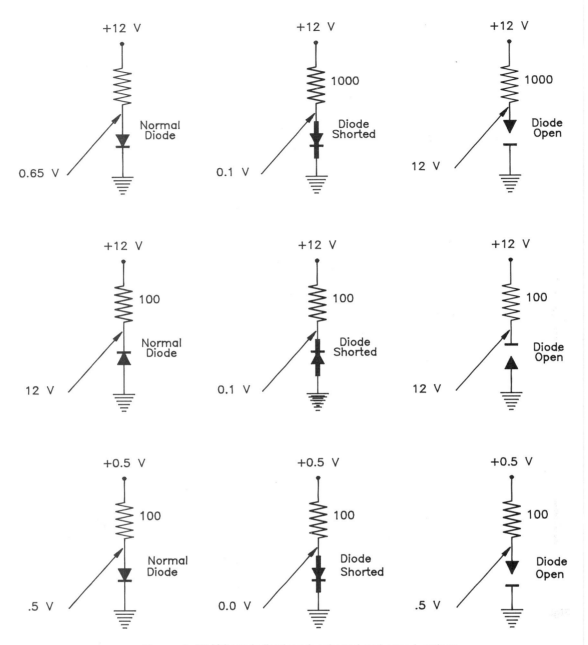

Figure 4-38 Voltage indications for shorted and open junctions

an *open circuit,* and can be eliminated from the circuit for the purposes of estimating circuit DC voltages. See Figure 4-38.

In summary, including a diode into DC circuits does not complicate matters to a great extent. If the polarity of the applied voltage is such that the diode is reverse-biased, the diode should be an open and "does not exist" for that circuit. If forward-biased, it should have either a 0.3 V (germanium diode) or a 0.7 V (silicon) DC voltage drop, regardless of the current applied, assuming the circuit has at least the 0.3/0.7 V required to bring the diode into forward conduction. Excessive voltages across a forward-biased resistor indicate an open diode.

EFFECTS OF THE BIPOLAR TRANSISTOR IN DC CIRCUITS — Estimating DC voltages in transistor circuits makes it necessary to *first determine the conducting state of the transistor.* Is it conducting across the emitter-collector leads, how much, or is it not conducting at all? In order to simplify this question, the bipolar transistor can be thought of as a diode and a resistor, as shown in Figure 4–39.

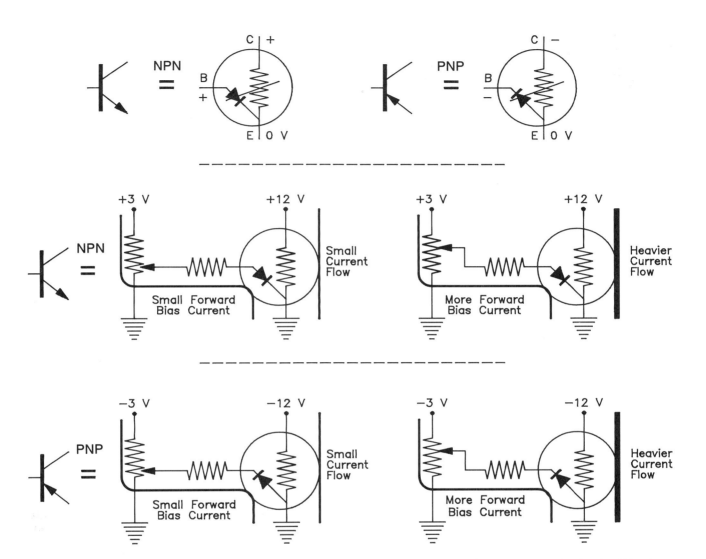

Figure 4–39 The transistor can be thought of as a diode and a current-controlled resistor for the purpose of estimating DC voltages

The technician's interpretation of the transistor as a diode and a resistor linked to it provides an easier way to interpret the transistor than to attempt recalling the electron and hole concept taught during transistor theory. The idea is simple: *If the emitter-base diode is not conducting, the emitter-collector resistor will be open; if the emitter-base diode is conducting, the resistance from emitter to collector will depend on the amount of current flow through the diode.* Be sure to remember that, once it is forward-biased, *the amount of forward current flow through the emitter-base diode cannot be determined by the voltage across that diode.*

DC transistor circuit analysis must begin with deciding which class of operation a transistor is using within the circuit. As a general rule, DC circuits use Class A stages; AC and low-frequency stages (including audio) use Class A for amplification and Class B for power; RF circuits use Class A for small signals and Class C for higher power stages; and pulse circuits use Class C. Look for a suitable resistor connecting the base of the circuit to the collector circuit, an indication of Class A or B operation. This statement applies to either NPN or PNP stages. See Figure 4–40.

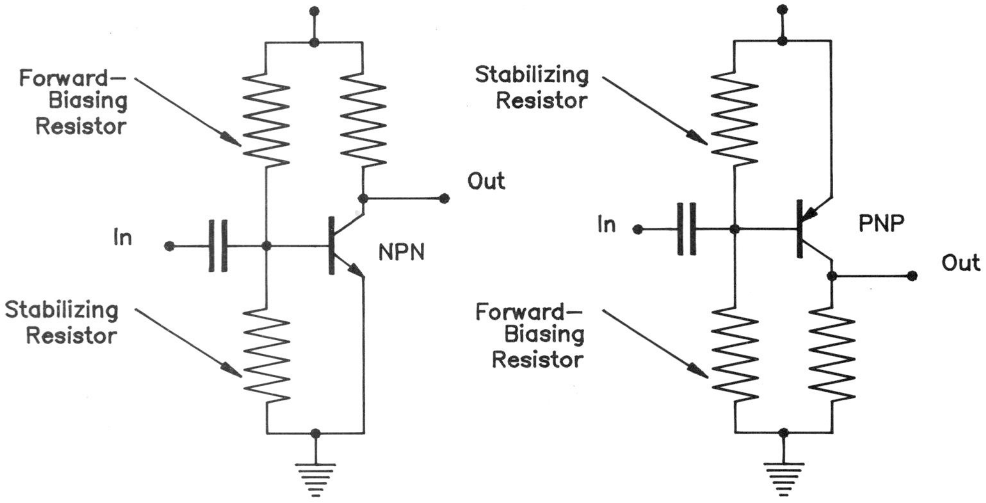

160 □ **Identifying the Defective Component**

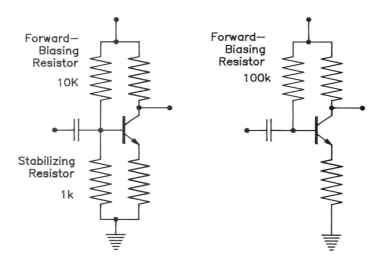

Figure 4–40 NPN and PNP stages with a resistor connected from base to collector, an indication of Class A operation

A Class A stage must be forward-biased from emitter to base, and the emitter-collector voltage should be approximately half of that available at the collector source. See Figures 4–41 and 3–67.

Figure 4–41 Typical Class A amplifier stage. Note the two ways of establishing the forward bias: with and without a stabilizing resistor. Note also the effect of this on the value of the forward-biasing resistor.

A class B stage will also be conducting, but very little. The emitter-collector voltage is typically high, almost that of the available voltage source, without an input signal. There will be a forward-biasing resistor to the collector circuit, but of relative high resistance value. Class B stages are frequently used with a second, complementary amplifier stage, to provide the opposite polarity of the input waveform. See Figures 4–42 and 3–68.

Operation of the Class B stage can be monitored very nicely by checking for the proper voltage across the emitter resistors. Without an incoming signal, there should be a small amount of voltage across the resistors, increasing dramatically as the stages are driven harder by an increasing input signal.

Class C stages do not have a forward-biasing resistor. See Figures 3–69, 3–80, and 3–82.

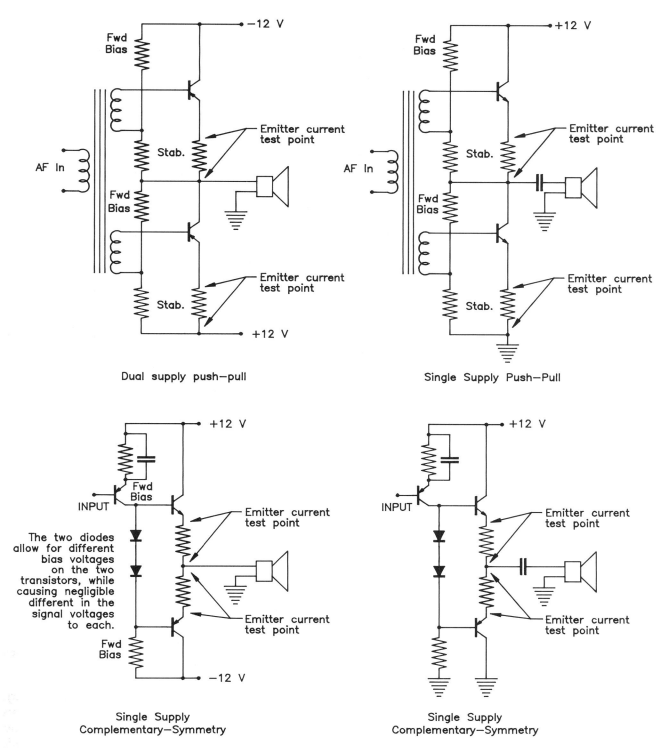

Figure 4-42 Typical Class B amplifier stages

The Class C amplifier should have no collector current in the absence of an incoming signal. Emitter-collector voltage should be that of the source. The emitter current should be zero also, as proven by a correct reading of zero volts across the emitter resistor if no input signal is present. As the stage comes into conduction, the emitter resistor provides a convenient place to monitor transistor current flow via a voltage measurement.

162 ☐ Identifying the Defective Component

In summary, the bipolar transistor in any static DC circuit will have an emitter-collector resistance of a value that is determined only by the amount of current flow through the emitter-base junction. Because of this, the status of the emitter-base junction must be determined before the emitter-collector resistance can be estimated. The forward voltage across the emitter-base junction is *not* an indication of the amount of current flow, nor of the amount of resistance to expect from emitter to collector.

The quickest test of a transistor that has forward-bias already applied (a Class A stage) is to short the base to the emitter. The resistance from emitter to collector should increase to an open circuit.

EFFECTS OF THE FET IN DC CIRCUITS — The field-effect transistor (FET) is available in two general types, the junction FET (JFET) and the insulated-gate FET (IGFET). Both transistors are characterized by having very sensitive input circuits. The FET is sometimes used in DC circuits where it serves as an amplifier, most often in instrumentation applications.

The JFET — The junction FET can be thought of as a resistor controlled by the voltage across an *open* circuit. See Figure 4–43.

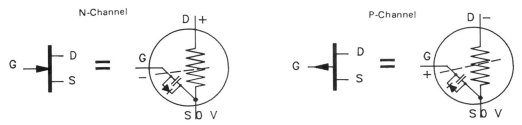

The internal diode is always reverse—biased, and therefore an open circuit

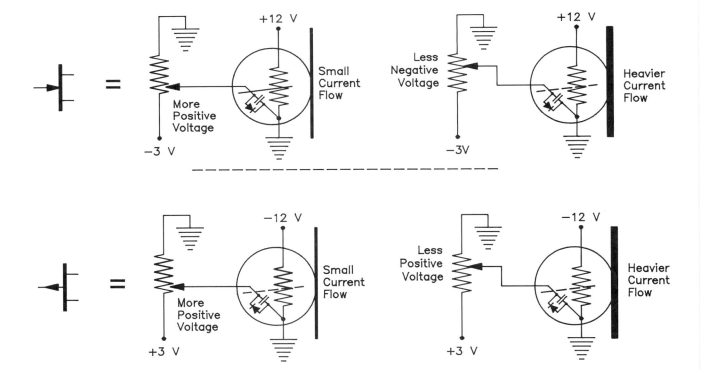

Figure 4–43 The JFET can be visualized as an open circuit input and a voltage-controlled resistor

The DC biasing of the JFET determines the operation of the output circuit. Because of the very high input resistance of this component, the connection of a voltmeter, even a digital multimeter with an input of 10 megohms, may completely upset the voltages of the input circuit. An easy way to provide bias for the JFET is shown in Figure 4-44.

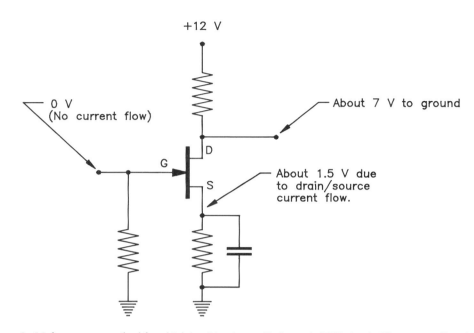

Figure 4-44 A common method for obtaining bias for an N-channel JFET circuit. The current flow through the source resistor provides a positive voltage. Using this as a reference, the gate is therefore negative by about 1.5 volts

The output circuit can be tested with the DMM because the output circuit of the JFET has much lower resistance than the input side.

Since the JFET is often used as a Class A amplifier in DC circuits, any disturbance of the input circuit should be reflected as a change in the resistance of the component from source to drain. Because of the high resistance of the typical JFET input circuit, sometimes just placing a finger on the input circuitry is sufficient to make a substantial change in the output of the JFET. If no change occurs, this is a good indication of a defective JFET.

The IGFET — Static troubleshooting of an insulated-gate FET (IGFET) circuit is similar to that used with the JFET. The resistance of the source-to-drain output circuit varies in response to the voltage of the input circuit. See Figures 4-45 and 4-46.

As with the JFET, the IGFET is often used in Class A applications. Any disturbance of the high-impedance input circuit, such as touching it with the hand, should produce a change in the resistance of the source-to-drain circuit. This can be monitored by watching the drain voltage with respect to ground while the input circuit is disturbed.

EFFECTS OF OPERATIONAL AMPLIFIERS IN DC CIRCUITS — The operational amplifier is sometimes used in DC circuits that can be analyzed under static conditions.

The operational amplifier is used in two basic configurations: with negative feedback or with positive feedback. The visible difference between them is the polarity of the feedback. See Figure 4-47.

The operational amplifier is often, but not always, operated with a single input, the remaining input being tied to a voltage reference.

The operation of operational amplifiers under static DC conditions depends mainly upon the kind of feedback, positive or negative. *With any degree of positive feedback, the op-amp output should be very near either of the two supply voltages.* These supply voltages, sometimes

164 ☐ Identifying the Defective Component

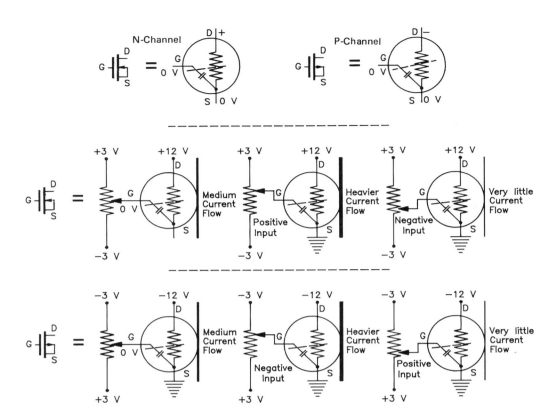

Figure 4-45 The depletion channel insulated gate FET varies the resistance of the source-to-drain channel in response to a small change in the gate-to-source voltage

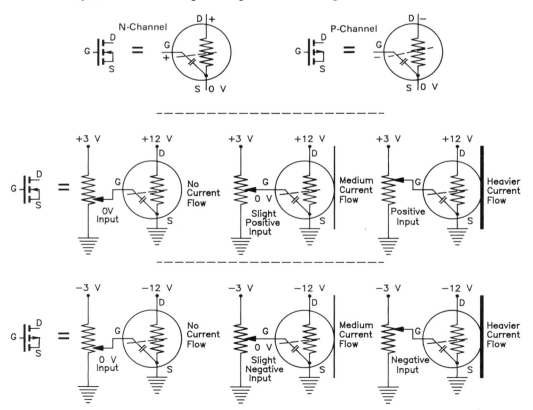

Figure 4-46 The enhancement channel insulated gate FET also varies the resistance of the source-to-drain channel in response to a small change in the gate-to-source voltage

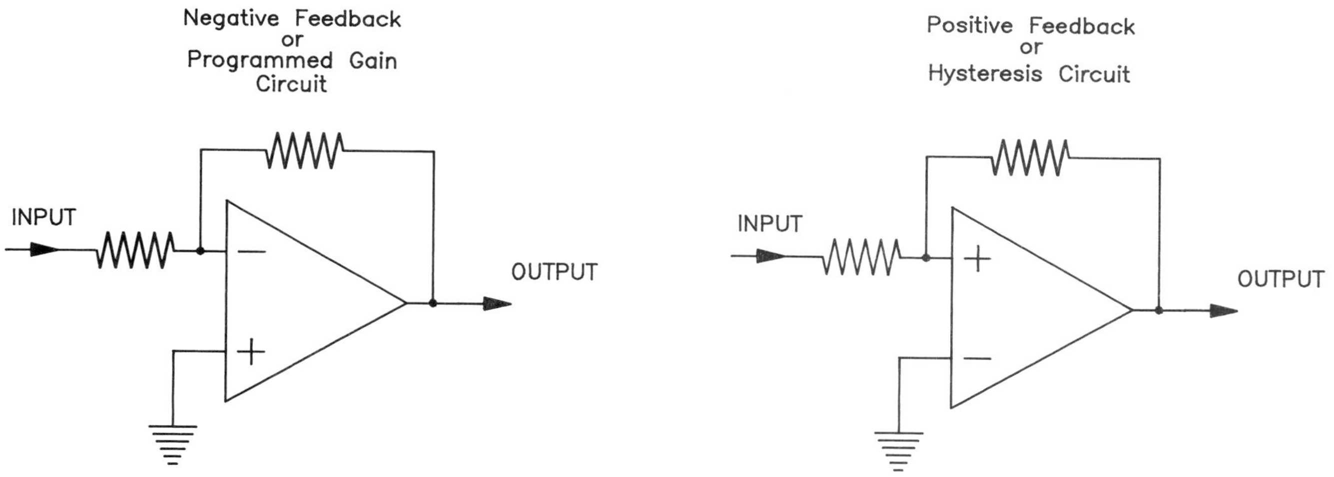

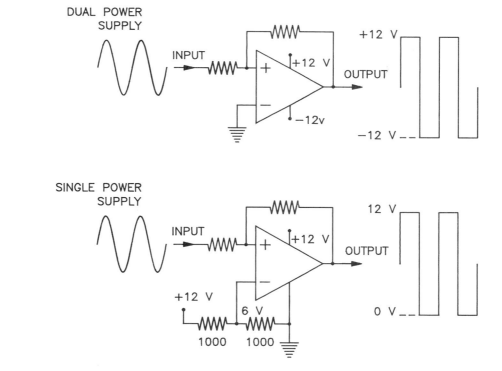

Figure 4-47 Basic operational amplifier circuits using negative and positive feedback configurations

called *rails,* can be separate positive and negative voltages, or often a single voltage and ground. See Figure 4-48.

Figure 4-48 Positive feedback means the output should be very near either one of the supply voltages because of the "snap action" of the circuit

There will be internal voltage drops of a small amount that prevent the output of the IC from becoming exactly the same as the supply voltages or ground.

On the other hand, *negative feedback will mean that the output voltage can be anywhere between the rails.* The exact output of the operational amplifier in a static circuit is determined by the *difference* in the voltages applied to the two inputs. The factors that determine the exact output voltage include:

1. The op-amp circuit with negative feedback and no difference in voltage between the inputs (shorted inputs) should match the output voltage to the same voltage as that on the non-inverting (+) input to the chip. See Figure 4-49.

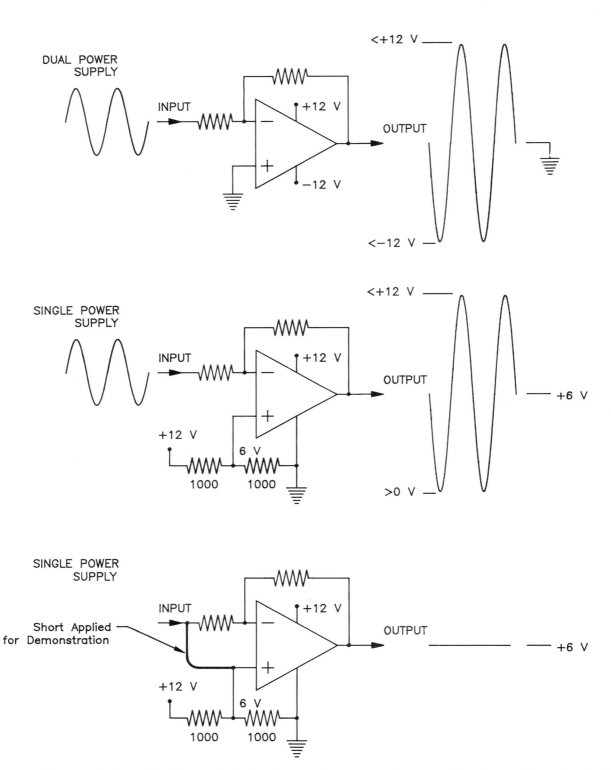

Figure 4-49 The operational amplifier with negative feedback centers the output waveform around the voltage on the non-inverting input

2. Any input voltage to the chip's inverting (−) input causes that pin's voltage to deviate farther away from the voltage on the (+) reference pin. It is this *difference in voltage* between the two pins that the amplifier uses as an input voltage. See Figure 4–50.

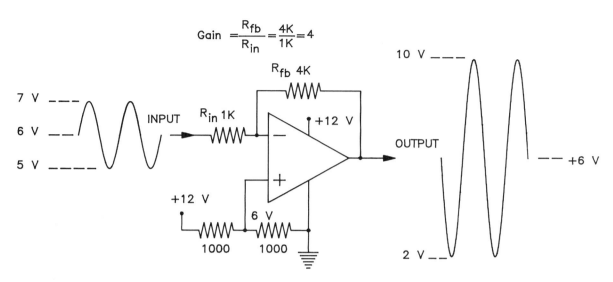

Figure 4–50 The difference between the (−) and the (+) inputs is the voltage that the op-amp amplifies

3. The input voltage is now subjected to an amplification factor determined by the ratio of two resistors, the input resistor and the feedback resistor. See Figure 4–51.

The feedback of the output to the input of this op-amp configuration produces a side effect that could be confusing to a troubleshooter. This effect is called the *"virtual ground."* See Figure 4–51.

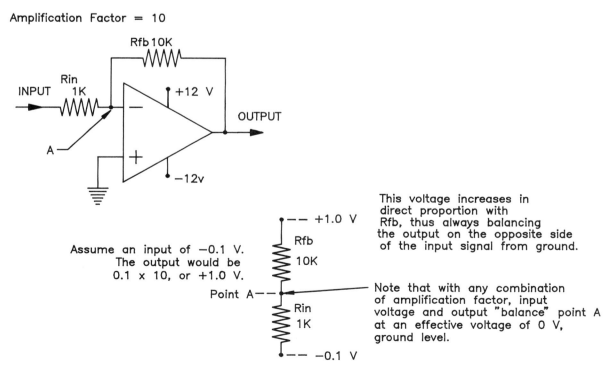

Figure 4–51 A DC operational amplifier circuit demonstrating the effect of the "virtual ground"

Since the operational amplifier with negative feedback amplifies only the difference between its two inputs, shorting the inputs together should result in the output assuming the same voltage as that on the (+) input or nearly so allowing for slight manufacturing differences within the op-amp. Be sure to allow the input resistor to be in the circuit, since it is used with the feedback resistor to control the gain of the circuit. This is a fast test to see if the input is responsive. See Figure 4–49.

Static DC troubleshooting of an op-amp circuit, particularly one with negative feedback, will have to take into account the operational amplifier characteristics. The virtual ground can be especially misleading unless it is understood and anticipated.

SILICON-CONTROLLED RECTIFIERS AND TRIACS IN DC CIRCUITS — Rectifiers and triacs usually operate only on AC voltages, typically power-line frequencies. Either of these components must be fired by auxiliary circuitry and, once fired, remain conducting until the AC cycle goes through zero voltage. Thus, they are not circuits that lend themselves to DC troubleshooting techniques. Refer to the end of paragraph 4.12 for troubleshooting tips on these components.

OPTICAL ISOLATORS IN DC CIRCUITS — Be aware of the fact that *an optical isolator can have dangerously high voltages connected to it.* One of the primary purposes of the isolator is to isolate a high-voltage circuit from another operating at a lower voltage.

The optical isolator is typically a light-emitting diode (LED) and a solid-state receptor of that light. The receptor can be another diode, a transistor, an SCR, or a triac. See Figure 4–52.

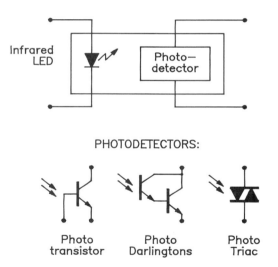

Figure 4–52 Schematic symbols for several kinds of infrared optocouplers/isolators

Optical isolator inputs act the same as any LED. The only difference between an LED and the isolator input is that no light is visible from the LED, being contained entirely within the opaque body of the component. The input LED port of the isolator can be tested under static conditions if there is a DC voltage limited to about 10 milliamperes steadily applied to it. There should be about a 1.0 V drop across an infrared LED, the typical LED chosen for this application. Note that this is not the familiar 0.3 or 0.7 voltage drop of a rectifier or signal diode.

The output component used inside the isolator can be tested as though it were a normal, discrete component. In other words, if the output component is a transistor, one would expect it to act in a circuit just like an ordinary transistor. Remember, of course, that there will be no normal input element, however. The light striking the component's sensitive input element, rather than a physical wire, becomes the input "lead."

UNIJUNCTION TRANSISTORS IN DC CIRCUITS — The unijunction transistor is a three-lead device that is almost always used as an oscillator. Troubleshooting this component in a static DC situation is not applicable. See the end of paragraph 4.12 for tips in troubleshooting this special component.

4.8 Make Static DC Voltage Readings in the Defective Stage

— Troubleshooting with a voltmeter usually requires several readings to find the problem. It is a good idea to mark down the voltage readings as they are taken. A good place to note these readings, in pencil, is right on the schematic.

Keeping in mind the various components and their effects on DC readings, make static DC voltage readings in the defective stage. Be sure you have the circuit common as the voltage reference. In other words, put the negative lead of the voltmeter where it is supposed to be. There are many pieces of electronic equipment that do not connect the power supply common to the chassis of the equipment. When in doubt, look for the filter electrolytic capacitors on the PC board. Is the negative lead connected to a trace that goes to the chassis? If so, then the chassis is probably OK to use as the voltmeter negative lead. If, on the other hand, the negative PC board trace avoids the mounting screws, it is a good indication that the chassis is not grounded to the power supply. See Figure 4–53.

Figure 4–53 A printed circuit that does not connect one side of the power supply to the chassis is often obvious about avoiding grounding screws

Another tip that the circuit does not use the chassis for ground is the use of nylon or plastic screws for mounting the circuit board to the chassis.

Many circuits will require pre-setting front panel controls to specified values before the DC readings in the circuits will be as specified on the schematic. If internal adjustments have been moved, the same principle applies. The controls must be set properly before the associated DC voltages will be correct.

4.9 FIND ANY BAD DC LEVELS?

— Differences between observed DC voltages and those expected either by estimation or by reference to a published value on a schematic should be carefully considered for several things:

1. Is the published value reasonable? Books are not always right. If not reasonable, compare the given values on a good piece of equipment, if possible.
2. Is your estimate correct? Re-evaluate the circuit and watch for paralleling paths of which you may not have been aware the first time through.

3. Is your reading taken correctly? How about the ground connection? Is it the correct one to use?
4. If the readings you take have suddenly dropped to zero or have become erratic, check the condition of your test leads. They may have developed an open or intermittent connection.
5. Do you have a *positively* good connection? A voltage reading of zero or one that is erratic can easily be caused by corrosion, sometimes invisible, on any connection made with the test probes. This is another good argument for *sharp* test probes. And don't forget the likelihood of a poor ground connection for the voltmeter.

The majority of circuit problems can be detected in analog circuits with the intelligent use of a DC voltmeter under static conditions. Perhaps 90 percent of all failures can be detected using this technique. Because of this, DC static voltage troubleshooting is the technician's first method of attack. There are exceptions, however. See Figure 4–54.

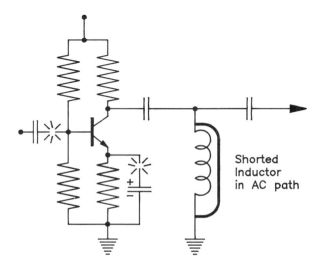

Figure 4–54 Several examples of component failures that will not affect the DC voltage readings

The fact that the DC readings are all normal should indicate to the technician that *those components that could have shifted DC readings but didn't shift them must be all right!* Look again at one of the circuits just shown and eliminate the components that must be good. See Figure 4–55.

Suppose your indicated voltage readings are close? How close is close enough? One clue can be taken from the resistors in the equipment. If they are 10% and 20% resistors, circuit voltage can probably vary as much as that percentage, too, without seriously affecting operation of the circuit. But if the resistors are precision, 1%, or wire wound precision 0.1% resistors, then tolerances on the indicated DC voltage tolerances should be tightened by about the same amount. For another indication of whether the voltage reading is close enough, review the original symptoms. If the circuit isn't working at all, it is seldom caused by a voltage reading that is only a few percentage points from normal.

If this approach still hasn't isolated the defective component, signal tracing is the second method that can be used to find the bad part. Signal tracing can also be called dynamic troubleshooting, because the signal entering the circuit causes constant changes of the voltages in the circuit.

4.9.1 Use Base-to-Emitter Test to Verify — If the suspected part happens to be a bipolar transistor operating Class A, there is a very quick check that can be made to see if the transistor is operating normally. Simply short the *base to the emitter* of the transistor. Be sure that you have identified the leads of the transistor correctly, as even

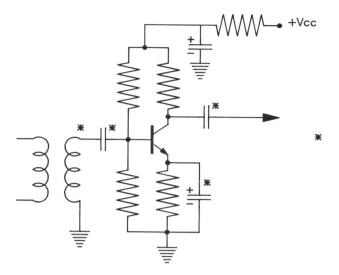

Figure 4-55 If this circuit should completely fail, yet not show a DC voltage out of normal, only the components marked could cause the problem

momentarily shorting the base to the collector could instantly destroy the transistor. The voltage at the collector should go from a value of about half the Vcc supply to full Vcc when the base is shorted to the emitter. See Figure 4-56.

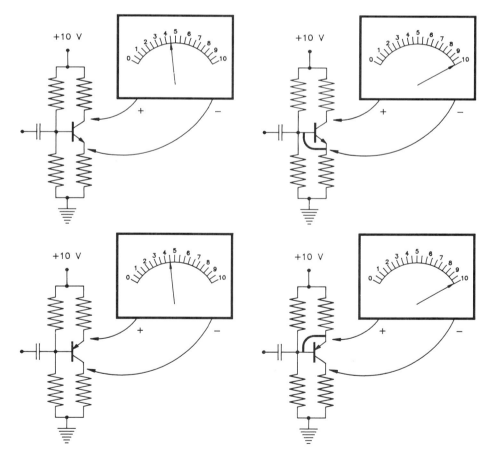

Figure 4-56 A quick test of the amplifying capability of a Class A amplifier stage

This test stops the forward conduction of the transistor. If the transistor is in good condition, the emitter-collector circuit should immediately become an open circuit, causing a full Vcc reading in the output circuit.

4.9.2 TRACE CAUSE USING THE PEROZZO 5-STEP METHOD — At this point in the troubleshooting procedure, the technician should have one or more DC voltage readings or signal levels that do not match those required for the circuit. *Pick the voltage level that is the farthest percentage from normal and concentrate the following efforts on that single point in the circuit.* Shift this point of concentration only if a more abnormal reading is found.

It takes experience to develop a "feel" for troubleshooting using a DC voltmeter or AC signal reading levels. The time required to develop this feel can be greatly reduced if the technician memorizes 5 simple steps to use at this stage in troubleshooting. These steps will reveal the offending component quickly and efficiently. Two cases must be considered: when the indicated voltage is less than it should be and when it is more. The first to be covered is the more common case where the indicated voltage is less than is required.

Voltage reading is too low or zero

1. Is the voltage source normal?
2. Is the meter operating normally?
3. Could the meter be loading the circuit excessively?
4. Is there an overload condition?
5. Is there too much resistance between load and source?

Explanation of these questions is based upon a generic circuit. This circuit represents *any* electronic circuit so far as voltage measurements are concerned. See Figure 4–57.

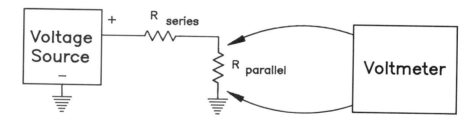

Figure 4–57 This circuit represents all circuits in which voltage measurements are made

Quite naturally, if the voltage source for a circuit is too low or too high, the voltages throughout the following circuits will be low or high by about the same percentage. One cure for unusually high or low AC line voltage sources is to bring the circuit voltage back within normal limits by changing the winding taps sometimes provided by power transformers primary windings.

IS THE VOLTAGE SOURCE NORMAL? — The voltage source referred to in Figure 4–58 is the last *good* voltage reading that is the source for the defective reading. It may or may not be an actual source of voltage, such as the power supply or a battery. See Figure 4–58 for a clarification of this concept.

Once a good voltage source is identified, the problem is already narrowed down to a specific area of the circuit. Circuitry from this point back toward the ultimate source, such as the wall outlet or batteries, must be operating normally.

IS THE METER OPERATING NORMALLY? — Checking for a good voltage source also is a check on the voltmeter or AC signal-detecting instrument, such as an oscilloscope. A defective instrument will not test a good voltage source successfully. Substitution of a good instrument will verify if a suspected instrument is bad.

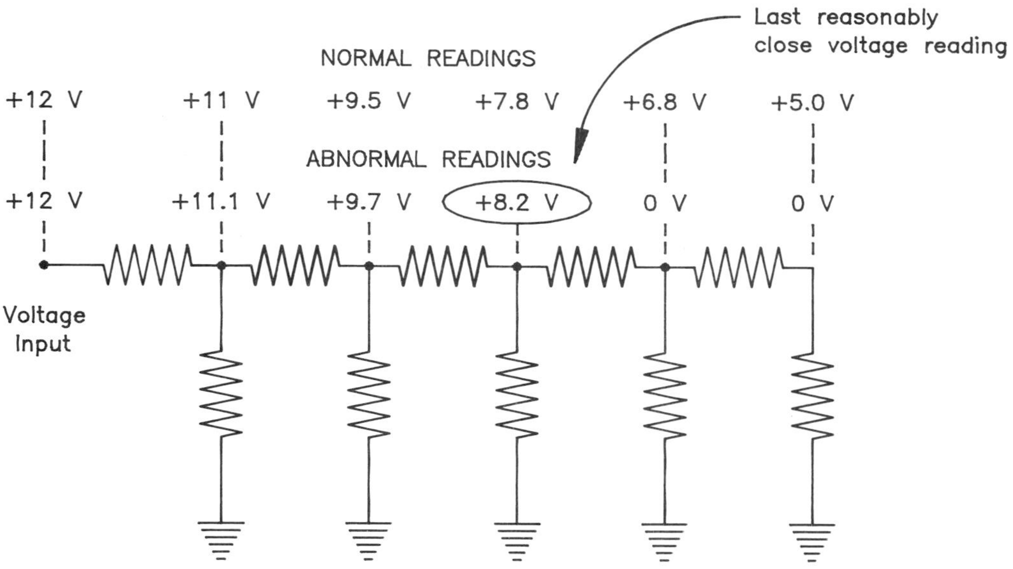

Figure 4-58 The voltage source for the 5-step method of voltage troubleshooting is the last reasonably close voltage that supplies the defective circuit

COULD THE METER BE LOADING THE CIRCUIT EXCESSIVELY? — The digital voltmeter endorsed throughout this text is adequate for perhaps 99.8 percent of the voltage measurements the technician might ever have to make. The input impedance of a meter, such as the one shown in Figure 4-21, is very high. A high input impedance simply means that the resistance of the meter is so high that it will not disturb the voltages in the circuit under test. Almost all circuits can tolerate an additional load of 10 megohms. About the only circuits that cannot tolerate this loading are the input circuits of some FETs. These will be obviously unusual circuits because they will contain resistances that are somewhere around the same magnitude as the 10 megohm input of the digital multimeter. As a general rule, if the circuit under test has resistors of less than one megohm of resistance, the digital multimeter will not load the circuit appreciably.

IS THERE AN OVERLOAD CONDITION? — Now an abnormally low voltage reading has been identified. The definition of "load resistance" needs to be clarified. *The load resistance is the circuitry in parallel with the voltmeter leads as the voltage reading is taken.*

If the resistance of the load upon the voltage source is too low, there will be too much current flow. This current flow will cause an abnormal distribution of the voltages between the series resistance and the load resistance. Some examples of overloads and short circuits with the voltages produced by these defects are shown in Figure 4-59.

Extremes of low voltage may go so far as a reading of zero volts across the load. This could be caused by an absolute short circuit rather than the partial overload with its accompanying lower-than-normal voltage. This overload condition often occurs as a result of component failures. Transistors and capacitors frequently fail by shorting.

IS THERE TOO MUCH RESISTANCE BETWEEN LOAD AND SOURCE? — The presence of too much resistance between the load and the voltage source can also cause too little voltage to be available at the load. See Figure 4-60.

Voltage readings of zero across the load can also be caused by a completely open circuit between source and load. An open circuit or excessive resistance between load and source is commonly due to an open pass transistor in a voltage regulator or possibly one of several resistors in RC filters intended to filter noise and ripple from the voltage supply line. Excessive resistance can also be due to poor ground connections, a fact often overlooked on a first pass of troubleshooting.

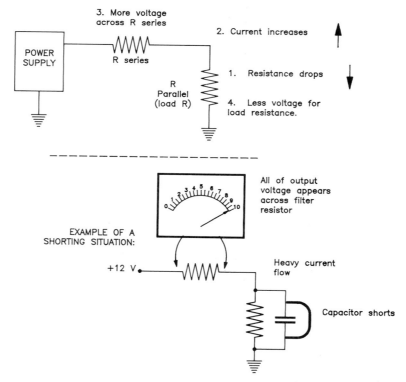

Figure 4–59 Why an excessive or shorted load has too little voltage drop across it

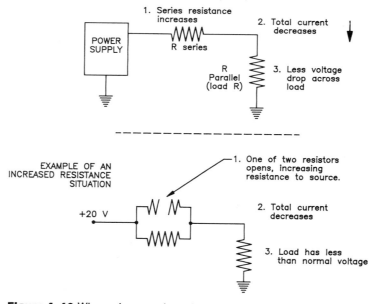

Figure 4–60 Why an increase in series resistance causes less load voltage

Voltage reading is too high — The second case of improper voltage readings concerns the voltage that measures higher than normal. Here are the questions which relate to this case:

1. Is the source voltage too high?
2. Is the meter operating normally?
3. Does the reading require a voltmeter load?
4. Is there too little load?
5. Has resistance between load and source decreased?

Using the same generic circuit as in the previous example, the same five questions can be reworded to cover this new situation.

IS THE SOURCE VOLTAGE TOO HIGH? — New batteries will often cause circuit voltages to be slightly higher than normal. Remember that an unusually high line voltage may be compensated for, as mentioned on page 165.

IS THE METER OPERATING NORMALLY? — A meter defect could result in a reading that is too high. Substitute another meter to check this as a possibility. During normal troubleshooting this is rarely a problem. It can surface as a problem when using meters that the technician has to share with others, or when a meter is borrowed or rented for temporary use.

DOES THE READING *REQUIRE* A VOLTMETER LOAD? — Some old circuits specified voltmeter readings must be taken with the volt-ohm-milliammeter (VOM). This instrument caused a substantial loading effect, particularly on the low-voltage ranges. The loading effect of such a meter on the 3 VDC range, for instance, is a mere 60,000 ohms, sufficiently low to affect low-voltage, sensitive circuits. To compensate for this loading effect, the schematic specifically states that readings were taken with the VOM. That comment might read something like "Voltage readings taken with a 20,000 ohm per volt meter." The digital meter, with its very high impedance input, does not load the circuit appreciably, and thus may produce a reading that is higher than expected. Although this is a rare case of higher-than-normal readings, the technician should be aware that this is a possibility when troubleshooting older equipment.

IS THERE TOO LITTLE LOAD? — If the resistance of the load has increased or opened, there will be too much voltage across it. See Figure 4–61.

HAS RESISTANCE BETWEEN LOAD AND SOURCE DECREASED? — The last possibility is that the resistance between the source voltage and the load has somehow decreased.

If the load voltage is the same as that of the source, the resistance between must be zero ohms, a short. Common causes of such failures are shown in Figure 4–62.

In summary, when using the Perozzo 5-step method, the technician should:

Step 1. Mark on the schematic the voltage reading that deviates farthest from normal.

Step 2. Identify the voltage source for the circuit and check for proper voltage. This step also checks the accuracy of the meter.

Step 3. Determine that the meter is not affecting the circuit. Circuit resistances are an indication.

Step 4. Note that the load resistance may have decreased or increased, resulting in lower- or higher-than-normal voltage readings, respectively.

Step 5. Note an alternative, that the resistance between the voltage source and the load may have increased or decreased, resulting in lower- or higher-than-normal voltage readings, respectively.

Memorizing these 5 steps will make the beginning technician a much faster troubleshooter and far more efficient than when attempting to "feel" the way through a circuit without really understanding what the instrument is indicating.

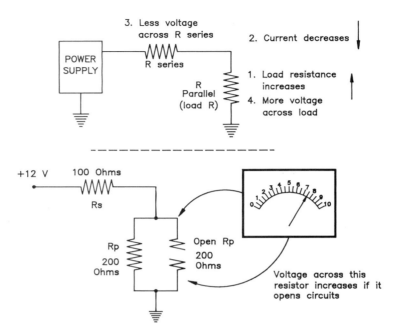

Figure 4-61 Why the load voltage goes up when the load decreases

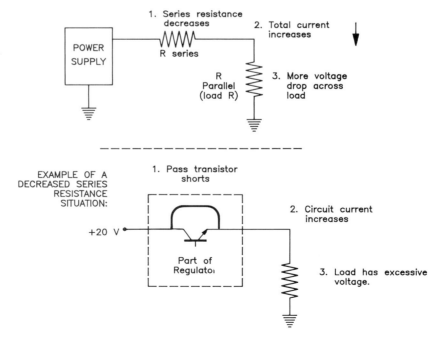

Figure 4-62 Why a decreasing resistance causes excessive load voltage

When troubleshooting transistor circuits, remember that the source for the base voltage is usually a resistor from the base to the collector source voltage. The source for the emitter voltage is principally the current flow through the collector circuit. See Figure 4-63.

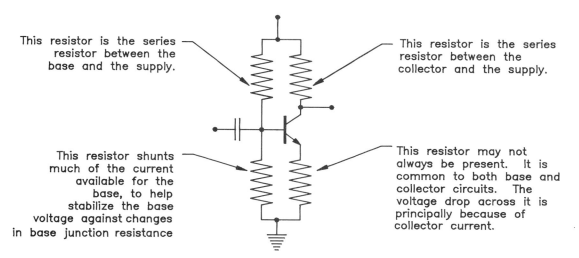

Figure 4-63 The operating Class A bipolar amplifier and its voltage sources

4.9.3 Consider the Odds of Component Failure — Most real world troubleshooting problems covered by the 5-step method boil down to either a defective series resistive component between source and load or a defective load. There are several possible components that can cause the problem. It often occurs that, during troubleshooting at this level, the technician is given a choice in selecting one of two components that may have failed. See Figure 4-64.

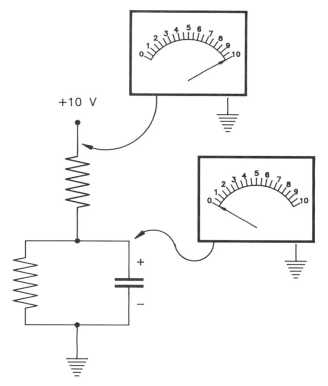

Figure 4-64 A troubleshooting choice. Is the capacitor or the resistor more likely to be the problem?

Such choices are not, as they might first appear, a 50/50 selection. Some components are much more failure-prone than others. The following list is the author's estimation of the failure rates of average components in actual circuits. When given a choice, as in Figure 4–64, it is most likely that the first component encountered in this list is the one at fault and that is the first one to be removed for out-of-circuit verification as the bad part.

Component	Failure Pattern
Batteries	Wear out
Fuses	Often open
Capacitors	Usually short, sometimes leaky, seldom open or out of tolerance
Semiconductors*	May short or open
Resistors	Sometimes open, occasionally out of tolerance, never short
Switches	Permanently open or closed
Relays	Contacts often erratic, coils rarely open
Vacuum Tubes	Wear out, break
Microphones	Occasionally inoperative
Speakers	Occasionally open
Inductors	Occasionally open, seldom partial short
Transformers	Occasionally open, seldom partial short
Meter Movements	Occasionally open, physical damage
Crystals	Seldom open, circuit usually at fault
Neon Indicators	Get dim only

*Includes bipolar transistors, FETs, integrated circuits, UJTs, SCRs, triacs, diodes, and LEDs

4.9.4 NOTES ON LIMITATIONS OF IN-CIRCUIT OHMMETER TESTING —

ALWAYS TURN OFF ALL POWER AND DISCHARGE ALL HIGH-VOLTAGE OR HIGH-VALUE CAPACITORS BEFORE USING AN OHMMETER OR SOLID STATE TESTER!

Once the technician has a good idea as to which components are bad, it is time to remove power from the circuit. Further testing with the power on is not needed.

At this point in the troubleshooting procedure there probably will be, as mentioned above, only one or two components that can produce the DC or AC voltage discrepancies. Considering the likelihood of failure rates, there may be more to learn about the failure now, before removing suspected components. Perhaps a quick check with the ohmmeter will verify which of the components are bad without the necessity of removing both in the case of a near 50/50 chance of failure.

By their very presence, existing resistances in the circuit can be used to help track a shorted component. A short circuit measured at one point cannot be located at another point if there is a large resistance in the path to it. See Figure 4–65.

Making resistance checks of a circuit before anything is removed is only a partially effective method of troubleshooting. The reason for this is that the circuit has, in all likelihood, many obvious parallel and series paths for the ohmmeter to detect plus some subtle ones, too. Here are some of the pitfalls to avoid.

Measuring semiconductor junctions in-circuit — Semiconductors, whether they are bipolar transistors, SCRs, triacs, LEDs or simple diodes, usually have paralleling components around the junctions of the device. Analyzing the schematic will show these parallel paths in most cases.

When using the ohmmeter to test these junctions in-circuit, readings will often indicate resistance where one would expect an open, because of paralleling paths. The junction, however, can still be tested *if* the ohmmeter can produce sufficient voltage across the measured

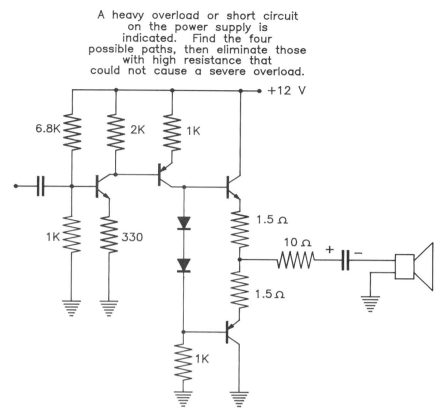

Figure 4–65 Use the resistances in the circuit to help eliminate those components that cannot be the cause of a high-current short

circuit to reliably forward-bias the junction. Most digital ohmmeters *do not* have sufficient voltage. It would not be unusual, for instance, to read 500,000 ohms in the forward-biasing direction of a diode when using such a meter on the ohmmeter scales. This reading might convince the technician that the diode is defective. This is seldom the case. The solution to this problem is the use of a special *diode scale* on the meter. This special scale attempts to put a reasonable current through a semiconductor junction, one milliampere being a common value.

The diode scale of a digital ohmmeter usually reads out in units of volts, not ohms. The resistance of a junction is immaterial since the resistance depends entirely upon the amount of current flowing through the junction. The voltage across the forward-biased junction, on the other hand, is a very pertinent reading. See Figure 4–66.

Using the diode test scale, in-circuit readings of the junctions can be made, but often with considerable interpretation necessary on the part of the technician. Low resistances normally in the circuit can upset the reading. If parallel resistance is sufficiently low, the diode scale will see that low resistance, thereby generating insufficient voltage to forward-bias the junction. See Figure 4–67. The usual indication of a good junction with this circuit is that the diode scale may read higher voltage in one direction than when the test leads are reversed. If the paralleling resistance is too low, the ohmmeter will "see" only the resistor.

Measuring resistors in-circuit — Resistors are often connected in parallel to something else in the circuit. These parallel paths will make the resistor reading measure less than the resistance of the resistor itself.

One particular case might be very misleading. If a resistor is in parallel with an open resistor of the same size, the resistance reading across the remaining good resistor will still be read on the ohmmeter. This would make the ohmmeter seem to indicate that the circuit

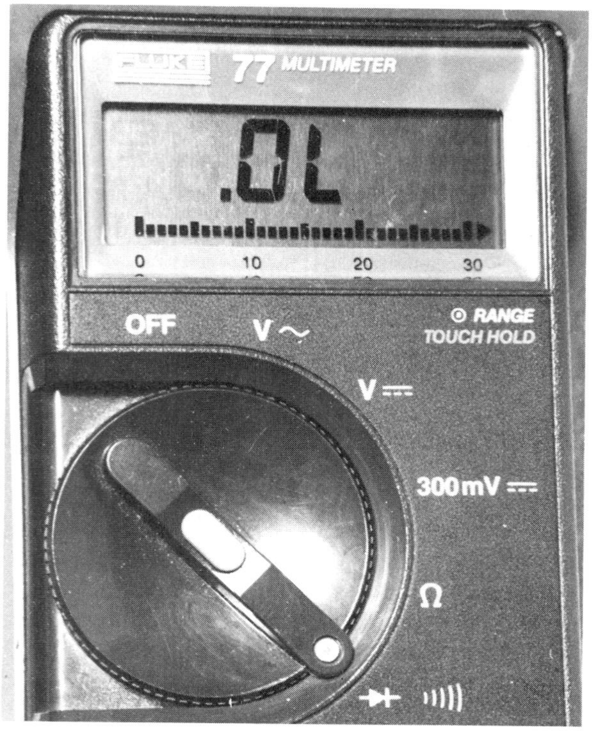

180 □ Identifying the Defective Component

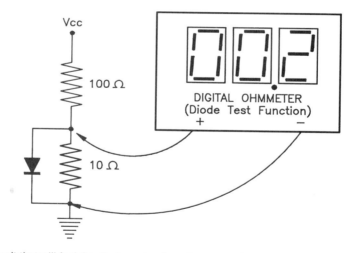

Figure 4-66 The digital multimeter may have a diode scale that reads out the conduction junction voltage—instead of its resistance

Figure 4-67 A circuit that will fool the diode scale of an ohmmeter when taking a reading from anode to cathode

is fine, where if it actually were, the reading would be half that of one of the resistors alone. See Figure 4-68.

Resistors in parallel with other components can sometimes be good indicators of where the problem lies. See Figure 4-69.

This is a common circuit configuration. Since the resistor "never" shorts, the zero ohmmeter reading is an excellent indication that the capacitor is shorted, not the resistor.

One of the biggest errors introduced into in-circuit ohmmeter readings is the paralleling paths of the rest of the stages and the power supply itself, when measuring across this bus. See Figure 4-70.

Electronic Troubleshooting □ 181

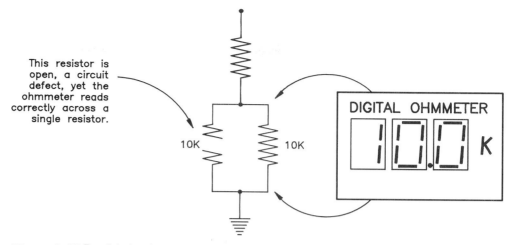

Figure 4-68 Paralleled resistors can be misleading. If one is open, an in-circuit ohmmeter check will show a value that might be mistaken for a good reading

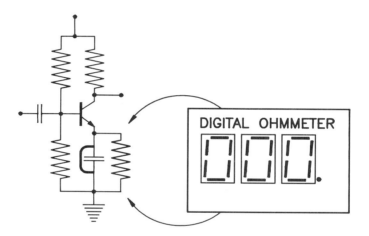

Figure 4-69 The resistor in parallel with a shorted capacitor will also appear shorted

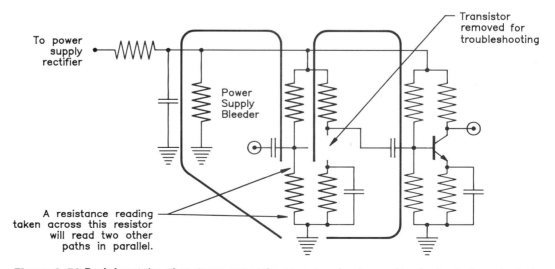

Figure 4-70 Don't forget the other stages across the power supply when making in-circuit ohmmeter tests. The power supply bleeder resistor is also easy to forget

Measuring inductors in-circuit — Inductors consist of turns of copper wire on some sort of core, depending upon the frequency at which they are used. Since copper wire has relatively low resistance when compared with other electronic circuit materials, the inductor will measure a very low resistance. The actual resistance will depend upon the size of the wire used in the inductor winding and the wire's total length. If a very small inductor has a hundred ohms or less, it is probably good. High-frequency inductors consisting of a few turns of heavy wire should have essentially zero ohms of resistance. Large power inductors may have values averaging from a few ohms upward to perhaps a hundred ohms for small inductors.

Inductors are almost always insulated from the case ground. An inductor with leakage or a short to the case is very likely to be defective.

The technician should carefully study the circuit being tested to see if a resistance reading across an inductor is a valid reading to take. Paralleling resistors or other components can provide undesired alternate current paths for the ohmmeter.

4.9.5 NOTES ON USING THE SOLID STATE TESTER FOR IN-CIRCUIT TESTS

**ALWAYS TURN OFF ALL POWER AND DISCHARGE
ALL HIGH VOLTAGE OR HIGH VALUE CAPACITORS
BEFORE USING AN OHMMETER OR SOLID STATE TESTER!**

The ohmmeter provides, at best, a two-point look into a circuit: one look with the ohmmeter leads with a given polarity, and a second look with the leads reversed. A solid state tester (SST) on the other hand provides two full dimensions of information about the components in the circuit. See Figure 4–71.

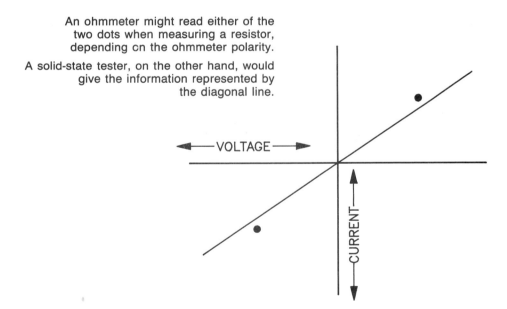

Figure 4–71 The ohmmeter can provide only two points of information where the solid state tester provides two dimensions, vertical and horizontal

Keep in mind the frequency and impedance ranges of the SST while testing. The displays shown are the circuit's response to a specific frequency of sinewave that the SST is providing as a stimulus.

As a starting point, the SST should be used on the lowest range and at 60 Hz (if this can be selected on the model being used) for testing single junctions. This is the case when testing bipolar transistors, diodes, and SCRs. When testing integrated circuits, one of the medium ranges should be used. This will provide a higher voltage to test the multiple junctions in series that these ICs contain.

The highest range should be used mostly for testing zener diodes and any other high-voltage device or semiconductor for voltage breakdown. The higher frequencies of excitation come in handy when working with small value capacitors and inductors.

Connect the test probes to the common and one of the two input terminals provided on the front panel of the SST. Switch to the correct input terminal for testing.

Test the instrument for operation by shorting the test leads. The display should then show a vertical line. When the probes are open the line should become horizontal. A permanent vertical line is an indication that someone has overloaded the instrument by applying power to the test leads, a definite "no-no." The internal fuse will have to be replaced before the instrument can be used. Once the instrument is working well, center the line horizontally and vertically by alternately shorting and opening the test leads and making the appropriate positioning adjustments.

The vertical amplifier of the SST shows the current flow produced by the introduction of a low-level, current limited AC signal into the circuit under test. The horizontal amplifier provides voltage information.

Learning to use the SST is best accomplished with a box of surplus parts, each tested out of circuit. Compare the displays shown with various components with those of paragraph 5.5. Once the basic waveforms of the resistor, capacitor, inductor, and semiconductor junction are recognized when viewed with the various ranges provided on the instrument, it is ready to be used for in-circuit testing. In-circuit tests can be done in several ways as shown in the Phase 6 flowchart. Briefly, these tests consist of comparing the displayed waveform with what the technician knows to be typical waveforms, comparing to another known good board, or by comparing using the methods of automatic test lead switching or using the accessory called the switcher to compare ICs. All of these methods of using the instrument are explained in Phase 6.

The SST will be used at this point in troubleshooting to verify what is already strongly suspected: that a specific part or parts are defective. The waveforms of defective components will deviate from normal indications and are clearly shown with this instrument. It is particularly good for spotting such items as open electrolytic capacitors and bad junctions that may not show at all using an ohmmeter. Similar limitations exist when using the SST, just as when using an ohmmeter. Refer to paragraph 6.0.1 for further details.

4.10 Apply Normal Input Signals — Simple signals can be generated by general-purpose test equipment such as audio, radio frequency, and function generators.

One or perhaps more input signals may have to be applied at this point in the troubleshooting techniques. The stage must be brought to *normal* levels of signal inputs.

Although usually a single input will suffice, some circuits require more than one input. If the stage is to be effectively analyzed, the signals must be correct. This means that they must have the proper amplitude and frequency. Some circuits require further signal criteria, such as two or more inputs of the proper phase relationship. Depending upon the circuit, it may or may not be brought into sufficiently normal operation with simple signal generators for troubleshooting. *If at all possible, the actual waveforms with which the circuit normally operates are preferred for troubleshooting.* Specialized circuits may require specialized test equipment to provide the unique inputs required.

Applying a signal to a circuit for test purposes can impose a problem for the technician. If the circuit is already set up for the application of a signal, such as an input connector to the first stage of a group of circuits, then the problem is easily solved. Simply apply the signal at the proper amplitude and frequency for the circuit under test.

Applying a signal into a circuit that is normally fed by a preceding stage imposes some problems. Any DC in the circuit must be blocked from entering the signal generator. The correct amplitude for the injected signal may not be easily estimated. The best solution to this problem is to use the preceding stages to condition the signal so that it appears as it should to the defective stage. Inject the signal at the normal input whenever you can. Refer back to Figure 3-10.

4.11 Select Instrument
— There are two common instruments that the technician might use to detect and provide a readout of a signal progressing through an analog circuit: the multimeter and the oscilloscope.

4.11.1 USING THE DIGITAL MULTIMETER TO DETECT SIGNALS
— The multimeter might be the only instrument at hand for troubleshooting, so it is good to know its limitations. Two operating scales apply, the DC and AC scales, each of which has its uses.

4.11.2 USING THE DMM DC SCALES
— The average voltage of an AC signal is zero. Using a DC meter to read AC should result in a value very near zero. This is because the DMM responds to the average value of a waveform, and equal + and − amplitudes of the sinewave will therefore cancel.

AC signals that pulse on and then return to zero are not the same. See Figure 4–72.

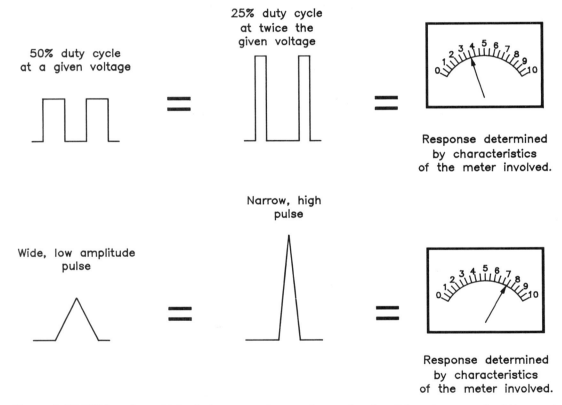

Figure 4–72 DMM readings obtained when measuring waveforms other than DC or sinewaves should be used for relative, comparative purposes only, as the voltage indications are not accurate

Some circuits operate by pulsing on and off. These can be partially analyzed with a DC meter. Such stages are those that are operating Class B or Class C. See the text accompanying Figure 4–42 for the Class B amplifier circuits and Figure 3–80 for Class C stages.

Using the emitter resistor of a transistor Class C amplifier is a common method of tracing signals through radio frequency circuits. The pulsations of RF are averaged out to a DC level by the bypass capacitor. The stronger the pulses, the higher the average voltage across the capacitor.

4.11.3 USING THE DMM AC SCALES
— The digital multimeter (DMM) can also provide a limited amount of information about AC signals, but only up through low audio frequencies, perhaps to 1000 Hz. The actual frequency limitations will depend upon the manufacturer and model of the instrument to be used. The AC scales of most DMMs are intended to be used

with 60 Hz waveforms and are calibrated to read accurately only at that frequency. Use of the instrument at other than this frequency may require a correction factor if an accurate voltage measurement is to be made. Most use of this instrument for troubleshooting, however, does not require extreme accuracy, and the instrument may be safely used up through the low audio frequencies, typically to 1000 Hz, without concern. Consult the multimeter specifications to be sure that it can be used at audio frequencies in the rare case that the AC accuracy is important.

The DMM provides readings without appreciably loading the circuit under test. The typical loading effect for a modern DMM on the AC scales is 10 megohms. This value is essentially an open circuit for most test purposes.

4.11.4 NOTES ON USING THE OSCILLOSCOPE —
The oscilloscope comes into its own in analyzing analog waveforms for troubleshooting. Beyond showing the mere amplitude of the waveform as the DMM can do, the oscilloscope can also show frequency, waveshape, and even the phase relationships between signals. Even when the oscilloscope is used for measuring simple DC voltages, it offers the advantage of showing unsuspected AC and/or pulse signals.

A disadvantage of the oscilloscope is that there is more room for operator error than when using a digital voltmeter. The vertical gain settings of the oscilloscope can be misinterpreted; when a 10× probe is used, for instance.

Whenever the oscilloscope is used, readings must be taken with respect to a common ground point within the circuit. It should not be used to measure the voltage across a component that is not connected at one end to ground. See Figure 4–73 for an explanation.

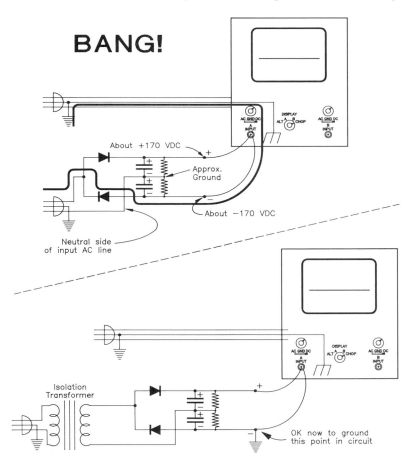

Figure 4–73 Connect the oscilloscope ground lead only to those points in a circuit that are at ground or which are completely unreferenced to ground. Unthinkingly connecting it to a negative point, assuming that it is a grounded point, can cause serious circuit damage

In order to get a "feel" for the signal frequencies and voltages being measured, the sweep time per division (sometimes marked horizontal frequency) and the vertical amplitude settings of the instrument should be carefully noted. *Without an awareness of the settings of the instrument, it is difficult to get any idea of the significance of the information on the screen.* An example might help point out the importance of this statement.

Suppose a technician is looking for audio signals as they progress through an audio amplifier. The oscilloscope is set for 0.1 V per division and the sweep time per division is set to 5 milliseconds. With these settings, about three cycles of a 60 Hz waveform would be displayed, and it would require about 2 VAC to give a pattern that extended three divisions up and three down. These would be good average settings to use in following low-frequency audio signals through a circuit. Signals appearing very small in amplitude on the screen should be investigated to see if this is normal for that point in the circuit. *If the technician cranks up the gain of the instrument to look at the signals, mental reference levels may be lost.* Under such circumstances it is easy to begin "chasing" signals through the circuit without coming to concrete decisions about the oscilloscope displays given.

The whole point is that the oscilloscope should not be operating at extremes of sweep time or voltage sensitivity unless there is a reason to do so. Start with common settings and modify them intelligently for other uses. For instance, looking at a higher frequency will make it necessary to increase the sweep frequency. If the waveform is too high on the screen, decrease the vertical gain setting, and so forth.

The most common mistake made by beginning technicians using an oscilloscope is to arbitrarily twist the instrument knobs in the hope that somehow the pattern will improve. Instead, observe the picture, *think,* and then reach for the proper control.

Learning to use the oscilloscope requires time. There are some very sophisticated instruments available these days that have a bewildering array of controls. Of these many controls, however, only the basic controls are used frequently. The additional controls are "frills," almost never used during routine troubleshooting work. It is a good idea for the beginning technician to put small removable labels or marks at the normal positions of the lesser-used controls and simply leave them alone until the instrument is thoroughly learned. Take these special controls one at a time, read the instrument operating manual, and experiment with their uses as time or workload permits.

The settings of an oscilloscope that are frequently changed during troubleshooting include the following items. An instrument with two input channels is assumed, 10× probes, and periodic signals such as audio or RF are being observed.

1. Vertical gains set to about 0.1 V per division for most troubleshooting.
2. Sweep speed set to about 5 milliseconds per division to observe audio frequencies. Faster (less time per division) as necessary to observe higher frequencies.
3. Vertical mode set to A only or Alternate if both probes are being used.
4. Vertical centering to put baseline at the vertical center.
5. Horizontal centering to start trace at the left edge of the screen, at the first graticule line.

These simple settings will take care of initially adjusting the instrument for most jobs. All other controls should be set at "normal" settings and left alone unless specifically needed for a reason.

Controls that are set once and not often manipulated are:

1. Intensity
2. Focus
3. Astigmatism
4. Beam finder
5. Delay time
6. Horizontal sweep mode
7. B Channel invert
8. Horizontal sweep vernier ("Cal")

9. Vertical amplitude vernier ("Cal")
10. Vertical mode switch
11. Triggering level

4.12 NOTES ON HOW COMPONENTS AFFECT AC DYNAMIC READINGS

A typical schematic may have a few DC voltage readings specified for the circuits at hand, but will have few, if any, AC signal readings noted. It will generally be left to the technician to measure the existing AC signals and determine if they are normal or not without benefit of any additional information. Once a stage is repaired, it is a good idea to mark the schematic with the normal AC signal readings for the next time.

Estimating AC voltages — Estimating the signal voltages within an operating circuit requires that the technician understand how various components change the signals as they pass through the circuit, just as under DC conditions. There is a great deal of effect on the circuit signals as they pass through the various components of the circuit.

EFFECTS OF CAPACITORS IN AC CIRCUITS — The effect of capacitors in any AC circuit depends on two basic factors: the capacity value of the capacitor and the frequency of the circuit. A special case arises when a capacitor is used with an inductor; this is covered separately in a later paragraph.

The series capacitor — The capacitor in series with the signal flow passes that signal on when the capacitor has a low reactance at the frequency in use. See Figure 4–74.

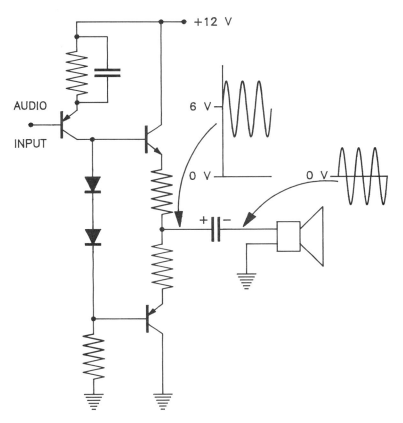

Figure 4–74 The series capacitor passes the AC signal but blocks any DC component from passing

The ability to stop any DC from passing along to the following stage is the principle reason the capacitor is used in most series applications. The capacitor is chosen so that it is relatively low in reactance at the lowest frequency to be passed, compared to the resistance it must feed in the following stage.

The paralleling capacitor — The parallel capacitor is used by itself to "load down" or bypass any AC signals, by providing a low reactance path for these signals to ground. At the same time, the capacitor is an open to DC. The parallel capacitor is most commonly used to smooth out short term variations and ripple in DC power supplies. See Figure 4–75.

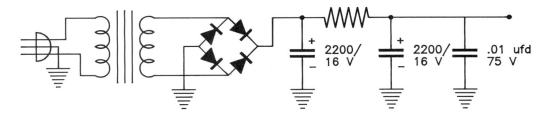

Figure 4–75 The electrolytic capacitor is often used to smooth out the voltage pulsations of a rectifier. Note the smaller high-frequency bypass capacitor also in parallel with the electrolytic capacitor

The electrolytic capacitor has a certain amount of unavoidable internal inductance. This inductance makes the electrolytic ineffective at high frequencies. The paralleling of another, smaller capacitor of the mica, monolithic, or ceramic type filters out these higher frequencies.

THE CAPACITOR AND RESISTOR COMBINATION WITH SINEWAVES — The reactance of a capacitor decreases as both the capacity and frequency increase. Typical values for practical circuits might be 25 ufd and up for power supply filter capacitors, 0.1 to 10 ufd for coupling in audio circuits, and 0.001 to 0.01 for radio frequency bypassing and coupling circuits.

Capacitors can fail by shorting, leaking, opening, or decreasing in value. The most common failure is the internal short circuit. A shorted capacitor is easily detected in most circuits by a drastic shift in the DC voltages within the stage. Leaky capacitors may cause the DC voltages of the circuit to shift, but not to the extent that a shorted capacitor might affect the circuit. An open capacitor is more difficult to detect. The open capacitor is usually found most effectively using a solid state tester or an oscilloscope and tracing signal voltages. The capacitor that has changed value is best verified as defective by substituting a known good capacitor of the proper value.

Using the capacitor and resistor combination with sinewaves causes changes in the amplitude and phase angle of the output signal in comparison with the input signal. As the capacitor increases in value or as the frequency increases, the voltage across the resistor will approach that of the input voltage because the reactance of the capacitor is decreasing, shifting more of the signal to the resistor.

The capacitor and resistor combination can assume two different configurations. See Figure 4–76.

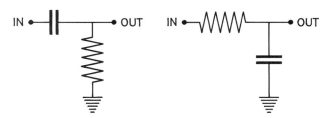

Figure 4–76 The capacitor and resistor in two different configurations

Coupling capacitors — The capacitor in a coupling configuration, in series with the signal flow, is usually very low in reactance at the circuit operating frequency and can be considered a signal "short," passing the signal without appreciable change, to the following stage.

A seldomly used effect of a capacitor used in combination with a resistor is phase shifting. There will be a particular sinewave frequency, for instance, at which the capacitor reactance and the resistance of the resistor will be equal. At this one frequency, the signal across the resistor will have about a little over half (70 percent) of the amplitude of the input signal. In addition to changing the amplitude of the signal, the output waveform across the resistor is a sinewave that is shifted 45 degrees and that *leads* the input signal in phase. See Figure 4-77.

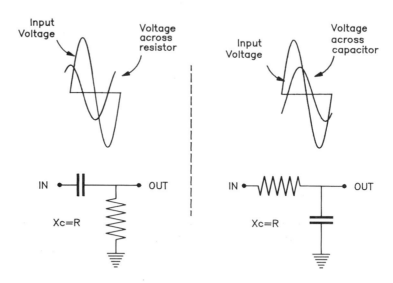

Figure 4-77 At the frequency where the capacitive reactance equals the resistance, the voltage between the resistor and the capacitor is shifted in phase by 45 degrees

A few circuits, such as the phase-shift oscillator, use the phase shifting principle for its operation.

Filtering capacitors — Consider the configuration in which the resistor feeds the capacitor. This circuit is most often used to eliminate higher frequencies and to pass DC and low-frequency signals. The choices of both the resistor and the capacitor determine the amount of loading effect on the high frequencies. See Figure 4-78.

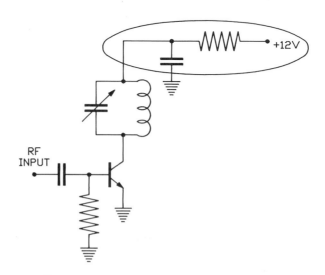

Figure 4-78 Typical circuit where the resistor/capacitor combination is used to filter out high frequencies, passing DC only

With a sinewave input, the output waveform at the frequency where the reactance equals the resistance will again be about 70 percent of the input signal, but now the output waveform across the capacitor will *lag* the input waveform by 45 degrees. See Figure 4-77.

In this configuration, increasing the value of the capacitor or increasing the frequency will result in the signal output voltage decreasing further in amplitude as the "shorting" action of the capacitor becomes more pronounced.

THE CAPACITOR AND RESISTOR COMBINATION WITH PULSES AND SQUAREWAVES — Using the capacitor and resistor in circuits with pulses and squarewaves produces effects entirely different from using them with sinewaves. The waveforms are often drastically changed when passing through the capacitor and resistor combination. The amount of change depends on the values of the resistor and capacitor used and the waveform of the input signal. Two representative examples will be considered to illustrate the changes that may occur to a pulse and to a squarewave. The first example uses a short input pulse and the second a 50% duty cycle squarewave.

Using the capacitor and resistor combination with pulses — The resistance and capacitance effect can be consolidated into a single factor called the *time constant* of that circuit. Simply multiply the capacity in farads times the resistance in ohms. This yields a timing factor that can be used to compare the speed of the RC circuit to the speed of the waveform applied. An example to show this relationship:

A capacitor of 1 microfarad is to be used with a resistor of 10,000 ohms. The time constant for this combination is:

$$T = RC = (1 \times 10^{-6}) \times (1 \times 10^{4}) =$$
$$1 \times 10^{-2} \text{ or } 0.01 \text{ second}$$

A signal with the same duration of 0.01 second would see this combination of parts as a medium time constant. A much slower signal with a duration of 0.1 second would see the same combination as a short time constant circuit when compared to itself. Of course, a fast signal that had a duration of 0.001 second would see this resistor and capacitor combination as a long time constant compared to its own period. Thus, time constant relates how an RC circuit affects the input waveform.

If the technician is troubleshooting pulsed circuits such as are used in radar, sonar, loran, television, and many other fields, the ability to predict waveform changes will be valuable. With the input waveform known, the output waveform can be approximated without schematics by comparing the time constant of the circuit to the duration of the voltage changes on the input. Refer to the waveforms that follow to get an idea of the proper waveform that should be at the output of a particular stage and compare that to the observed waveform as seen on an oscilloscope.

Pulsed circuits often use a capacitor to pass a pulse to the following stage. The pulse will be passed on essentially without change if the capacitor and resistor combination has a long time constant when compared to the duration of the signal. By the time the capacitor begins to charge, the pulse is gone and the input voltage drops to zero. The capacitor returns to its original voltage between pulses by discharging through the circuit resistances. Then the cycle repeats. See Figure 4-79.

If the capacitor and resistor combination has a relatively short time constant (an unusual case with pulses as the input), the waveforms of Figure 4-80 would result.

Note that the use of the series capacitor will block any DC voltage on the input from the output circuit.

Using the resistor to feed a long time constant circuit, on the other hand, is a different application altogether. This circuit is sometimes used in practical circuits to *integrate,* or average out, the voltage value of a string of pulses. See Figure 4-81.

The voltage that appears across the capacitor depends on the amplitude of the pulses, their width in time, and how often they occur. The capacitor voltage thus provides a measure of the energy of the pulsed waveform. The most common use for this circuit is the filter

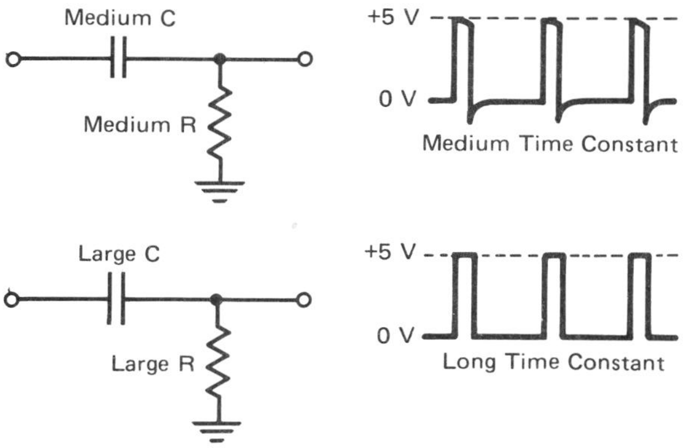

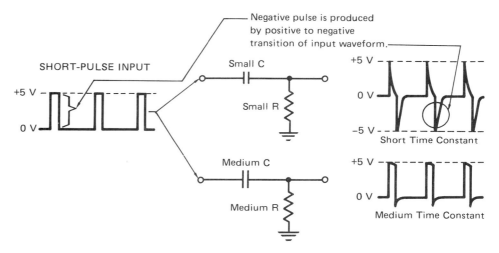

Figure 4-79 Typical circuit where a capacitor feeds a resistance a pulsed voltage

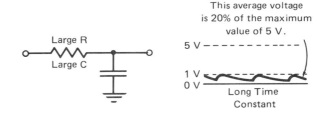

Figure 4-80 Using a capacitor and resistor with a medium and a short time constant with a pulsed input signal

Figure 4-81 Resistor used to feed a large capacitor integrates, or averages out, the input pulses

capacitor in a power supply. The capacitor is supplied half sinewave pulses and because of the very large values of capacity used, the capacitor averages the output voltage. A different way to look at the same circuit is to say the capacitor filters out high-frequency components by shorting them to ground. At the same time, the capacitor is an open circuit to the DC component of the power supply.

It is less common, but this same circuit could be built with medium or short time constants. The circuit would then couple the signal without integrating it, becoming mere coupling circuits with a very small amount of high frequency filtering as the time constant became very short. See Figure 4-82.

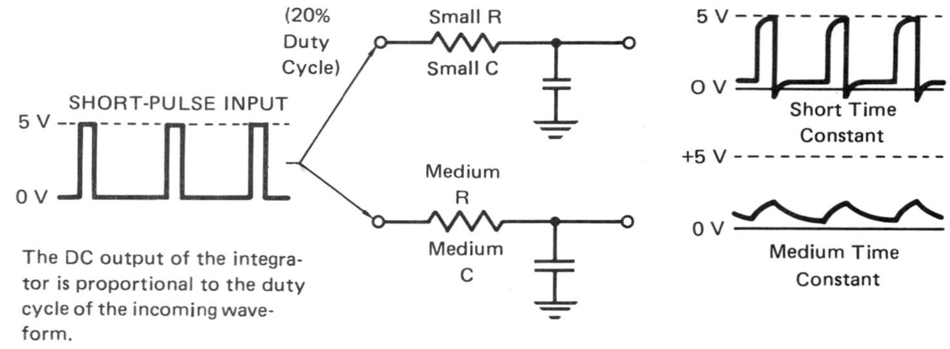

Figure 4-82 Effects of medium and short time constants with a pulsed input signal

Using capacitor and resistor combination with squarewaves — A squarewave applied to a capacitor that feeds a resistor results in a *differentiator* circuit. The output of a differentiator circuit will vary depending on the time constant for that circuit. The effect of the time constant on the output waveform when the input is a 50% duty-cycle waveform is shown in Figure 4-83.

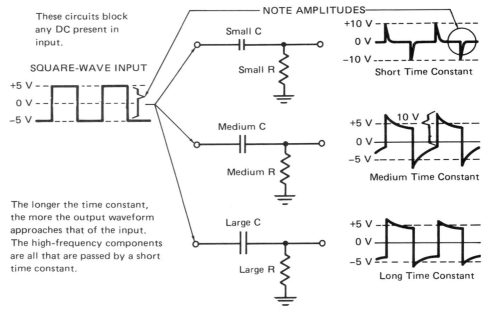

Figure 4-83 Effect of time constant on an RC differentiator circuit

Two of these circuits are common, the circuit with the short time constant and the one with the long. The short time constant circuit is often used to produce a single, short pulse for each transition of the input waveform. One polarity of the output waveform is generally not used in this application. The long time constant circuit is used to couple the squarewave signal to the next stage with minimal changes.

Because of the series capacitor, each of these circuits will block any DC that might be present in the input signal from appearing in the output circuit.

Reverse the two components to an input resistor and capacitor to ground once more. Feeding this circuit with a squarewave results in a circuit that will *integrate* the input signal to a DC voltage across the capacitors explained for pulses. It will also couple across any DC in the input circuit. This averaging of a DC and an AC signal is sometimes used in practical circuits. Most generally it is desirable when doing this to provide large time constants that essentially result in a DC voltage from the circuit. Medium and short time constants are less often used. See Figure 4-84.

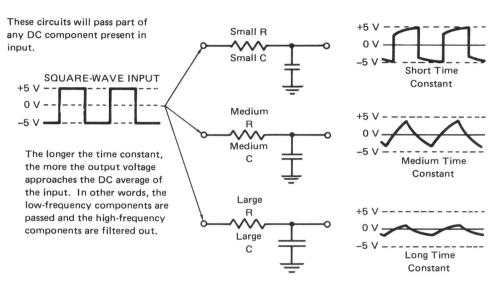

Figure 4-84 Using the averaging characteristics of a resistor and capacitor circuit to average a squarewave and DC input together. Note the DC level of the input waveform and the output average voltage

EFFECTS OF INDUCTORS IN AC CIRCUITS — The effect of inductors depends upon the value of the inductor and frequency in use. The effects are similar but opposite to those of the capacitor. Where an increase in operating frequency causes the reactance of the capacitor to decrease, the inductor increases in reactance.

Inductors are more costly to manufacture than capacitors. By using the capacitor in "opposite circuits," the same effect as an inductor can sometimes be obtained at less expense. See Figure 4-85.

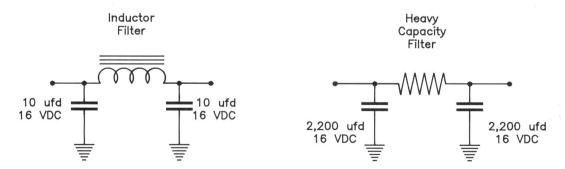

Figure 4-85 How the capacitor can do the job of the inductor for less money

The inductor in series — The inductor is most commonly used in series with a signal path to pass DC and to reject power-line AC signals, audio, or RF signals depending on the size of the inductor. See Figure 4-86.

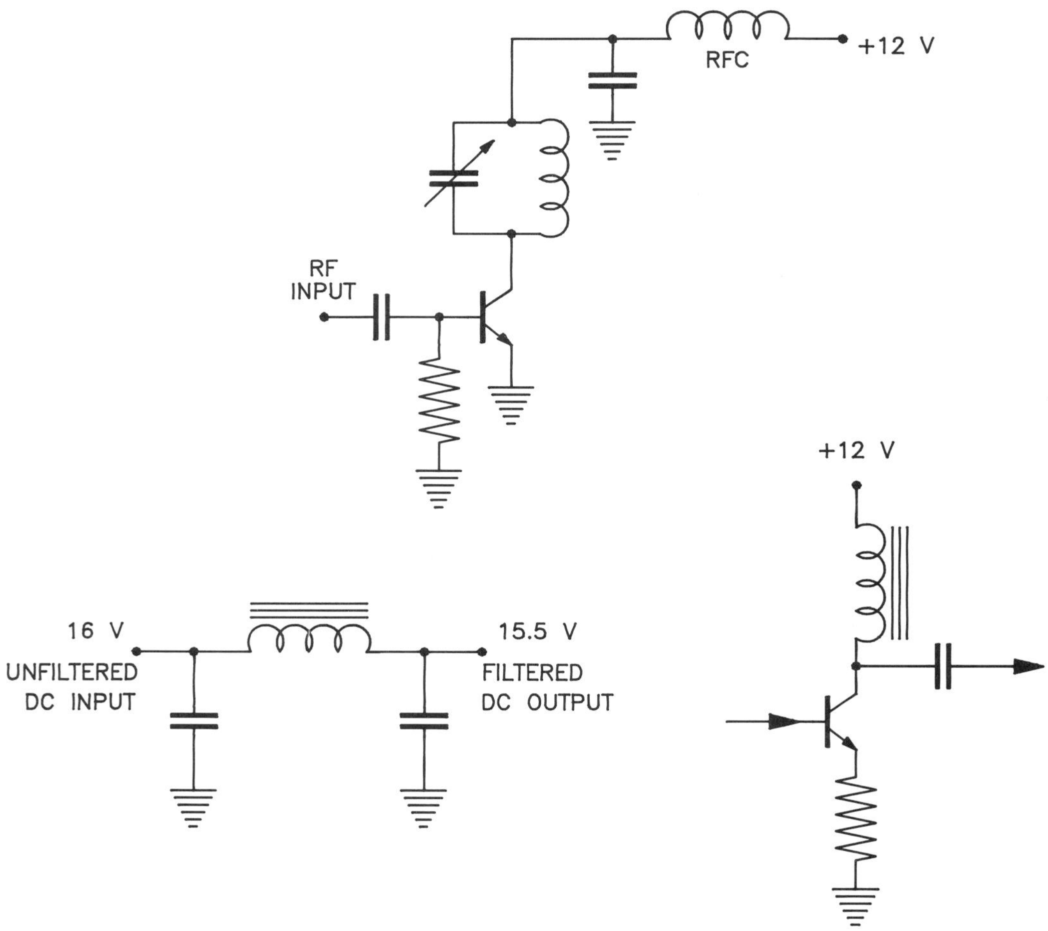

Figure 4-86 The inductor is often used to pass DC and to block power line hum, audio or RF signals

The inductor in parallel — The effect of an inductor in parallel is to allow all DC and low frequencies to pass through the inductor, but to look like a high reactance to higher frequencies. It is seldom used in this capacity, though, since a capacitor in series with the signal can be made to accomplish the same result and is less expensive to build.

THE INDUCTOR AND RESISTOR COMBINATION WITH SINEWAVES — Inductors in circuits other than power supply filters are rare. On those occasions where they are used, their failure is often easily detected using DC static readings. See paragraph 4.7, "Notes on How Components Affect DC Static Readings." On those occasions where DC readings do not conclusively prove a bad inductor, substitution should isolate the problem.

The normal series inductor should block signals for which it is wound. The RF inductor should block RF, the audio inductor should block audio frequencies, and the power supply filter should block power line hum frequencies of 60 or 120 Hz. *The abnormal inductor will not block the signals* and yet may pass DC normally. These symptoms are characteristic of shorted turns within the coils of the inductor. If a small percentage of the turns are shorted out, the total resistance reading for the coil may not be far enough from normal readings to be able to say definitely that the inductor is bad. Substitution is the best method in such a case.

THE INDUCTOR AND RESISTOR COMBINATION WITH PULSES AND SQUAREWAVES — The inductor and resistor combination has a time constant relationship with the applied waveform, similar to a resistor and capacitor circuit. The inductor-resistor relationship is a bit different, however. In order to find the time constant of an inductor and resistor, divide the inductance in henries by the resistance in ohms.

$$\frac{1 \text{ henry}}{100 \text{ ohms}} = 0.01 \text{ second}$$

A time constant of 0.01 second would be long when compared to an input signal period of 0.001 second. The same circuit would seem short if the input signal had a period of 0.1 second. The LR time constant is very rare in practical circuits. Capacitors and resistors are cheaper and used instead.

Using inductance and resistance with pulses — If the inductor is used to feed the resistor, the output waveforms will be identical to those of the resistor feeding a capacitor. See Figure 4–87.

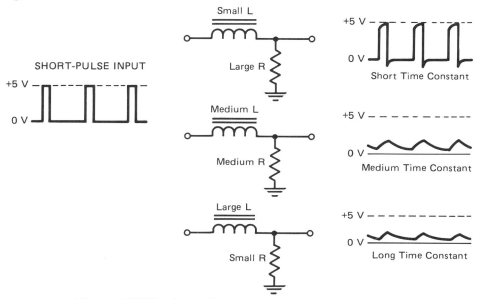

Figure 4–87 Waveforms of inductor feeding a resistor short pulses

If the resistor is used to feed the inductor, the output will be the same as the capacitor feeding a resistor. See Figure 4–88.

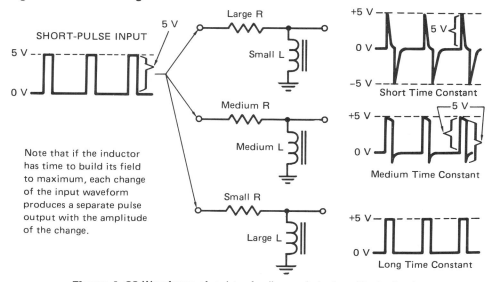

Note that if the inductor has time to build its field to maximum, each change of the input waveform produces a separate pulse output with the amplitude of the change.

Figure 4–88 Waveforms of resistor feeding an inductor with short pulses

Using inductance and resistance with squarewaves — As the technician might suspect, using an inductor to feed a resistor has the same result as feeding a capacitor with a resistor. See Figure 4-89.

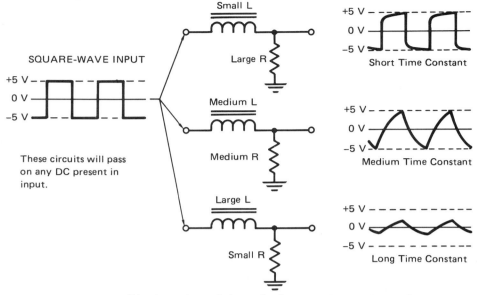

Figure 4-89 Waveforms for an inductor feeding a squarewave to a resistor

Last, but not least, is the possibility of feeding an inductor a squarewave through a resistor, much the same as feeding a resistor the same signal through a capacitor. See Figure 4-90.

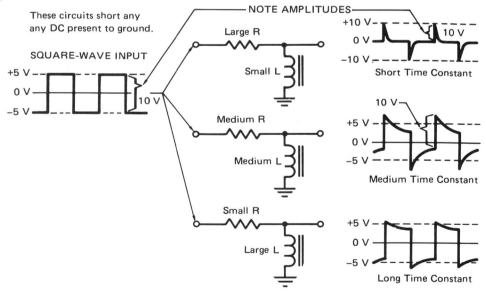

Figure 4-90 Waveforms for a squarewave fed by a resistor to an inductor

EFFECTS OF CAPACITORS USED WITH INDUCTORS IN AC CIRCUITS

The series capacitor and inductor with sine waves — The capacitor and inductor in series can look like one of three different circuits depending upon the frequency applied. There will be a single frequency where the reactance of the inductor and the capacitor are the same. Since the two reactances are in series and exactly opposite each other, the net result at this *resonant frequency* will be a *short circuit* from one end of the circuit to the other. Frequencies less than resonant frequency will see only some of the reactance of the capacitor in series with the signal. Frequencies higher than resonance will see only some of the reactance of the inductor. In summary, the series LC circuit looks like a higher series reactance the farther the input frequency departs from resonance. See Figure 4–91.

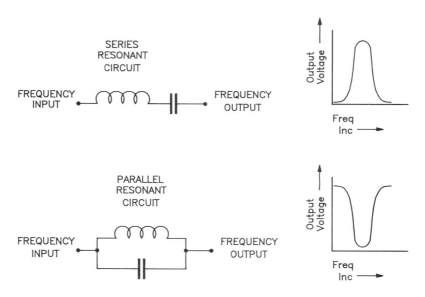

Figure 4–91 Graph of the effect of frequency on series and parallel resonant circuits

The parallel capacitor and inductor combination — Using the capacitor and inductor in parallel with one another also results in a single frequency where the two reactances are equal. At this resonant frequency, the parallel combination becomes "transparent," an open circuit. Frequencies below resonance see only the reactance of the inductor and those above resonance see only the capacitive reactance. Again, see Figure 4–91.

Effects of inductance, capacitance, and resistance with sine waves — Adding resistance to a series or parallel resonant circuit (this includes the normal resistance of the inductor wire) will reduce the sharpness of the tuning curve. This technique is sometimes deliberately used to broaden the frequency response curve of a tuned circuit. See Figure 4–92.

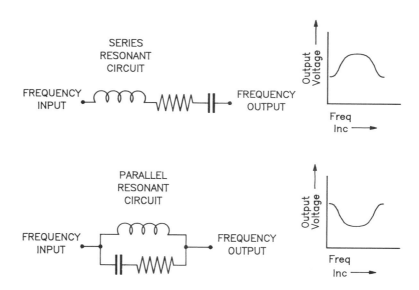

Figure 4–92 Effect of adding resistance to the tuned circuit. Compare to Figure 4–91

The capacitor and inductor combination with pulses and squarewaves — A capacitor and inductor in series or parallel with each other makes a resonant circuit. Application of a pulse or squarewave causes an LC circuit to oscillate with a sinewave shape, back and forth a few times with decreasing strength at each cycle. This process is called "shock exciting" the resonant circuit. The circuit "rings," much like the mechanical analogy of a bell being struck by a hammer. Adding resistance to such a circuit increases the losses within the circuit, causing the oscillations to decrease in amplitude more rapidly. The resistor that is installed across such a circuit for damping oscillations is called a "swamping" resistor.

Broadening frequency response in RF circuits — Adding resistance to an LC circuit will cause the bandwidth of the circuit to increase.

Another way of widening RF bandpass is to put several stages in a string, each tuned to a slightly different frequency. This is called stagger-tuning. See Figure 4-93.

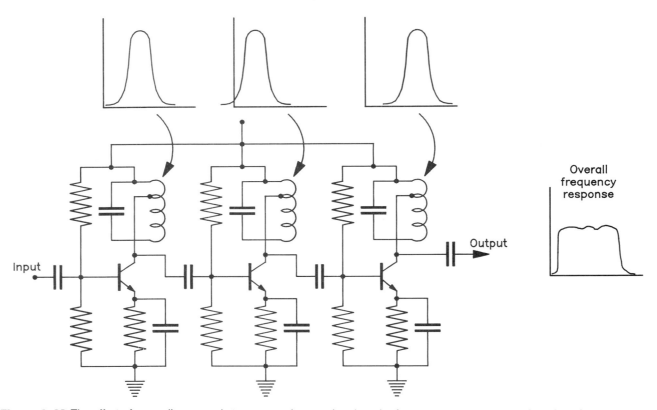

Figure 4-93 The effect of cascading several stagger-tuned stages is a broader frequency response curve than that of a single stage

Another way a stage might be designed to have a broad bandwidth is to over-couple the primary and secondary windings of an interstage coupling transformer. See Figure 4-94.

THE SPECIAL CASE OF THE DIODE AND CAPACITOR PLUS A LARGE SIGNAL — There are several circuits where a large signal is applied to a diode and a capacitor. The half-wave rectifier is the most common case. A second, less obvious circuit that will do the same thing is an ordinary bipolar transistor input circuit. See Figure 4-95.

The large signal causes the capacitor to accumulate a charge because of the diode action of the transistor's base junction. The "mysterious" appearance of a DC voltage apparently without a source is really not so mysterious after all. It is merely the rectification of a large input signal by the not-so-obvious base-emitter junction.

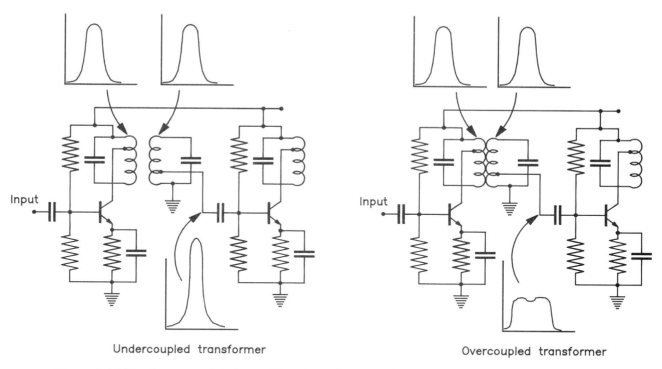

Figure 4-94 Transformer coupled stages with under- and over-coupling, and the effect on frequency response

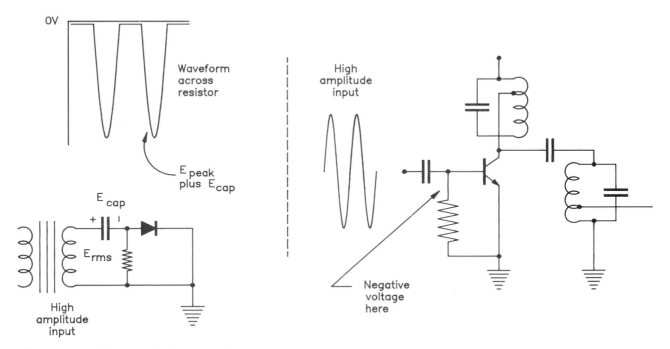

Figure 4-95 A high amplitude input, a diode, and a capacitor equal a new voltage. The new voltage is DC, resulting from rectification of the AC signal

SCRS AND TRIACS IN AC CIRCUITS — The SCR or triac requires an input pulse to trigger into a shorted condition. Removal of the voltage across the output terminals, decreasing the voltage level to zero, or reversing the polarity of the voltage applied to the main terminals will be required to make the component cease to conduct once it is triggered.

The triac conducts in both directions, whereas the SCR conducts only in one direction, much like a diode.

The input pulses for operation of these components are almost always sent through some isolating component, such as a small pulse transformer or an optical isolator. This is because the controlling circuitry of the triac or SCR usually must be operated with a different ground reference than the output circuit of the device. See Figure 4-96 for an example of a triac circuit.

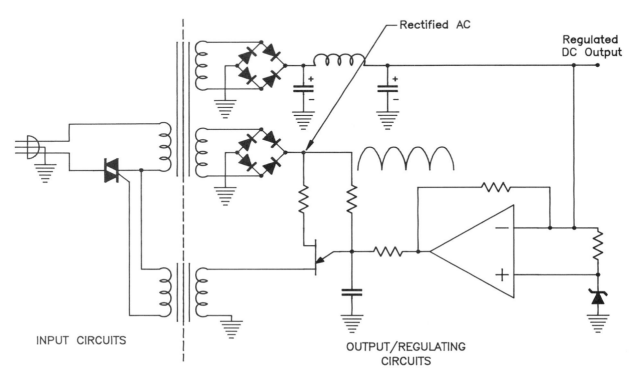

Figure 4-96 Sample triac circuit showing isolation of input and output circuits

Because the SCR and the triac are usually used with relatively high AC voltages on the output elements, there is a hazard for the technician and the test equipment. Besides the safety precautions required to prevent electrical shock, the output circuit of the device is probably not at earth ground, thus preventing the safe application of an oscilloscope probe to the circuit. For these reasons, troubleshooting SCR and triac circuits is often best done by turning off the power and troubleshooting the circuit dead as covered in Phase 6.

THE UJT IN AC CIRCUITS — The UJT should have a sawtooth waveform at the emitter and very short, sharp pulses produced at either of the two bases. See Figure 4-96.

Since the UJT is used almost exclusively as a pulse oscillator, troubleshooting the circuit should be easy. As cautioned earlier however, be sure that connecting an oscilloscope will not cause circuit damage.

4.13 Make Dynamic AC Voltage Readings in the Defective Stage
— The reason an AC signal must now be traced through a circuit is that the problem, whatever it might be, has not shifted the circuit's DC voltage readings sufficiently to indicate where the problem is, thereby rendering the easier DC voltage troubleshooting methods ineffective.

The technician must now shift the mental view of the circuit to that which AC would see: capacitors and inductors now assume reactance and are no longer pure shorts or opens.

The size of these components in microfarads and henries must be considered in view of the frequencies of the circuit.

Almost without exception, dynamic signal readings in analog circuits are taken with reference to the circuit common or ground. This means that an oscilloscope can be used, referencing it to ground.

Signal tracing often depends on the settings of certain controls within the circuit. Be sure that these controls are at normal or specified settings. See Figure 4–97.

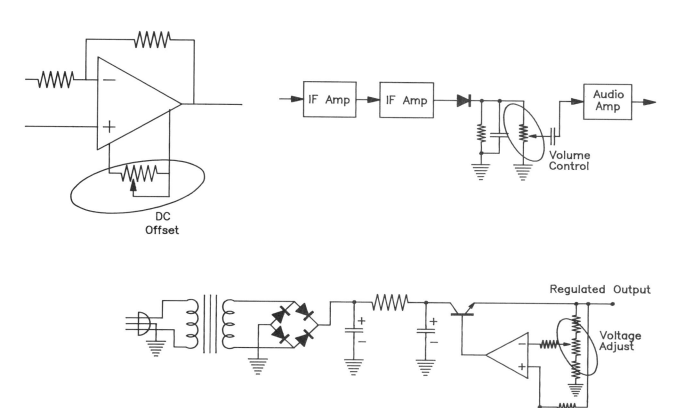

Figure 4-97 A few examples of internal and external adjustments that must be set correctly before tracing DC or AC signals

Consult the schematic notes or elsewhere in the maintenance manual for the required settings for each control before expecting signal tracing voltages to be correct.

4.14 FIND ANY BAD AC SIGNAL LEVELS? — Bad signal levels may be signals that are too low or too high in amplitude in comparison to what those levels should be in a normal working circuit.

As recommended in troubleshooting DC levels, it is a good idea to physically mark the schematic with the exact location of the reading that is farthest from normal. Concentrate on the possibilities that might cause the bad reading. Details of how this leads directly to the problem are covered in "The Perozzo 5-Step Method," covered in paragraph 4.9.2. Just substitute "AC readings" for "DC readings" in that paragraph.

4.15 NOTES ON DISTORTION AND WAVESHAPE PROBLEMS — Once in a while a problem shows up as a distortion of the proper waveshape. These problems are difficult to troubleshoot because almost any of the components in the circuit can affect waveshape. Resistors, capacitors, inductors, and amplifier circuits, including diodes and transistors, are

possible causes of waveform discrepancies. Here are some possibilities for the technician to consider if waveform distortion is defined as the problem.

Class A amplifier waveform problems — Amplifiers must have the proper voltages to operate. Verify that the proper bias, Vcc, and Vss voltages are present.

CLIPPING — A Class A amplifier output voltage has limits. If the incoming signal is too high in amplitude, the output voltage of the stage will attempt to follow but will stop when reaching the power supply and/or ground voltage limitations. Refer back to Figure 4-20 noting the upper right corner.

The cure for this kind of distortion is to simply reduce the amplitude of the incoming signal. It is too high for the amplifier to handle.

AMPLIFIER DRIVEN INTO SATURATION — A Class A common emitter amplifier stage will produce distortion on the "bottom" of the output waveform if that stage has too much forward bias and is not operating at the center of its operating curve. Refer to Figure 2-20, noting the bottom part of the figure.

The cure for this problem is to find the reason the amplifier is not operating at the midpoint of the operating curve. This is almost always an excessive biasing current problem in the input side of the circuit. See Figure 4-98.

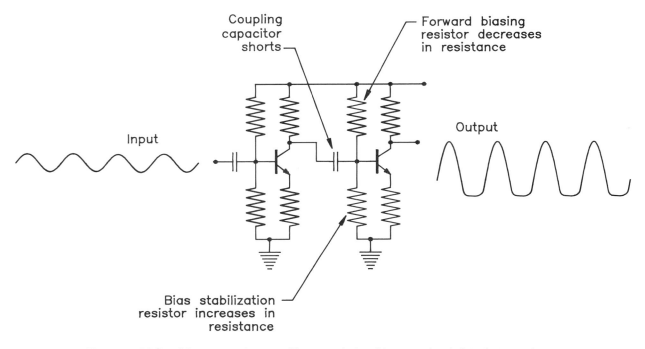

Figure 4-98 Possible causes of an amplifier stage being driven too deeply into the operating curve, causing negative clipping of the output

AMPLIFIER DRIVEN TO CUTOFF — On the other hand, a Class A common emitter amplifier that is not biased "on" enough will clip of the top portion of the waveform. Again, see Figure 2-20.

The amplifier circuit producing this clipping also needs proper bias to be restored. See Figure 4-99.

Class B amplifier waveform problems — The Class B amplifier is supposed to work at relatively high power levels. Biasing problems with this kind of amplifier stage are generally due to too little or too much forward bias. The normal bias is such that the stage is slightly "on" at rest. Too little bias results in distortion as shown in Figure 4-100.

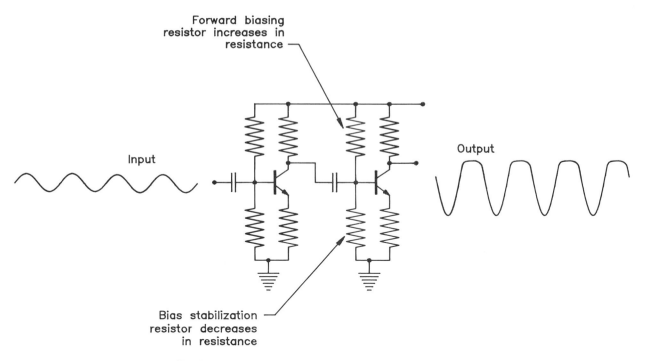

Figure 4-99 Possible sources of problems causing positive clipping of the output signal

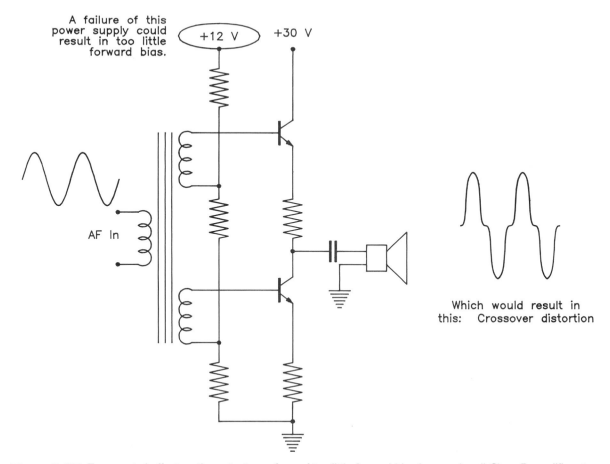

Figure 4-100 Exaggerated effect on the output waveform of too little forward bias in a push-pull Class B amplifier stage

This kind of distortion is called "crossover" distortion. It is cured by increasing forward bias to both the positive and the negative sides of the circuit.

Too much forward bias on the Class B stage will probably cause the amplifiers to overheat or the power supply fuses to blow because of the excessive current flow through the transistors. A second possible symptom of too much forward bias would clip either or both of the output waveform peaks. If the bias on one side of the circuit is normal, but the opposite side has too much forward bias, only one side of the output waveform will be clipped, much the same as described when the Class A amplifier is driven into saturation.

Class C amplifier waveform problems — The Class C amplifier stage requires a signal to drive the base into the conducting region of the operating curve. Too much drive signal can cause negative clipping of the output signal in a common emitter configuration.

Most applications of the Class C amplifier use RF or pulsed input signals. These applications are not usually sensitive to waveform distortion, however. RF circuits often use a tuned LC output circuit to remove any distortion produced by the stage. Many pulsed circuits do not depend heavily on waveform, and require the mere presence of a signal of sufficient amplitude to operate normally.

The diode as a waveform modifier — The diode is commonly used in circuits to shape waveforms. Its most common use is in clamping circuits. See Figure 4–101.

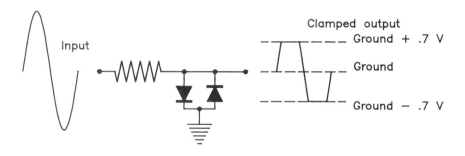

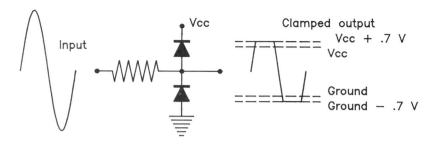

Figure 4–101 The diode finds common use as a clamping device, limiting voltages to specific levels

Waveform problems in circuits using diodes will often be caused by the diode. Diodes short or open, producing drastic changes in the resulting waveforms.

Another circuit that uses the diode is one in which the diode dampens the ringing of a coil when used in DC circuits. See Figure 4–102.

The voltage across the coil in this circuit can assume a very high initial peak value, followed by a damped oscillation waveshape if the diode should open. A shorted diode will prevent any appreciable voltage from appearing across the coil at any time.

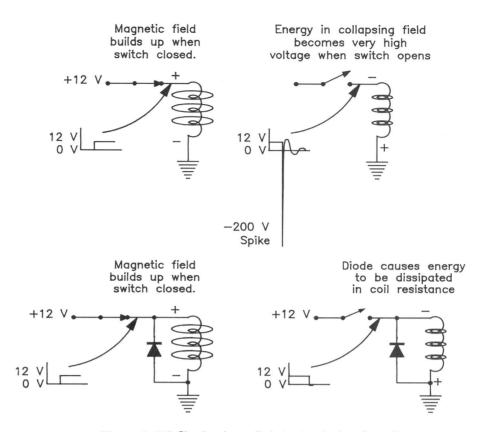

Figure 4-102 Circuit using a diode to stop ringing of a coil

In addition to stopping the ringing of the coil, the diode prevents the very high voltage spike from damaging other components. The diode in parallel with a coil, such as in a relay, is a required component when the coil is operated by solid state components. Without the diode, the high voltage spike produced by the collapsing magnetic field of the coil would very likely ruin the transistor or IC driving the coil.

4.16 Consider Degree of Original Problem, Possible Realignment — Troubleshooting is easy when the stage in question has completely failed. Real circuits do not always fail completely, however. Most analog circuits can be partially affected by the normal aging of components. When this kind of failure occurs, performance may be less than what the specifications require, but the circuit is still working.

One of the more subtle yet a common failure resulting in reduced performance is the opening of the emitter bypass capacitor in the common Class A audio amplifier. See Figure 4-103.

If low gain of a Class A stage is suspected, a quick check of the capacitor can be made by watching the output signal amplitude while paralleling the emitter capacitor with a good electrolytic of equal or more capacity than the original. If the gain increases substantially, the original was probably open.

The presence of one or more internal adjustments labeled calibration, balance, or zero is an indication that the circuit may be sensitive to any changes in the circuit including aging of components. The unavoidable drifting of component values over time may require periodically checking or adjusting these controls. Lack of maintenance and calibration can produce symptoms that make the equipment appear to need repair.

The degree of the original failure may give the technician a clue as to whether the circuit has indeed failed and will require component replacement, or whether an adjustment, internal or external, is all that is required.

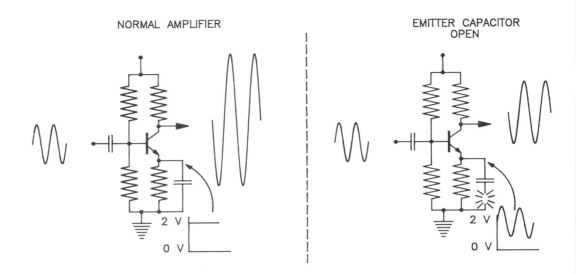

Figure 4-103 The open emitter capacitor results in normal DC readings but low AC signal gain

4.17 COULD REALIGNMENT CURE THE PROBLEM? — If the
stage is not quite measuring up to normal standards and there are adjustments labeled calibration, balance, or zero within that stage, there is a good argument for checking the alignment of the stage. Checking alignment is a matter of going step by step through the manufacturer's adjustment procedures. This information must be provided to the technician as lack of calibration procedures is a good reason to set the job aside until they are made available.

Accomplishing the realignment routines may point to a component that needs replacement. Alignment procedures may proceed normally up to a certain point if there is a failure; then further calibration may not go as planned. This may be used to help indicate where the problem lies. Marginal operation of a circuit is in itself a good reason to realign the circuit. It is also an acceptable method of troubleshooting for a partial failure.

4.17.1 Realign as Appropriate — Realignment, as a general rule, means to supply
precise inputs (DC or AC or combinations of both) to a stage and to adjust that stage for an exact output. For example, an amplifier stage in a calibrated test instrument would probably be required to provide a very precise amount of amplification. Calibration of such a stage would probably require the application of a very precise amplitude of input signal and careful monitoring of the amplitude produced at the output of the stage. The internal adjustment, in this case probably called a "calibration" adjustment, would be adjusted carefully until the output voltage matched that specified for the stage.

Realignment procedures must follow the manufacturer's recommendations *to the letter.* An apparently small deviation from the set methods or the order in which adjustments are made can make the entire alignment process invalid.

4.17.2 IS THE PROBLEM GONE NOW? — If the problem appears to have
cured itself, there is a possibility that the problem is an intermittent. In this case, it is necessary to re-determine if this stage is still defective. Go back to the beginning of Phase 3 and be sure that the problem still exists in this stage. If it doesn't exist, go back to Phase 1, checking for an intermittent problem as covered in paragraph 1.2.

Sometimes a technician may realize that the problem is not defined positively, and there is doubt as to whether the stage in question is really bad after all. In this case go back to the beginning of Phase 3 and verify it.

4.18 Verify this Stage Is Still Bad – Intermittent? — If the problem
seems to be coming and going during the course of trying to track it down, it may be an intermittent situation. Intermittents are categorized and covered beginning with paragraph 1.2.

☐ Phase 4 Summary

It should be evident by now that troubleshooting to the component level is an art requiring a broad background of circuit theory, together with an intimate knowledge of test instruments. There is a great deal more to know and do to repair a circuit to the component level other than simply changing circuit cards to find the bad one. Because of the multitude of today's throw-away circuits, the technician will often be doing component-level work only on new, expensive circuits. High-volume, mass-produced circuits will be cheaper to discard and replace as a unit.

Troubleshooting analog circuits to the component level begins with DC readings and reverts to dynamic signal tracing as a second choice.

By this point in the job, one or two components should be suspected as being defective. Verification and replacement takes place in Phase 5. Troubleshooting digital circuits generally involves two chips, the source and the load ends of the defective signal. Making a choice between them involves some additional troubleshooting if the chips are not easily replaced.

☐ Review Questions for Phase 4

1. You suspect that a certain chip is not performing its job. What is the first check you should make on the chip?
2. What is the normal voltage and tolerance for a TTL circuit?
3. You are fortunate to find that the source and the load chip for a "stuck" signal are in sockets. What is a quick way to find out if the problem is in either of these chips?
4. You find that a logic gate has input signals but no output signal, being always "low." What problems could cause these symptoms?
5. What are the indications of a logic probe when troubleshooting an open trace between logic gates?
6. If the logic probe is used to troubleshoot bipolar transistor circuits, will the probe normally indicate activity if the transistor is operated on 5 V in a TTL circuit?
7. You have confirmed that you have a "stuck bus" between two ICs of a logic circuit. What are the two kinds of shorts?
8. You have confirmed that you have a "stuck bus" between two ICs of a logic circuit. How can you test the circuit to see what kind of short is present?
9. What test instrument will indicate which of 10 load ICs connected to a single output chip is causing an indication of a "stuck bus"?
10. What is a common name for transistors meant to be either saturated or cutoff in normal operation?
11. A logic gate is noted on the schematic to have a resistor from the output pin to Vcc. What might this indicate about the chip?
12. Name two interface circuits that can clean up switch bounce.
13. What is the danger of using dull test points on test lead?
14. Why should all input signals be removed from a circuit when DC troubleshooting?
15. Name four disadvantages of the VOM as opposed to a good DMM when used to service solid state circuits.
16. What is a typical input resistance for a high quality DMM?
17. Comparing known or estimated circuit voltages with what the circuit actually has can reveal discrepancies between good and bad DC voltage measurements. Which should come first, the estimation of a voltage or the measurement of it?
18. When is a calculator used by a technician in estimating voltages?
19. You find that two DC voltage readings are not as they should be. One is 50% of normal, the other is 80% of normal. On which one should you concentrate?

20. Is the chassis of equipment the point from which to take all DC voltage readings?
21. Supply voltage is 12.0 VDC. A resistor has 12.0 VDC across it. Does this prove there is current flow through the resistor?
22. Supply voltage is 12.0 VDC. A resistor has 0.0 VDC across it. Does this mean the resistor is defective?
23. Supply voltage is 12.0 VDC. A resistor has 5.22 VDC across it. Does this mean the resistor has current flow through it?
24. A schematic shows two electrolytic capacitors in series across a relatively high DC voltage. There are also two high-value resistors across the same voltage, with their midpoint connected to the midpoint of the capacitors. What is the function of the resistors?
25. A schematic shows two transistors sharing the heavy output current of a regulated power supply. Each of the transistors has a very low-value resistance in series with it. What is the purpose of these low-value resistors?
26. How should a capacitor look to DC?
27. Generally speaking, how would an inductor look to DC?
28. Transformers are used only in _____ or _____ circuits.
29. How should a fuse look to DC?
30. A good switch should always look like a _____ or an _____ in a circuit, never like a _____.
31. If there is sufficient forward voltage and current, the voltage drop across a normal diode junction will be about _____ V or _____ V, depending on the type of diode material.
32. If the forward-biasing current through a normal diode junction increases, how would you expect the diode voltage to change?
33. How does a reverse-biased diode junction look to DC?
34. What determines if a given diode is forward- or reverse-biased in the circuit?
35. The bipolar transistor can be thought of as a _____ and a _____.
36. Can the amount of current flow between the emitter and the collector of a bipolar transistor be estimated by the voltage drop across the emitter-to-base junction?
37. There are two common ways of providing forward bias for a transistor. What are they?
38. Does a Class B audio amplifier transistor require a forward-bias resistor?
39. What kind of biasing does the JFET require for linear operation?
40. Draw the most common way of biasing the JFET transistor in a Class A application.
41. What is the principal difference in the biasing of a depletion-type FET and an enhancement FET?
42. What are the two basic circuit configurations for an operational amplifier?
43. What will the output voltage from an operational amplifier be when it is connected in a positive feedback circuit?
44. What will the output voltage from an operational amplifier be when it is connected in a negative feedback circuit?
45. What is the voltage that is amplified by an operational amplifier?
46. Discuss the phenomenon of the "virtual ground" as it relates to the operational amplifier.
47. What caution should come to mind when a circuit is seen to have an optical isolator?
48. What voltage drop should be expected across the junction of an infrared emitting diode?
49. Name two ways that should alert the technician to the fact that the circuit common lead is probably not connected to the chassis of the equipment.
50. Discuss: Identify the components of Figure 4–33 that could 1) open or 2) short, yet not disturb the DC readings of the circuit.

51. Discuss: Describe the failures, open or short, of the marked components that would not disturb DC readings.
52. What could the technician use as an approximate indication as to how far DC voltage readings might vary from normal without seriously affecting the circuit?
53. If a particular stage has ceased operating entirely and you have found a voltage reading off by about 5%, are you near the problem?
54. You have a small Class A bipolar transistor amplifier that you wish to test to see if it is amplifying. What simple check could you make?
55. You have decided to concentrate on finding the reason that particular DC voltage reads only 10% of normal. What are the five possibilities that can cause a reading to be too low?
56. Discuss the differences in the Perozzo 5-step method item checklist when considering a circuit voltage that is abnormally high rather than one that is abnormally low.
57. What is the failure pattern of a capacitor?
58. What is the failure pattern of a resistor?
59. Which is more likely to be the cause of a particular problem, a defective capacitor or a bad resistor?
60. Which is more likely to be the cause of a particular problem, a fuse or a switch?
61. In-circuit use of an ohmmeter or a solid state tester requires a single, important caution. What is that caution?
62. How effective are in-circuit resistance checks with an ohmmeter?
63. Why does a digital ohmmeter provide poor information when testing a diode junction?
64. You wish to test a diode junction. What scale of a digital ohmmeter must be used to get a valid reading?
65. You have removed the mixer transistor of Figure 4–33. Using an ohmmeter, you wish to check the resistor that provides forward bias for the transistor. Identify all, if any, paralleling resistances that would influence the reading.
66. How would most inductors appear to an ohmmeter test?
67. Can the solid state tester be used for in-circuit testing?
68. What caution should be remembered when introducing an input signal to defective equipment?
69. The typical digital AC voltmeter is calibrated to read RMS at what frequency?
70. What is the best instrument for tracing and analyzing signals through analog equipment?
71. What is the severe limiting factor when using an oscilloscope so far as grounding is concerned?
72. Why is it a good idea to leave the vertical gain and horizontal sweep settings of an oscilloscope alone once they are set?
73. How can you identify inexperienced technicians by their use of an oscilloscope?
74. The effect that a given capacitor will have in an AC circuit is dependent upon what two factors?
75. A capacitor is meant to pass _____ voltage and to block _____ voltage.
76. There will be a phase difference of _____ degrees between the voltage across a capacitor and the voltage across a resistor in series with it, at a frequency where the capacitive reactance is equal to the resistance.
77. What is the time constant of a capacitor of 0.01 ufd and a resistance of 100K ohms?
78. An RC circuit is allowed to charge, using a source voltage of 100 volts. What voltage will be across the capacitor after the time represented by a single time constant?
79. An integrator circuit passes the _____ component of the input.

80. A differentiator circuit passes the _____ component of the input waveform.
81. What circuit will average an input signal to a DC level?
82. An inductor and capacitor connected in series or in parallel exhibits the phenomenon of _____.
83. What is the effect of adding resistance into a resonant LC circuit used in a pulsed circuit?
84. What is the effect of adding resistance into a resonant LC circuit used in an RF circuit?
85. Name three ways of broadening the frequency response of a series of RF amplifiers.
86. A large input signal and a diode and capacitor can give rise to an unexpected _____.
87. What will be the effect of too much forward bias in a bipolar amplifier?
88. What common test instrument will best show distortion?
89. Why would a diode be used across an indicator if operated on DC?

PHASE 5
REPLACING DEFECTIVE COMPONENTS

☐ **PHASE 5 OVERVIEW**

There are several ways the technician might reach the point of replacing components when repairing something electronic. One way is to go the whole route of troubleshooting a problem from the system to the equipment, then down through the card, stage, and component levels. Another possible route to this point might be the isolation of a problem using only the techniques described in Phase 6, bypassing all live circuit troubleshooting. Of course, if the problem was visible without troubleshooting, then the next logical step is to replace the bad component. Replacing the bad component is only part of what has to be done, however. A single bad component may have either caused another component to fail, or it may be defective because another failed. More troubleshooting must yet be done.

Figure 5-1 A soldering operation

5.0 ELECTROSTATIC DISCHARGE AND OTHER PRECAUTIONS — Be sure that all power is off before doing any component replacement work on electronic circuits.

The technician must be aware of and take special precautions to avoid static damage to FETs and ICs during installation of new parts. Review the ESD precautions in paragraph 2.0.

Equipment using capacitors with very high capacity — Capacitors do not have to have a high voltage rating to be dangerous to your test instruments. Low-voltage electrolytic capacitors can store a great deal of energy at low voltages. These capacitors will appear physically large compared with other electrolytics. See Figure 5–2.

Figure 5–2 These large electrolytics store considerable amounts of energy at low voltage

These capacitors must be discharged when working on unenergized circuits. Contact with their terminals may not cause a shock, but the energy stored can damage test instruments. An ohmmeter or solid state tester, for instance, could be damaged by the current flow resulting from a charge retained by these components.

Discharging these large-value capacitors is best done using a resistor of a few ohms. This prevents the extremely high current that would result if discharged through a short circuit. Once they are discharged it is a good idea to keep the short connected for a few seconds to let the capacitors drain.

5.1 NOTES ON SOLDERING AND SOLDERING IRONS — The tip of any kind of soldering iron may be connected to ground via the power cord. Do not attempt to solder circuitry that is still energized because there is the danger of shorting the circuitry to the soldering iron ground.

There are three general types of soldering irons: the soldering gun, the simple resistive soldering iron, and the temperature-regulated soldering iron. See Figure 5-3.

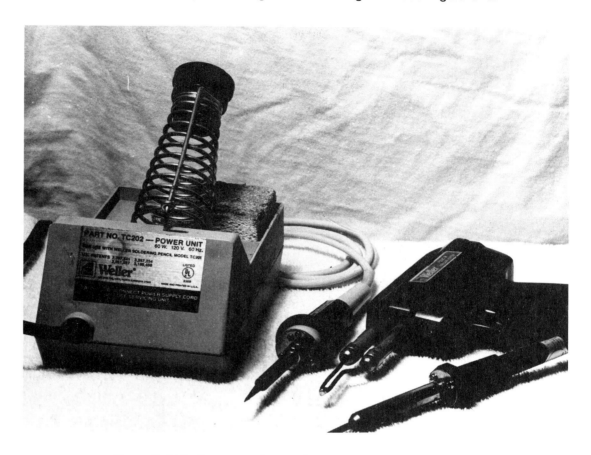

Figure 5-3 The three general types of soldering irons used in electronics

The soldering gun — The soldering gun is recommended for soldering heavy wiring and those components that do not involve semiconductor materials. It is *not* recommended for use on any PC board.

The soldering gun operates on the principle of resistive heating of a low-resistance copper tip. Heavy currents flow through the copper tip. Because of the currents involved, a bit of normal corrosion of the attachment nuts will result in slow or no heating of the tip at all. A quick fix for this problem is often just a loosening and retightening of the tip nuts.

Never use a soldering gun that has a burned-open tip. Replace the tip when it becomes pitted. Be sure to keep the soldering gun tip clean and well-tinned.

The resistive soldering iron — The resistive soldering iron is the cheapest soldering iron in use. Some stores offer them for as little as $3.00. Using them for precision electronic work carries a big risk. The heat produced at the tip is far in excess of that needed to melt solder. These irons must be designed for high temperatures to make them usable on reasonably large connections. This excessive heat can be very damaging to PC boards and to semiconductors of all kinds. The simple resistive iron should be used only in emergencies, and when used should be applied for the bare minimum of time to prevent overheating damage.

The temperature-regulated iron — The temperature-regulated iron operates much cooler than the simple resistive iron. These lower temperatures are much better than any other soldering iron to use on PC boards and semiconductors.

The regulated iron can use lower temperatures because the regulating mechanism can increase the duty cycle of the heating current to compensate for heavier thermal loads such as physically large connections. Once the iron is producing the proper temperature, the duty cycle stabilizes to maintain the correct heat.

The heating curves of the simple resistive and the temperature-regulated irons are shown in Figure 5-4.

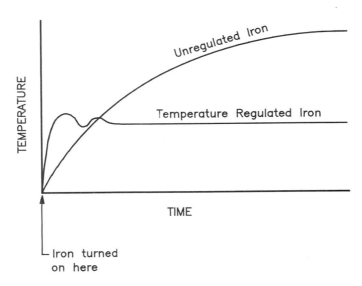

Figure 5-4 Superimposed curves of the temperatures of simple and temperature-regulated irons show marked differences

Temperature-regulated irons are also available with an adjustable temperature control. This control sets the temperature at the tip of the iron, ensuring that the work is made no hotter than necessary. See Figure 5-5.

Work on CMOS circuits and other static and transient sensitive circuits may require the use of an iron that produces no transients during the temperature regulation cycle. This kind of iron is available, using an SCR and special circuitry to ensure that the iron never cycles on or off during the time when the power line voltage is other than or close to zero. This feature is called a "zero current switching" attribute.

Care of the soldering iron — Heating the connection to be soldered is best done with a clean, tinned soldering iron. A dirty tip will not heat the connection or melt the solder properly, as the dirty, crusty tip will prevent the conduction of heat.

The tinned portion of an iron is "wetted" with the molten solder. A tiny amount of solder applied to a tinned iron will provide a good heat transfer and will result in the intended connection being heated to proper temperature in a minimum of time. Be sure to apply only a small amount of solder to conduct the heat, however. If the tip of the iron is merely tinned without a bit of excess solder, it will be considerably more difficult to heat a connection properly. Besides, it is better for the iron if it is kept tinned with a bit of excess solder at all times.

A damp sponge is good for keeping the soldering iron tip clean. The technician does not need to keep the tip shiny all the time. To strip the tip of all solder and allow it to sit heated in the holder for long periods will result in oxidation of the tip, and an oxidized tip will not transfer heat. The iron should be will tinned at all times. Use the sponge to clean the tip of any burned solder bits, then immediately re-tin it before using or allowing the iron to stand heated for any length of time.

An oxidized soldering iron tip can often be rejuvenated by alternately cleaning with a damp sponge and applying tiny amounts of rosin core solder over as much of the tip area as will accept it. When the tip becomes dripping wet with solder, remove it with the sponge

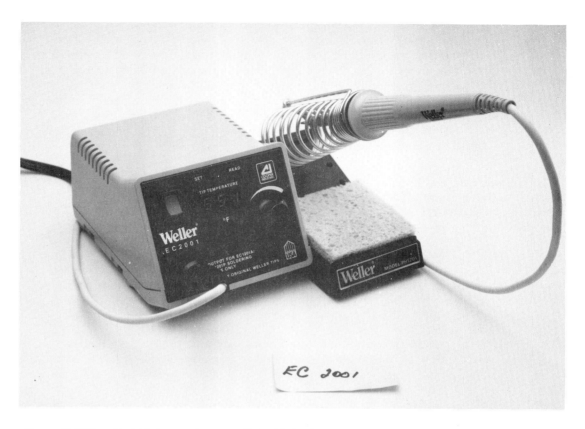

Figure 5-5 The adjustable temperature-controlled soldering iron makes it possible to use the minimum temperature for any soldering application

and repeat the tinning process. It may take several minutes to reclaim a burned tip. Stubborn cases may require replacement of the tip.

Never file the tip of a plated iron. Filing of the tip will remove the plating, ruining the tip.

Rosin is available in a paste for heavier cleaning jobs. It should not be used on a temperature-controlled iron, however. Excess flux works its way up into the sleeve of the iron and into the internal parts. Use only the flux available in cored solder.

Desoldering operations are discussed in detail beginning with paragraph 5.2.1.1. Soldering suggestions are given in paragraph 5.11.

Solder — There are two common kinds of solder. One has a rosin core and the other an acid core. *Only rosin-core solder should be used for electronic work.* The rosin flux serves the purpose of providing a cleaning agent that makes the solder flow readily into the connection to be made. Under the heat of the iron, rosin does not last long. The burning of the rosin is what makes the smoke during a soldering operation. Without a little puff of smoke, there is no cleaning action and the connection will probably not be as good as it could be. Simple application of a fraction of an inch of fresh solder to an old joint is often all that is needed to clean it up.

Rosin-core solder is available in several different ratios of tin to lead. Electronic work is best done with a 60:40 ratio. This combination melts at the lowest temperature and hardens quickly without an intermediate "mushy" consistency.

Acid-core solder is for use on plumbing where corrosion is heavier and a more powerful cleaning agent is needed. Acid core residues can badly deteriorate PC boards, however, and solder containing acid flux should *never* be used on electronic equipment. Leave acid-core solder for the plumbers.

Soldering tools — Several soldering accessories are recommended for soldering operations. They include at least one pair of hemostats for manipulating tiny parts and holding them for soldering. The solder sucker is used to suck melted solder from circuit board holes. When this tool is not effective, solder wick can be used for a similar purpose, mostly for removing solder from flat surfaces. A powerful magnifying glass is handy to inspect joints before and after soldering. Denatured alcohol is used to clean the connection after completion; this makes the joint look better and allows a thorough inspection of the finished joint. A small, stiff brush is useful for cleaning the joint; the alcohol is used as a cleaning agent. See Figure 5-6 for examples of these soldering accessories.

Figure 5-6 Soldering accessories recommended for routine solder jobs include magnifying glass, hemostats, solder sucker, solder wick, denatured alcohol, and a stiff brush

Adequate lighting is also necessary to do a good job. A goose-neck lamp is recommended for the detailed work of soldering.

5.2 IS THE BAD PART AN IC?

There are different options available if the failed part happens to be an IC. These options are valid for both digital and analog ICs. Troubleshooting discrete component analog circuits, however, suggests different options. If the part to be replaced is not an IC, proceed to paragraph 5.3.

5.2.1 SAVE THE IC?

The answer to this question will determine how the IC is to be removed from the circuit: either quickly, but in the process destroying the IC, or slowly and carefully on the chance that the IC might prove to be good after removal and thus may be reinstalled.

The choice of removal method depends upon several factors. The availability of the replacement IC must be addressed; if the IC is a common one and there are a dozen in stock, this would probably make the suspect IC expendable. The cost of the IC must also be taken into account. Some ICs cost less than a quarter; these are not worth the time it can take to remove them when the technician costs the company perhaps $35 an hour. Pure economics dictates that if the replacement IC is both available and inexpensive, the old one should be removed as quickly as possible.

5.2.1.1 Remove the IC by Careful Desoldering — If the IC must be saved, the desoldering must be done with care to prevent damage to either the board or the IC. The first step is to remove as much of the solder as possible from the underside of the board, using a solder sucker. This IC desoldering operation can be done two ways.

The first way is to lay the PC board flat on the bench, heating the connection with the iron in one hand and holding the solder sucker ready in the other hand. When the connection becomes molten, remove the iron, place the solder sucker over the connection, and press the trigger, all in one smooth movement. This method requires speed and a bit of practice. Hesitation in getting the sucker onto the connection and pressing the trigger will allow the connection to cool and the whole operation will have been ineffective. The only result will be that both the board and the IC will become heated, which is something that must be kept to a minimum. See Figure 5–7.

The second way to desolder an IC is to work with the board held vertically. See Figure 5–8.

Figure 5-7 Desoldering on one side of the board requires speed and timing to accomplish solder removal with minimal damage to the board and IC

Figure 5-8 Desoldering an IC with the board held vertically

Place the solder sucker over the connection to be worked on, on the underside of the board. Without changing the position of the sucker, shift your eyes to the component side of the board. Hold the soldering iron gently against the IC pin where it enters the board. When the solder melts, simply press the release on the solder sucker.

If one or more of the pins still refuse to give up all the solder, it is time to try the solder removal braid. The braid should be placed between the tip of the soldering iron and the work. As the work heats, the solder becomes molten from the heat transferred through the braid. Capillary action then absorbs the molten solder, much like a terry cloth towel absorbs water. Braid is particularly effective on the top of the board when the IC has been soldered on both the top and bottom of the board. See Figure 5–9.

Connections that still refuse to come loose may require the use of the suggestion given at the end of paragraph 5.2.3.

218 ☐ **Replacing Defective Components**

Figure 5-9 Using solder removal braid to remove solder from the top of the board

5.2.2 Clip Pins at IC Body — If a replacement IC is available and costs very little, the logical choice is to remove the old IC quickly and easily, and with as little heating of the PC board as possible.

The quickest way to remove an old IC is to clip off the legs right at the body of the component. See Figure 5-10.

A sharp pair of cutters is necessary to make these cuts. The tips of the cutters must be very narrow for reaching between the pins as they are clipped. A pair of cutters could easily be modified to perform this task by grinding down the sides of the cutting tips.

Once the body of the component is out of the way, the pins still remain in the circuit board.

5.2.3 Remove Pin Stubs — Most of the pin stubs can be easily removed by using a solder sucker.

Use a solder sucker on the bottom of the board. Remove solder from the top of the board, using desoldering braid if the IC has been soldered on both sides of the board. Remove all of the solder from the pins using these methods. After all available solder is removed, most of the pins can easily be removed with a pair of small needle-nose pliers. *Do not* pull hard on the pins! A bit of a turn and they should literally fall out if all of the solder is gone.

A few of the pins may not cooperate entirely with this method. This may be because one or more of the IC pins are connected to the internal ground plane of a multi-layer board. This internal layer is a large heat sink and prevents easy removal of the solder. Refer back to Figure 3-24 to get an idea of this heat path.

The solution to this problem may be to use the desoldering braid as in Figure 5-9.

Another suggestion may help. Once most of the solder has been removed from a connection, further application of heat may or may not cause the last bit of solder to be melted.

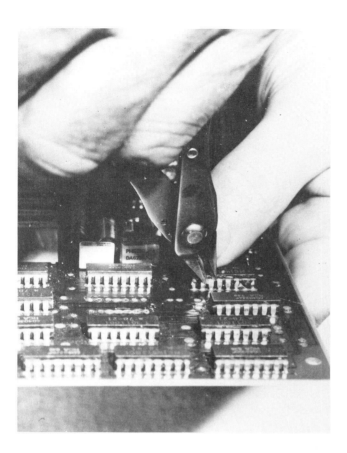 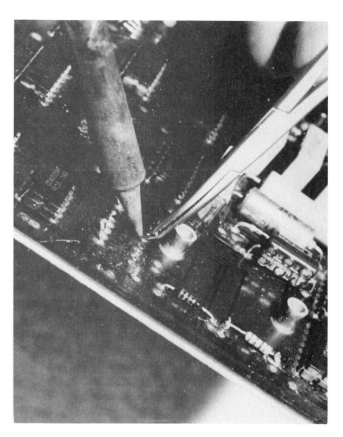

Figure 5-10 Removing an IC by clipping the legs at the component body and removing the remaining stubs

One result of this is that the board can become overheated. If this happens, *reapply solder* to the connection, just as though you were going to solder the pin back into the board. This new pool of solder will provide the heat transfer necessary to begin the desoldering operation on that pin all over again.

5.2.4 Clear the PC Board Holes
— The holes in the PC board should be inspected carefully after the removal of an IC. Hold the board to the light and look through it. Each of the holes must be clean and round before attempting to install a new IC. The solder sucker is particularly helpful in cleaning up partially cleared holes. Heat from one side while using the solder sucker from the other.

5.2.5 IS A TEMPORARY CHECK DESIRED?
— It may be desirable to make a check of the circuit to see if a replacement IC will work without actually soldering the IC into the circuit. This might be the case if the replacement chip is expensive. This is seldom the case, however, and the replacement chip can be installed with or without the installation of an optional socket.

5.2.5.1 The Toothpick Test
— An IC can be temporarily installed directly into a circuit board without the installation of a socket. This method will work only if the holes of the PC board are plated all the way through the board. Put the IC into the circuit (Watch the orientation of the chip!) and push toothpicks into the holes to jam the pins against the sides. See Figure 5–11.

Power and signals can be applied to test the circuit with the new IC temporarily "installed." The IC can be permanently soldered into the circuit if the circuit works properly. If not, further troubleshooting must be done to find the bad IC, and the new toothpicked IC has not solved the problem. If this is the case, return to Phase 4, "Identifying the Defective Component."

Figure 5-11 The preparation of an IC for a quick test without a socket or the necessity of soldering the IC into the circuit

5.2.6 Install Optional IC Socket and New IC — An IC socket may be installed in the circuit board if there is any chance that this same problem may come up again. If not, it is just as well to solder the new IC directly into the board. In either case, keep the soldering operation short to avoid heat damage to the board. Be sure that the IC is properly oriented (pin #1 is where it should be) before soldering. Too much or too little solder should be avoided.

5.3 Remove the Old Component — Now that the suspected component has been identified, it must be removed from the circuit. The removal of most components is a matter of common sense. Be sure all power is removed from the circuit. Minimize the possible damage to the circuit board by using a suitable iron and by applying heat no longer than necessary. The solder sucker is a help in removing most components, as detailed in paragraph 5.2.1.1.

It is generally a good idea to keep the original component until the replacement confirms that the old one was indeed the bad one. It is not unusual to find that the removed component was a good one after all.

There are some special arrangements for mounting components that bear mention at this time. Power transistors and regulator ICs are often mounted so that their normal dissipation of heat is carried away by a mass of metal; this is often the chassis of the equipment. Whatever is used as a heat sink, there may be a need to insulate the transistor electrically from it. There may also be a greasy compound applied to the mounting surface to help conduct the heat from the component to the heat sink. See Figure 5-12.

Figure 5-12 Typical mounting arrangements for power transistors. Note the white compound used to help in the heat transfer from transistor to heat sink

Installing one of these new parts will require the application of new heat transfer compound. Be sure to use the old or similar new hardware to insulate the component electrically from the heat sink. This hardware includes various special "shoulder washers" and insulators.

Some components are polarity sensitive. Electrolytic capacitors, transistors of all types, ICs, and some transformers must be replaced with due regard to the proper connection of the new components. *Be very careful to note or write down the original component orientation and connections before removing it.* It is embarrassing to remove a component, only to find that somehow the original orientation of the component is now lost. Here are some suggestions to be heeded *before* removal of the old component:

Transistors	Note which way the original component "faces" in the circuit.
ICs	Note the position of the marking on pin #1.
Electrolytic capacitors	Note which PC board hole is indicated as the "+" or "−" lead.
Transformers	Note wiring color coding (write it down) or the orientation of pin #1.
Diodes	Note the orientation of the banded end of the component.

Soldering static-sensitive components — CMOS devices and field effect transistors (FETs) are very prone to static discharge damage, as pointed out in paragraph 2.0. Soldering operations involving these components should include additional protection against possible leakage currents from the heating element of the soldering iron. These currents can be passed harmlessly to chassis ground by attaching a ground lead from the metal part of the iron to the chassis of the equipment under repair. Soldering irons using a three-wire power cord should be checked with an ohmmeter to be sure that the tip is actually connected to the ground pin of the cord.

Components should *never* be removed or installed in a circuit while the power is on! Input signals should not be applied while power is off. Observing these precautions will help prevent transient voltage damage.

Keep these sensitive components in their original packing until actually ready to install them. Keep any shorting wires on the leads of the component in place until the part is installed.

When approaching a piece of equipment to work on it, it is a very good idea to touch the chassis first, before touching any of the internal components. This procedure will relieve any static buildup, especially if the chassis is grounded via the power cord.

5.4 **NOTES ON OUT-OF-CIRCUIT TESTING WITH OHMMETER** — There is a great deal of difference between using a modern digital ohmmeter with a special diode testing function, and the analog volt-ohm-milliammeter. The analog meter produces up to several hundred milliamperes of current when using the R × 1 scale and can produce up to 30 VDC across the probes on the highest ohmmeter ranges. Both of these conditions are not healthy for the microscopic spacing of modern semiconductors. Therefore, all of the following tests will assume a high-quality digital ohmmeter that has an additional range for testing semiconductor diodes. The special diode range is usually identified by a diode symbol on the range switch.

The next step after removal of a suspected component from a circuit is to verify whether or not that component was really defective. While instruments such as capacitor bridges or transistor testers could be used as appropriate, the ohmmeter will show most problems and is the instrument of first choice for general purposes.

Testing resistors — Testing a resistor with an ohmmeter is simple. Connect the leads of the ohmmeter to the suspected component. Keep your fingers off the metal tips or the meter will be influenced, if reading higher resistances, by the paralleling resistance of your fingers and body.

Resistors usually fail by opening. Partial failures can be caused by a resistor changing value. The older carbon resistors such as those shown in Figure 5–13, were prone to changing value, particularly if they were overheated.

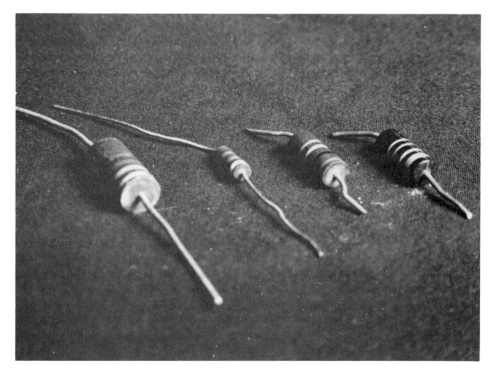

Figure 5-13 Carbon resistors, such as those shown above, can change value if overheated

The direction and magnitude of the resistance change of a carbon resistor is unpredictable. The resistance can increase or decrease drastically. If not burned open, other resistor types will probably be close to their marked value, and should be within marked tolerance.

Testing capacitors — Capacitors usually fail by shorting. The inside of the electrolytic capacitor, for example, amounts to two sheets of thin foil insulated from each other by a thin paper-like insulation. It is relatively easy for the insulation to fail and allow the two foils to touch.

The shorted capacitor is easy to verify with an ohmmeter. The capacitor will show less than an ohm and the resistance reading will be stable.

A good capacitor will act much differently. The good capacitor should accumulate a charge from the ohmmeter's internal battery over a period of time. See Figure 5–14.

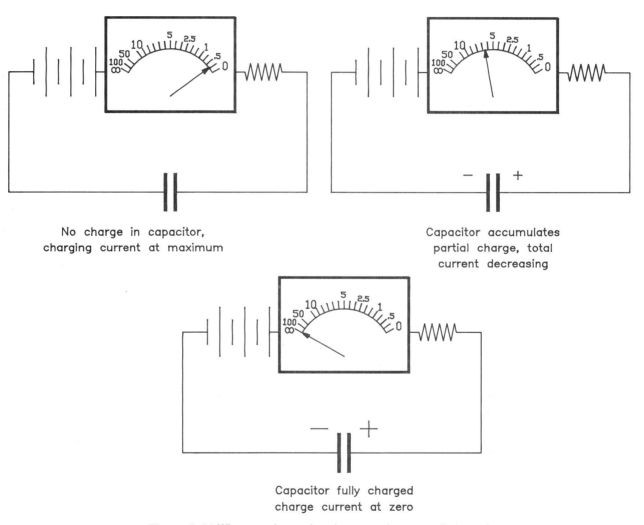

Figure 5–14 Why a good capacitor shows a resistance until charged

The larger the capacitor value, the more time will be required for the capacitor to reach full rated charge. The length of time required to reach a specified resistance, as shown on an ohmmeter, can be used as a rough indication of the capacity of the component. Comparing the capacitor's charge time with a known good capacitor of the same value provides a reference point. Be sure to begin the timing comparison on both from the moment a short is removed from the capacitor.

An auto-ranging digital ohmmeter may give apparently erroneous measurements of the charge time of a large-value (filter) capacitor. The ohmmeter should be switched off the auto-range function and a single high resistance range used to test these capacitors.

Testing semiconductor junctions/diodes — The modern digital multimeter uses extremely small values of voltage and current on the ohmmeter function. Although this is no problem when testing components such as capacitors, resistors, and inductors, it makes testing of semiconductors unreliable. Semiconductors *require* as much as 0.7 volts to bring a junction into conduction. An ohmmeter that produces only a few tenths of a volt cannot give the correct status of a junction unless that junction happens to be shorted.

The answer to this problem is to provide a special scale on the digital multimeter that is used for testing semiconductor junctions. This scale tries to pass a current of approximately one milliampere with a maximum open-circuit voltage capability of perhaps 2 volts. This is ideal for testing junctions. The criteria for a good silicon junction is that it has about 0.7 volts across it during conduction. The less common germanium junction should have about 0.3 volts drop. *The readout of the diode test scale of the DMM indicates voltage, not resistance.*

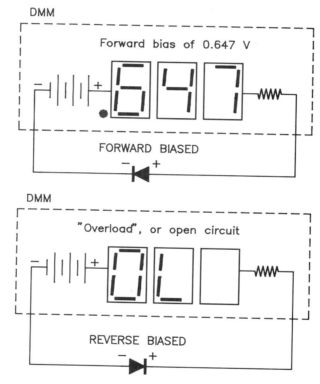

Figure 5-15 How to test diodes using the DMM diode scale. Two readings must be taken on the diode, reversing the leads from one reading to the other.

Forward conduction voltage of about 0.7 (or 0.3 in the case of germanium) should be indicated when the DMM black (− or negative) lead of the DMM is on the banded (cathode) end of the diode and the red lead is on the opposite end. Reversing the leads should show an open circuit. Remember to keep your fingers off the test leads to avoid influencing the ohmmeter readings with the leakage through your hands.

The common failures of a diode are to open or to short. Rarely, a diode will be leaky, and will show some resistance in the reverse-biased direction.

Testing bipolar transistors — For the purposes of testing with the ohmmeter, the transistor should be considered to be two diodes back-to-back. Large power transistors may have a third junction and resistor from collector to emitter. See Figure 5-16.

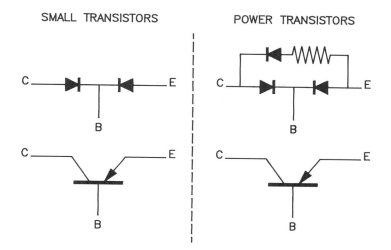

Figure 5-16 To the ohmmeter, the bipolar transistor looks like at least two diodes. Large transistors may appear to have a third diode

Testing the bipolar transistor is very similar to testing semiconductor diodes. Test each internal junction of the transistor separately. *Be sure to use the diode scale* to test transistors for the reasons explained under the section on "Testing Diodes." Check the base-emitter junction and take readings both forward and with the ohmmeter leads reversed. If one reading is not 0.7 volts (0.3 for the germanium transistor) or if the opposite reading is not an open, the transistor is definitely bad. Then check the base-collector junction for the same results. Last, check the emitter-collector and reversed, collector-emitter, junctions. Small transistors should show an open in both of these directions. Large transistors may show some leakage in one emitter-collector direction. Comparison with a known good transistor of the same type may be necessary to determine if the leakage value reading is normal or not.

Testing inductors — Inductors are basically just long wires to the DC produced by the ohmmeter. Since wire has a relatively low resistance, the reading can be expected to be low for a good inductor. An open inductor will show an open or possibly thousands of ohms if the failure involved burning of some of the insulation along with opening the internal wire. Again, a resistance reading comparison with a good component is a conclusive test of a suspected open inductor.

Shorted turns within an inductor can be hard to detect with an ohmmeter because resistance readings may be very near normal. If the percentage of shorted turns is low, resistance readings alone may not be conclusive, even when comparing a good inductor to the bad one. In this case, substitution of a known good inductor into the circuit and operational testing of the equipment will be the only 100 percent sure way of determining the status of the original inductor.

A word of caution: Very large inductors in good condition may require a few moments to read properly. These inductors will initially show as an open circuit, but after a period of time (depending upon the value of inductance) the ohmmeter reading will approach zero ohms. Disconnecting the ohmmeter at this time will result in breaking the flow of current through the inductor. The magnetic field, which at this time is at maximum, will collapse quickly and *will produce a very high voltage* across the inductor. *This is a definite shock hazard.* This effect is particularly noticeable when using an analog Volt-Ohm-Milliammeter instrument. These ohmmeters may use several hundred milliamperes on the low ohms scales, quite enough to produce a nasty shock when testing large inductors. The cure for the problem of the high-voltage spike is to short the inductor leads together *before* disconnecting the ohmmeter. Almost all inductors are insulated from their case or frame metal structures. Be sure to test for an open between either end of the coil to the case or frame.

Testing transformers — Each winding of a transformer can be tested just like an inductor, covered above. In addition to testing each of the windings separately for continuity or opens, the transformer should be tested for possible shorts between any of the windings and for shorts between any of the windings to the case or core of the transformer. Again, comparison of the transformer with a known good one may be required in inconclusive cases. See Figure 5-17.

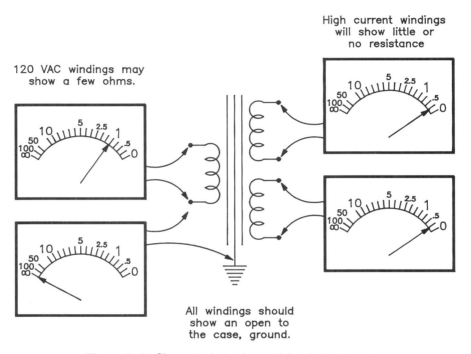

Figure 5-17 Ohmmeter tests of a multiple winding transformer

Tapped windings should be checked to be sure there is equal continuity reading between the taps and the ends of the windings. Symmetrical windings, such as those used with the center-tapped full wave rectifier, must have identical readings from the tap to either end of the winding. Not all tapped windings are symmetrical, however. The schematic diagram or parts list may help to determine if the transformer should be symmetrical.

Testing switches — Switches should be either open or shorted. Anything other than these extremes indicates a bad switch.

There are many types of switches. Regardless of the type, the ohmmeter must show the switch as an open or short as the position of the switch is changed.

An important detail to note when testing spring-loaded switches is to look for small abbreviations near the connections on a switch. NC, NO, and a C or COM mean normally a closed circuit (NC), normally an open circuit (NO), and the common connection of the switch (C or COM). "Normally" simply means in the normal resting position of the switch without anyone pushing or activating the switch. The positions COM and NC should have continuity or a short indicated under normal or "rest" conditions that should change to an open if the switch position is then changed. Likewise, there should be an open connection between the COM and the NO contact until the switch is changed. There must always be an open between the NC and NO contacts of the switch regardless of the position of the switch.

Testing SCRs and triacs

SCRs — A digital ohmmeter with a diode function should see a good SCR as an open circuit from cathode to anode in either direction of applied ohmmeter polarity. The gate to cathode should show a semiconductor junction. See Figure 5-18.

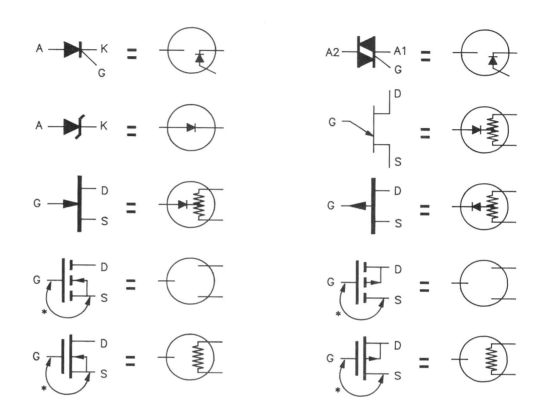

* Short Gate to Source lead

Figure 5-18 How various components look to an ohmmeter

SCRs often fail by shorting from cathode to anode, a simple thing to detect by measuring continuity between these two elements.

TRIACS (THYRISTORS) — An ohmmeter using the diode function should see a good triac as an open circuit from main terminal 1 (MT1) to main terminal 2 (MT2) in either direction of applied ohmmeter polarity. The gate lead to MT1 should show a diode junction. See Figure 5-18.

Since the triac will often fail by shorting from one main terminal to the other, the diode function of the ohmmeter can easily detect this failure by a very low reading in either or both directions between the MT1 and MT2 terminals.

Testing FETs

THE JFET — The junction field effect transistor comes in two types, the N-channel and the P-channel. In either case, only one diode junction should be evident, from the gate to either one of the two remaining leads. Checking resistance between the source and drain leads is inconclusive because a normal negative bias cannot be applied easily using only an ohmmeter. Most failures of the JFET will involve the gate junction and should be detectable as a diode having a short or having bad leakage when it should be an open. See Figure 5–18.

THE IGFET — The insulated gate field effect transistor (less precisely called a MOSFET) should be open between the gate and either of the two remaining leads (source and drain). Regardless of ohmmeter polarity, any leakage indicates a puncture of the gate insulating material and therefore a defective component. The ohmmeter scale should be used for this test. See Figure 5–18.

The resistance of an FET device from source to drain will depend upon the bias applied to the device. The practical problems of connecting a test jig to provide bias for comparison of two devices, suspected and good, is probably not worth the effort. It is simpler to install the devices in a normal circuit and test by substitution.

Testing UJTs — Unijunction transistors should be checked with the diode function of the multimeter. The emitter lead should show a normal diode junction to both of the two remaining leads, base 1 and base 2. A measurement of base 1 to base 2 in either direction should show an open when using this scale. In case of doubt about the readings obtained, compare readings taken on a known good UJT. See Figure 5–18.

Testing Zener diodes — The Zener diode can be tested as a normal diode. Using the diode function of the multimeter, one direction should show a normal junction and the reverse should show an open. Only a relatively high-voltage circuit can show the avalanche, or Zener voltage. The ohmmeter has a maximum voltage on the diode function as little as 2.0 volts, insufficient to test the Zener voltage of the diode. This test will, however, indicate the usual failures of an open or shorted Zener. See Figure 5–18.

The avalanche, or Zener point, of a Zener can be tested using a relatively high-voltage DC power supply and a current limiting resistor. See Figure 5–19.

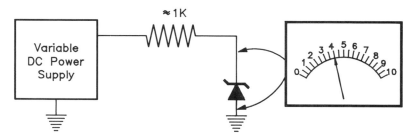

Figure 5–19 Testing the avalanche or Zener voltage of a Zener requires a DC supply higher than the Zener voltage, a current limiting resistor, and a voltmeter

Testing ICs — While it is not practical to test an IC with an ohmmeter for specific problems, the readings obtained on a suspected chip can be compared with those taken on a known good chip. Usually, however, the most conclusive test is to replace the chip and try the circuit for normal operation. If the circuit works with a new chip, the old one must have been bad.

OUT-OF-CIRCUIT OHMMETER TEST SUMMARY

— The ohmmeter is able to test most components for common failures, but it has definite limitations. *The final conclusive test for a bad part is to compare operation of the old with that of a new part.*

Testing batteries — Since we are considering here the testing of suspected components, this is the most logical place to discuss the testing of batteries. Yet they are, of course, not tested with an ohmmeter. The voltmeter must be used.

The only valid test of the condition of a battery is to measure its voltage while providing a reasonable current to a load. Turn on the equipment that is battery-operated and then measure the voltage that the battery is providing. If the battery voltage is within about 10 percent of normal voltage, the battery is working as it should, although it may be weakening if the voltage is near the low end of this range.

When testing batteries, it is good to know the nominal voltages to be expected of the various kinds of cells. In addition to this voltage, it is necessary to know how heavy a load to apply to a single cell *if* it must be tested without a normal load. The following table calls for heavy test loads. For detailed information on common cells, see Figure 5–20.

BATTERY TYPE	LOAD (R)	MINIMUM VOLTAGE
Cylindrical Cells and Batteries		
1.4-V hearing aid	60	1.2
1.5-V carbon-zinc (general use cell)	10	1.1
1.5-V alkaline	1	1.1
1.2-V ni-cad	1	1.1
6-V camera battery	1.2 K	4.4
6-V lantern battery (dry-cell type)	40 Ω	4.4
9-V carbon-zinc	250	6.6
9-V alkaline	15	6.6
Miniature Cells		
1.4-V mercury†		
watch, calculator		
high drain	200	1.0
low drain	3 K	1.0
camera	300	1.1
hearing aid	1 K	1.2
1.5-V general use manganese	200	1.0
1.5-V silver		
watch, calculator		
high drain	200	1.3
low drain	3 K	1.3
camera	300	1.1
hearing aid	1 K	1.2
1.5-V carbon-zinc	200	1.0
1.5-V lithium-iron	3 K	1.0
3.0-V lithium-manganese	600	2.0

Information courtesy Eveready-Union Carbide.
†*Caution: watch polarity as it seems reversed from other cells.*

Figure 5–20 Table of various cells and terminal voltages under heavy loads (Courtesy Union Carbide)

5.5 NOTES ON OUT-OF-CIRCUIT TESTING WITH THE SOLID STATE TESTER

— The ohmmeter generates a direct current and applies it to the circuit under test, reading the resulting current on a display. The solid state tester, on the other hand, generates a sinewave signal that is applied to the circuit by the test leads. The resulting current/voltage waveform is displayed on a cathode-ray tube. In-circuit component testing was covered in paragraph 4.9.5.

Using an AC signal and a CRT allows the performance of the component under test to be analyzed simultaneously over a range of voltages rather than at a single voltage as the ohmmeter provides. Of particular importance is the examination of non-linear components such as semiconductor junctions. These are much more thoroughly examined using this instrument. Small but important differences between a suspected component and a known good one are easily noted on the CRT.

Testing resistors — The resistor is the one component that might be best analyzed using the ohmmeter. The solid state tester will show resistors as a straight line, at an angle somewhere between the horizontal (an open circuit) and a vertical (a short circuit). See Figure 5–21.

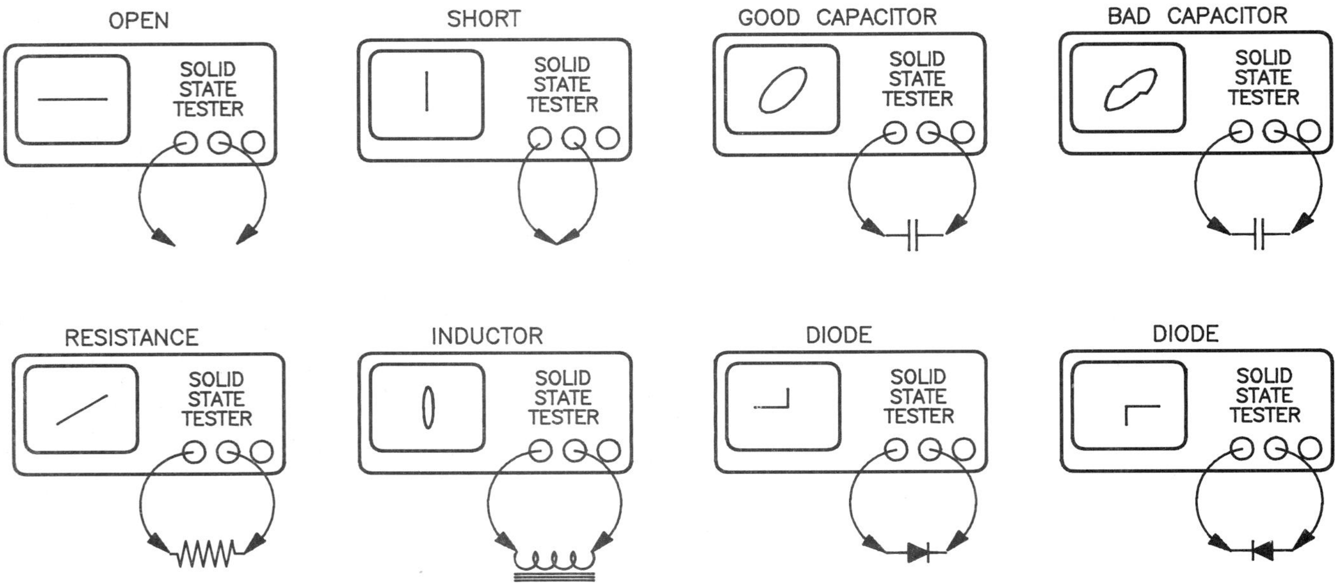

Figure 5-21 Solid state tester patterns for common components

    The angle of the line representing resistance will be affected by the value of the resistor and the range used on the SST. The high ranges use an internal reference resistor of a higher value and will show resistors as more a short than lower ranges.

    Potentiometers are tested quite adequately for wiper noise by using the SST. Noise produced by moving the shaft of a pot is the result of poor contact between the mechanical wiper and the fixed resistor against which it slides. Connect the SST from either end to the wiper and turn the shaft of the potentiometer. The display should make a smooth transition from one extreme to the other (depending on how you connected it) of somewhere near vertical to somewhere near horizontal. If the pot is noisy, the line will not make a smooth rotating tilt, but will "smear" while the control is being moved. When movement of the shaft ceases, the pattern may or may not stabilize into a solid line again.

Testing capacitors — Capacitors should basically show an elliptical pattern. The angle of the pattern provides a comparative indication of the amount of capacity in the circuit. As the pattern moves from a horizontal line into a circle and then into a vertical pattern on a given scale, the amount of capacity represented has increased. Filter capacitors with large amounts of capacity will look very much like short circuits, the vertical pattern becoming a barely open double vertical line. This is understandable because these capacitors are used as a "short" to power line frequencies of 60 Hz and above.

    As the size of capacitors decreases, the more the display will tend to lie horizontally, an open. The sensitivity of the instrument can be improved for these smaller capacitors by increasing the frequency of the voltage provided to the capacitor by the SST. Later models have this provision, enabling the operator to select the basic 60 Hz range, 400 Hz, or 2000 Hz. The 2000 Hz range will have the effect of increasing the capacitive reactance sensitivity by a factor of 33 times that of the 60 Hz range.

    While most capacitors should show an ellipse, those that show breaks or kinks in the pattern are defective. This is the result of the capacitor showing suddenly different reactance at certain voltage input levels. See Figure 5-21.

Testing inductors — A small RF inductor (choke) will look like a short circuit at 60 Hz. An audio inductor will have a discernable ellipse when viewed on an SST. Filter inductors, with their very large values of inductance, will look more like open circuits at this low frequency. A better comparison can be made of smaller inductors with known good components if the frequency applied is increased to the 400 Hz or 2000 Hz as provided on later models of the SST. As the test frequency is increased, the inductor display should slant more to the horizontal, indicating a higher reactance with increasing frequency. Incidentally, the ellipse may look like that of the capacitor, but in reality it is caused by the beam of the CRT rotating in a circle in the opposite direction from that of the capacitor. See Figure 5-21.

Be sure to look for a short to the case or core of the inductor. Measure from either end of the winding to the metal parts of the component. Only a relatively small amount of capacity should be evident. A vertical line indicates a short to the case, almost always indicating the part as defective.

Suspected inductors can be compared to known good inductors of the same value. This method is successful in detecting shorted turns in inductors. The defective inductor will look more like a short than a good one.

Testing semiconductor junctions and diodes — The SST really begins to show its value when testing solid state circuitry. It provides an actual curve of voltage to current for the device, automatically reversing in polarity to show both forward and reverse polarity response.

The basic semiconductor junction will show a pattern that looks like a check mark. The check mark can be right side up or inverted, it does not matter. Switching the test leads around will invert any pattern, but no new information is available in so doing. See Figure 5-21.

Testing bipolar transistors — Bipolar transistors are made up of two junctions back-to-back. Testing these components with the SST should show the existence of the base-to-emitter junction and the base-to-collector junction. The emitter to collector should be an open circuit, possibly with a small break at one end of the waveform. See Figure 5-22.

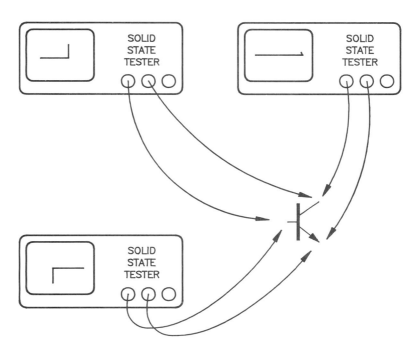

Figure 5-22 Normal indications when testing bipolar transistors with a solid state tester. Some good transistors may have a bit of a hook at one end of the emitter-collector pattern

Any gradual curving of the patterns given by a bipolar transistor indicates a defective device.

Testing transformers — Transformers are tested as inductors, each winding separately. In addition to testing the windings, be sure to test at least one end of each winding to the case or core of the transformer. While there may be a slight amount of capacity from the winding to the case, a short should not be found.

Testing switches — The SST must show switches as either shorts or opens, depending on the configuration of the switch. Any other indication is proof of a defective switch.

Testing SCRs and triacs — Silicon controlled rectifiers and triacs are widely used in electronic circuits, particularly those dealing with 120 and 240 VAC power. These components are best tested out of the circuit using the later models of SST. Not only will the SST test these components "dead," but some SST models have a triggering function provided for testing these components. See Figure 5-23.

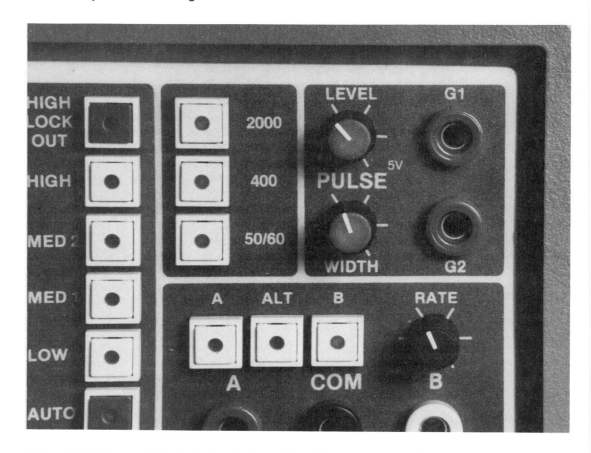

Figure 5-23 These controls of a Huntron tracker solid state tester operate a pulse generator used to test SCRs and Triacs

Using this feature, the component under test can be triggered and brought into full conduction. Detailed instructions for the testing of these components is found in the SST instruction manual.

Testing zener diodes — Zener diodes should test on an SST as a normal diode if using a range that provides too little voltage to bring the zener into avalanche. If the zener voltage is within the voltage sweep range of the SST, the pattern will have two hooks, one in the forward direction of current flow, and another at the opposite end, in the opposite direction.

Testing ICs — Integrated circuits can be tested very well with an SST. The medium ranges should be used because they have a slightly higher test voltage that is required to properly exercise the multiple junction architecture of these components.

The SST can be used to detect failures in the input and output circuitry of ICs. It cannot, however, detect failures deep within the chip, in the driver areas of a chip. Failures in these internal circuits are not evident by sampling of the input or output components of the chip.

Failures deep inside the IC are rare, however. Most of the stresses of normal operation as borne by the input stage, where incoming signals may be too high in voltage due to transients, and the output stage, where a small amount of power is normally involved and overloads are most likely to occur.

IC test patterns shown by the SST should be examined for clean, straight line segments without curved lines or breaks. Curved lines and small portions of lines that seem to flicker are indications of failures. Integrated circuits of the same generic number may show differing patterns if made by different manufacturers. The patterns should show only straight, stable segments. See Figure 5-24.

GOOD TRACE   GOOD TRACE   BAD TRACE
                                        (NOTE "FRACTURED"
                                        KNEE OF CURVE)

These indications are only a few of the possible traces obtainable in troubleshooting. The instruction book with each Huntron instrument should be consulted for specific waveforms.

Figure 5-24 Normal and abnormal indications as seen while testing ICs out of circuit

Impending failures of ICs can sometimes be detected by using an SST. Abnormal patterns, such as in Figure 5-24, may be detected on an IC that is still apparently working in the circuit. Since an IC cannot heal itself, the abnormal pattern can only get worse with time. Replacing such components may prevent a future failure. The SST is the only generic test instrument that can be used to predict a failure.

Other components — Waveforms for other components such as the JFET, IGFET, UJT, some zeners, and ICs are best studied by reference to the SST instruction manual. It is important for the technician to be familiar with the basic waveforms of Figure 5-21 without reference to the manual, however.

5.6 Check Component with Ohmmeter or SST (Compare Old/New)
— Sometimes tests of a suspected component are not conclusive. Something doesn't look as clear-cut as it should when using either an ohmmeter or an SST. One solution is to compare a suspected part with a known good component.

The bottom line in testing any component is proper operation in the circuit. The heating stresses of removal from the circuit may, at least temporarily, make a component seem fine when tested.

5.6.1 Leave It Out of the Circuit
— If a suspected component tests good when removed from a circuit, do not be in a hurry to replace it. Further testing might be easier if the old component is left out of the circuit. See Figure 5-25.

5.6.2 Look for Other Bad Components in Same Area
— If a component is left out of the circuit, testing other components in the stage in question is a bit more conclusive. Previous troubleshooting should have narrowed the problem to this one stage, and components in that stage can now be quickly checked using the ohmmeter or an SST.

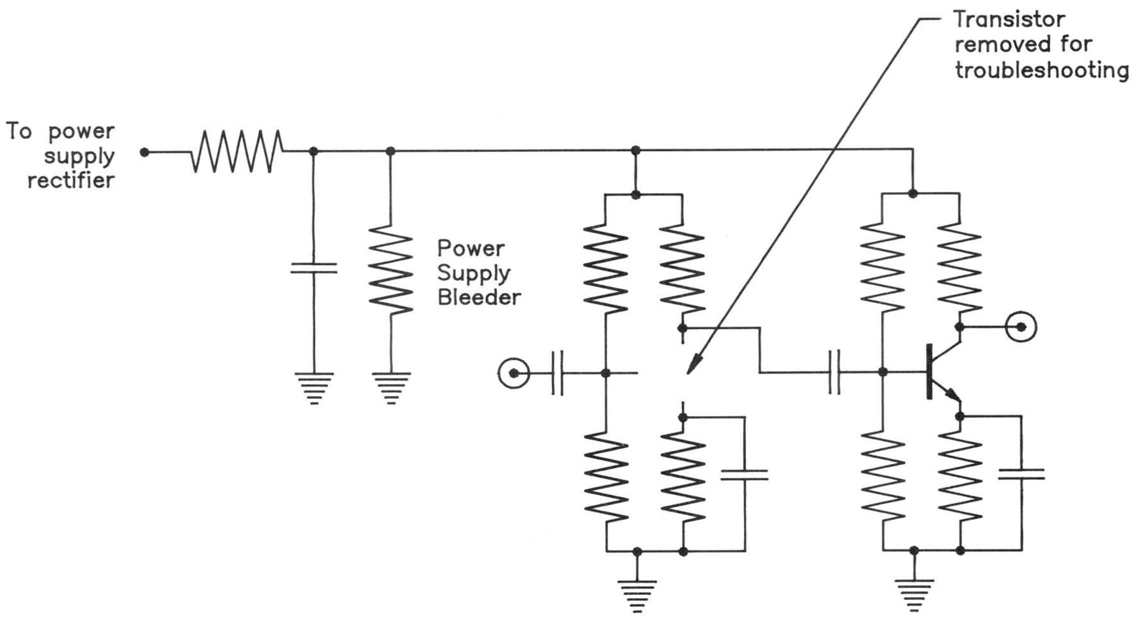

Figure 5-25 A suspected transistor was removed from this circuit. The transistor was left out after finding it tested good. This leaves several nearby components disconnected at one end, an ideal situation for further testing with an ohmmeter or solid state tester

5.6.3 Verify This Stage Is Still Bad —
If components in the defective stage all check out as good, then the stage may be operating properly. Perhaps an intermittent component is causing the problem, and the testing, soldering, and probing during troubleshooting have cured the problem, at least for the moment. If the equipment now works properly when tested, it leaves an uneasy feeling with the servicing technician. An intermittent problem may be indicated, which should be investigated using the methods in paragraph 1.2. Failing this, the technician must backtrack to verify that the stage is still not doing the job intended. Go back to the beginning of Phase 4.

5.7 Look for Primary or Secondary Component Failure —
This is a very important consideration. It is tempting to replace a definitely bad component and then immediately test the equipment for normal operation. It is an error to do so without considering the kind of failure that occurred. As an example, an open resistor is usually caused by the shorting of another component, thus causing excessive current flow through the resistor. See Figure 5-26.

Replacing the resistor in this case would obviously cause the new part to fail. Avoid this by doing some detective work. Consider what could have caused the resistor to fail.

Some classic examples of multiple failures are:

- Resistor is open. Cause: An electrolytic capacitor or transistor in series with the resistor has shorted, thus causing excessive current flow through the resistor.

- Transistor is shorted or open. Cause: If the transistor is in a DC-coupled amplifier, test previous transistors for failures. In any case, check biasing circuit for possible excessive base current. Be sure power transistors have good thermal bonding to heat sinks. Keep in mind when working on a complementary-symmetry circuit that the normal failure is finding both of the transistors overloaded and either shorted or open.

- Transformer is defective. Cause: Excessive current flow. Test all loads for shorts or current overload.

- Integrated circuit is bad. Cause: Possible electrostatic damage. Particularly susceptible chips are those connected to the "outside world," such as the card edge connector.

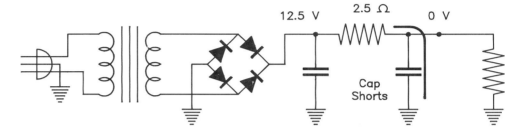

Normal Output Voltage: +12 V
Normal Load Current: 200 MA
Power dissipated by R: 0.1 Watt

Abnormal Output Voltage: 0 V
Abnormal Load Current: 5A
Power dissipated by R: 62.5 Watts!
(Instant Burnout!)

Figure 5-26 The common failure of an open resistor is a strong indication that another component failed first, causing the resistor to open

5.8 Install Optional Socket
— If there is a possibility that this component might fail again some day, by all means install a socket. There is no more stress to the PC board when installing a socket than the installation of the component. This act entirely avoids the heating stresses on the new component. Of course, physical considerations must be taken into account. Perhaps there is not enough room for a socket along with the new component.

5.9 SIMILAR PART AVAILABLE
— It often happens that an identical part is not available, but similar parts may exist. This brings up an important point in the political sense: Is it permissible to substitute? Military and aviation applications might demand an exact replacement with no substitutions of any kind permitted. If the exact replacement part isn't available, then the equipment remains inoperative until the new part arrives. Substituting anything but an exact replacement can carry heavy penalties in some cases.

Sometimes a manufacturer forces customers to use them exclusively for parts by using generic parts, such as transistors and ICs, by removing the original markings, replacing them with their own "special" codes. Perhaps a technician can recognize the functions performed by a bad chip in a circuit. A generic part might work in such a case, but this is doubtful.

Larger manufacturers make their own chips, and nothing except an exact replacement, available only from that manufacturer, will work in a given circuit. Hewlett-Packard is a good example of such a company.

One last part bears mentioning here, the PAL, or programmable array logic chip. This is a series of logic gates that is configured by the manufacturer to perform a specific and highly customized function in a circuit. Only a direct replacement, programmed by and available only from the manufacturer, will work in the circuit.

5.10 Check Electrical and Mechanical Specifications before Installing

— Applications of a less critical nature may allow substitution of similar or even better parts than the original.

The following list of components gives individual considerations when substituting similar but not identical parts in place of an original component.

Resistors

RESISTANCE VALUE — The resistance value given on the resistor is the value used by the designing engineer. The closer the substitute part comes to this value the better. Many circuits will work well when the actual value is substantially different, however. Analog circuits are more sensitive to resistor values than digital circuits as a general rule. Instrumentation and calibrated circuits must have replacement values exactly the same as the original value to perform properly.

Required resistance values can also be made up from series and parallel connected resistors. When connecting resistors in series, the sum of the individual resistors must be equal to that of the original resistor. In other words:

$$\text{Total resistance} = R_1 + R_2 + R_3 + \ldots \text{ etc.}$$

When connecting resistors in parallel, a special formula is needed to calculate the total value of the combination:

$$\text{Total resistance} = \frac{1}{\frac{1}{R_1} + \frac{1}{R_2} + \frac{1}{R_3}} + \ldots \text{ etc.}$$

It is often convenient to remember that two resistors of twice the original value will work if connected in parallel; three resistors of three times the original value will work if connected in parallel, and so forth.

RESISTANCE TOLERANCE — Some circuits require less precise resistors than others. As a general rule, replace with a resistor having the same or a better tolerance figure. For instance, if the circuit has a 10 percent tolerance resistor originally installed, a resistor of 5 percent or 10 percent tolerance can be substituted, but not one with a 20 percent tolerance.

POWER RATING — Resistors of more than approximately a half watt power rating will probably be expected to produce some heat. These resistors must be replaced with resistors of the same or greater power rating. Resistors of less than a half watt rating are usually run at substantially lower actual power dissipation and any resistor power size will work satisfactorily. In cases of doubt, calculate the actual power dissipation of the resistor as used in the circuit and use a resistor power rating equal to or greater than this value.

When using several resistors to substitute for an original, the power actually dissipated by each of the substitute resistors is additive. The total power dissipation of the substitute arrangement must equal or exceed that of the original resistor.

VOLTAGE RATING — A few circuits operating at very high voltage require resistors that have greater than normal separation of the input and output leads.

These resistors must be replaced with the same type of special resistor if a single resistor is used. Normal resistors might be used if several resistors are connected in series to provide the high voltage separation requirement of the circuit.

PHYSICAL SIZE, MOUNTING METHOD — There may be obvious limitations on the physical size and the method of mounting the resistor used as a substitute.

RESISTOR TYPES — There are a variety of kinds of resistors, described by the material of which they are made and the method of making them. Examples are the carbon composition resistors, wirewound and wirewound non-inductive resistors, and metal film resistors. When substituting resistors, it is best to stay with the same identical type if at all possible.

Capacitors

CAPACITY VALUE — Capacity value in a given circuit may or may not be critical. An indication of this is given in the kind of capacitor that is used. Electrolytics are most often used in applications where the substitution of more capacity than the original is usually quite acceptable. Filter capacitors, for instance, will filter even better if more capacity than originally installed is added to the circuit. Similarly bypass capacitors, like those used across emitter resistors in amplifier stages, also filter better with added capacity. Interstage coupling capacitors benefit when adding more capacity since the circuit will amplify low frequencies better than as originally designed.

RF bypassing often uses disk ceramic, mica, Mylar, or monolithic capacitors. This is another case where more is better. Replacing a 0.01 ufd bypass capacitor with a 0.02 ufd capacitor of the same type is quite acceptable.

Critical applications, generally speaking, include those where a capacitor is used in combination with other components. This includes timing circuits where the capacitor is used with a resistor, such as some multivibrator circuits. Other applications may use a capacitor in combination with an inductor for tuning purposes. As a rule, replace these components only with parts identical to the original. Capacitors used in these applications sometimes have special additional characteristics that would be difficult to duplicate, such as temperature coefficient.

Paralleling capacitors results in a total capacity that is the sum of the individual capacities. See Figure 5-27 for an example.

Capacitors are often connected in parallel, particularly in filter circuits. Using several smaller capacitors instead of a single large capacitor is often more favorable because of space limitations.

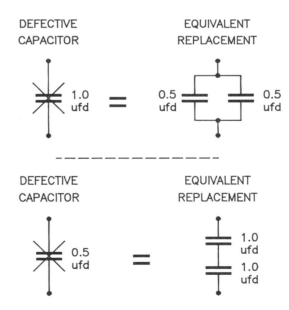

Figure 5-27 Examples of capacitor replacement with other than original values

Occasionally, it may be necessary to substitute several capacitors connected in series for a bad original capacitor. The trick here is to remember that if two capacitors are used, each must be at least *twice* the capacity rating of the original. See Figure 5-27 for an example.

VOLTAGE RATING — The voltage rating of a capacitor should never be exceeded by the circuit in which they are installed. Exceeding the voltage rating by only a small amount is inviting the capacitor to fail by shorting.

Circuits that operate at high voltages sometimes operate electrolytic capacitors in series to reduce the voltage across the individual capacitors. See Figure 4-30.

Note that when electrolytic capacitors are operated in series there must be equalizing resistors connected across them. These resistors force a more even distribution of DC voltage across the capacitors than would result if the internal leakage of the capacitors was the only factor involved.

CAPACITOR TYPE — Of the many types of capacitors available, it is generally best to use the same type that was originally installed. A disk ceramic capacitor should be used to replace an original disk ceramic, for instance. A listing of capacitor types would include the electrolytic, tantalum, oil, mica, ceramic, paper, glass, air Mylar, and monolithic capacitors.

CAPACITOR CURRENT RATING — A few applications require a capacitor to continuously pass substantial radio frequency current. These capacitors are usually made of mica, and may be physically large. Replacement of one of the capacitors will require another of not only the same capacity rating, but the same current rating as well.

SIZE AND MOUNTING — There may be obvious mechanical difficulties in substituting physically large capacitors. The different types of mounting may present a problem, along with the method of connecting them to the circuit. Individual cases must be considered on the basis of their unique problems.

Fuses

CURRENT RATING — The current rating of the fuse is the major factor in selecting a substitute fuse. *The substitute fuse should be no larger in value than that of the original, unless further operation of the circuit is more important than the probable damage that will result.* Probable damage estimates must include the possibility of fire damage. It is safe to use a smaller fuse than the original, but such a fuse will probably blow quickly, or at least will render the circuit less than normally reliable.

VOLTAGE RATING — The voltage rating of a substitute fuse should be the same or greater than the original. The voltage rating of a fuse is the safe maximum voltage at which the fuse can operate. In the event of a dead short in the load, the excessive currents that could flow could result in a terrific amount of energy being handled by the fuse alone. The voltage rating of the fuse is that voltage that the fuse can safely withstand while opening the circuit. Use at a higher voltage could cause the fuse to literally explode.

SLOW- OR FAST-BLOW — Sometimes a circuit has a normal surge of current during initial startup. The slow-blow fuse has been designed to tolerate this surge, yet provide protection against a sustained overload current well below the surge value. Using a fast-blow fuse in such a circuit will result in the immediate blowing of the fuse because of the initial surge current. Use a slow-blow fuse where one is indicated.

A fast-blowing fuse is used where the circuit must be turned off immediately when circuit current exceeds the value of the fuse. Such fuses are supposed to act very quickly, even if there is a small sustained overload. Using a slow-blow fuse of the same current value would result in removal of most of the protection offered by the fast-blow fuse. Circuit damage would probably result before the fuse opened.

PHYSICAL SIZE, STYLE — Fuses come in many different physical configurations.

While it is possible to change fuse styles and retain all of the protection of the original fuse type, this is not often done. When it is, physical mounting and layout of the new fuse holder is of major consideration.

Transformers

INPUT VOLTAGE — The voltage used by the primary winding is the first consideration when replacing a transformer with a similar but not identical type. The replacement transformer must be capable of operating with the given input voltage. Replacement transformers may have one or two primary windings for operation on either 120 or 240 VAC. These windings are intended to be wired in series for 240 VAC or operated with the two windings connected in parallel. It is also possible that this kind of transformer is intended to be operated on 120 VAC using only one of the two windings rather than operating them in parallel. See Figure 7-9.

FREQUENCY OF OPERATION — A transformer may be designed to operate over a range of input frequencies such as those used in a few audio interstage applications. These applications are rare, however. A failure of such a transformer would require a direct replacement.

Power transformers are specifically designed to work on 50, 60, or 400 Hz circuits. European power lines are commonly operated at 50 Hz. Power transformers designed for 50 Hz should work well with the 60 Hz standard of U.S. power lines. The reverse may not be true, however. A transformer designed to operate at 60 Hz may eventually overheat if operated at full current ratings on 50 Hz because of a lack of sufficient core material.

Aircraft applications commonly use 400 Hz for AC power circuits because transformers of a given power rating can be made much smaller and lighter in weight because less core material is needed. Only exact replacements must be used in aviation applications.

OUTPUT VOLTAGE, CURRENT — A replacement transformer will be required to provide the same output voltages as the original. Windings may be placed in series with phasing boosting or bucking each other. Refer back to Figure 1-15.

The output current of windings connected in series will be the current winding of the lowest rated winding. For example, if two 6-volt windings were rated at 3 amperes and 6 amperes, the maximum current that could be drawn with the windings in series must not exceed 3 amperes. See Figure 5-28.

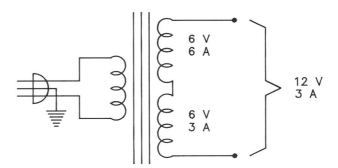

Figure 5-28 The current rating of windings in series is that of the lowest rated winding

SIZE, MOUNTING — Physical mounting requirements may vary widely. Replacement transformers must fit within the confines of both chassis area and within the cover. Be sure to place the transformer in position and try the cover on for size to avoid surprises later.

Be sure to observe the phasing requirements of the circuit. These are particularly important when transformers are used in pulse circuits where phase reversal of a winding could result in the complete failure of a circuit to operate. As an example, through improper phasing a pulse amplifier stage could get a pulse to turn off, rather than to turn on as required.

Power transformers must also have the proper phasing to operate if there are several windings in parallel on the input side of the transformer. If in doubt as to the proper phasing of a transformer, a soft fuse can be used to prevent overload damage to the transformer during initial wiring. Using the soft fuse, a shorting situation is harmless.

Switches

CURRENT RATING — One of the important specifications for switches is the amount of current that the switch can reliably make and break a reasonable number of times before the contacts fail. Replacing a switch with one of lower current rating invites an early failure.

VOLTAGE — A voltage rating is sometimes given for a switch. This is a measure of the insulating qualities of the switch. Be sure that the switch to be considered for a replacement is rated at equal to or more than the voltage involved in the circuit.

SPRING LOADING — All other factors being equal, replacement of a spring-loaded switch with one that is not, or the reverse case, may not be convenient for an operator, but the switch will work for temporary use until a proper replacement can be obtained.

NORMALLY CLOSED OR NORMALLY OPEN? — This question must be determined for any spring-loaded switch. Which contacts are closed without the switch being activated? These are the normally closed contacts. The contacts that close upon activation of the switch are called normally open contacts. A replacement switch should have the same arrangement of NO and NC contacts as the original. See Figure 5–29.

Figure 5–29 Spring-loaded switches should be marked Common, NO, and NC

NUMBER OF CONTACTS — A replacement switch may have more contacts than the original. Extra contacts may be left unused or, if the functions are identical, they may be paralleled with others to help in current-carrying capability. See Figure 5–30.

TOGGLE — There are a few switches that toggle. This switch changes the position of the contacts every time it is activated. A single push might cause the contacts to close, the next cause them to open. While this is a convenient switch for an operator, a simple switch will often do the same job.

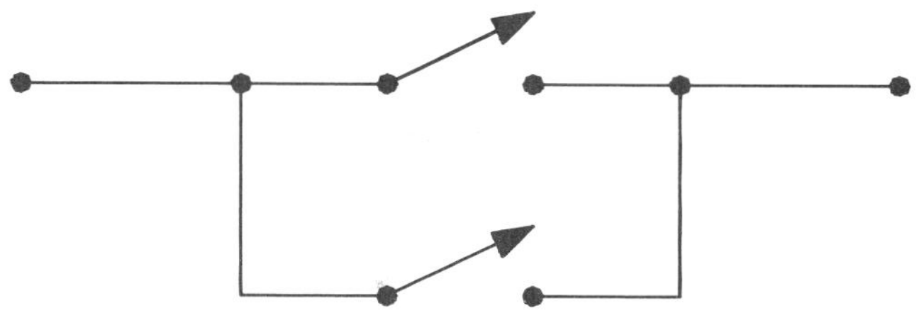

Figure 5-30 Extra switch contacts can be wired in parallel to help carry current through the contacts

ROTARY SWITCHES — Rotary switches are used when complicated functions need switching together or when many positions are necessary. When such a switch fails it is usually necessary to get an exact replacement. A single exception to this rule might be when the switch is a simple multiple contact switch with a single common connection. These switches are used in applications such as selecting different resistors within a circuit. See Figure 5-31.

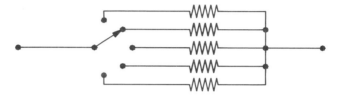

Figure 5-31 A simple multiple contact switching arrangement is often used in electronic circuits

Rotary switches may be made in either of two ways, break-before-make and make-before-break. In any case of rotary switch replacement, this is a major consideration. See Figure 5-32 for an explanation of this principle.

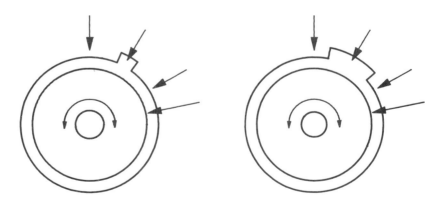

Figure 5-32 The break-before-make and make-before-break arrangement of a rotary switch

Semiconductors — Diodes, transistors, zeners, SCRs, triacs, FETs, and UJTs are among the many kinds of semiconductors made today. There are several major manufacturers of these components that have attempted to provide a line of relatively few semiconductors to substitute for the thousands of different semiconductors in use. A single transistor, for instance, might work very well in substituting for over 200 different transistors made by other manufacturers. Much time and effort has gone into *replacement guides* that match things such as the current, voltage, power, frequency and gain specifications of these components, to name just a few. Recommended replacements are supposed to match or exceed the specifications of the original component.

Replacement is simple when using such a guide. Just look up the old part number in the listing provided and look across to find the generic identification number of the recommended replacement part. Be careful to note any special codings given in the guide, as such coding may alert the technician to important things such as a difference in the designation of the leads of the replacement device.

If the bad part cannot be found in the guide, identification becomes much harder. It is possible that the manufacturer of the equipment in which the device is used may provide a replacement recommendation. It is equally possible that the manufacturer may want to sell you a direct replacement at an inflated price. See Appendix C for information that may be useful in identifying the manufacturer of a component.

The replacement guides can also be consulted to verify that two components are approximately interchangeable. Look up the original part and then a proposed replacement. Both entries in the replacement guide should refer to the same generic replacement part.

Good information on the considerations of replacing original parts with generic replacements is usually given in the front of the replacement guide. This is good material to review before using a generic replacement component.

Integrated circuits — Integrated circuits can be replaced using the same replacement guides mentioned for semiconductors. There are two families of digital ICs that are so common that manufacturers have made a real effort to standardize on their specifications: TTL and CMOS ICs. The same part number should be interchangeable from one manufacturer to another. A 7402 TTL IC made by GE will replace one made by RCA, for example. Beyond these two families, similarities of numbering systems are less common. The replacement guides should be consulted in all other cases.

Again, the appendices of this text may be of help in identifying the manufacturers of integrated circuits.

Batteries — Perhaps a direct replacement battery is not available or operating time needs to be extended by means of a different battery. Basically, operating time can be extended by using physically larger cells or by paralleling individual cells. Higher voltage can be obtained by connecting cells in series with each other. See Figure 5–33.

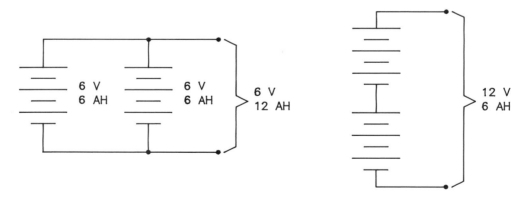

Figure 5–33 Connecting cells in parallel increases current capability and/operating time. Series connected cells result in higher voltages

Battery operating time is expressed as an ampere-hour rating. An ampere-hour rating of 4 AH implies that the cell or battery will deliver 4 amperes of current at normal voltage for a period of one hour. Other operating times can be approximately estimated by dividing the AH rating by the current required in amperes.

Replacing carbon-zinc cells with alkaline cells will result in longer battery life in a given application. Alkaline cells also have a longer shelf life and will provide higher currents than carbon-zinc cells.

RECHARGEABLE? — One major reason for changing the original kind of battery is to allow recharging of the cells rather than discarding the old ones and buying new ones.

Carbon-Zinc and alkaline cells are the standards of the throwaway cell line. They have cell voltages of 1.5 and 1.4 V respectively. Changing to nickel-cadmium (ni-cad) cells for their advantage of being rechargeable means that cell voltage must be considered. Four carbon-zinc cells will provide a nominal 6.0 VDC. Replacing them with four ni-cad cells will result in only 4.8 V because a ni-cad cell produces only 1.2 V. Five ni-cad cells will provide the same battery voltage as four carbon-zincs and slightly more than four alkaline cells. See Figure 5–34.

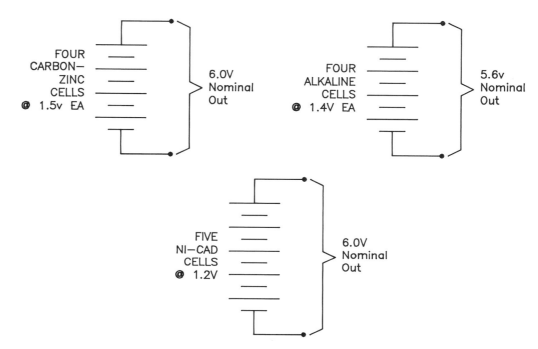

Figure 5-34 Voltages of carbon-zinc, alkaline, and ni-cad 6 volt batteries

Relays

COIL VOLTAGE — If a direct replacement is not available for a relay, substitutes can often be made. The first consideration is the operating voltage for the coil. Common coil voltages are 6, 12, and 24 V DC, and 12, 24, and 120 VAC.

If the old relay is operated from AC, the replacement relay can be an AC or DC relay. An AC relay will have a small copper attachment to the end of the core, under the armature. If the replacement has a DC coil, the addition of a series diode and a filter capacitor can be made to operate the relay. The diode is connected as a half-wave rectifier with the capacitor in parallel with the relay coil. See Figure 5–35.

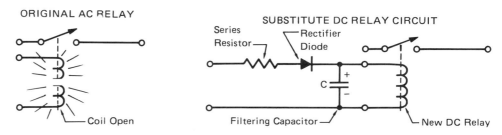

Figure 5-35 Operating a DC relay coil in place of an AC relay, using a diode and filter to convert the input power to DC

If the old relay was operated on DC, an AC relay should work fine with nothing more added than possibly inserting a resistor to reduce the current flow because the relay coil will now be operating without reactance.

If the relay driving circuit is a driver transistor feeding the relay coil, it may be necessary to consider the current requirements of the replacement relay to ensure that the driver can supply the current needed.

CONTACTS, NUMBER AND ARRANGEMENT — Contacts on relays are usually simple and straightforward. The replacement relay must have at least the same number and kind of contacts as the original, such as the double pole, double throw configuration. See Appendix D for examples. As previously stated in this paragraph (5.10), for switch replacements, the current and voltage requirements of the circuit must be met by the relay replacement. Additional current and voltage capability beyond that demanded by the circuit are desirable but not necessary.

PHYSICAL MOUNTING — The replacement relay can usually be mounted in the same space as the original. Although most relays can be mounted in any position, mercury-wetted relays must be mounted in accordance with the "this side up" markings on the relay.

Meters
Meters should be replaced only with the same type of meter as the original. Most meters are of the D'Arsonval movement type, but other types include the iron-vane and the dynamometer types.

CURRENT AND RESISTANCE — The principal ratings of a D'Arsonval meter are the full-scale sensitivity and the resistance of the meter coil. A basic meter movement can be wired externally to make it function as a voltmeter or an ammeter. Adding a high resistance in series with the movement makes it a voltmeter. Add a rectifier to this and the meter becomes an AC voltmeter. Instead of a series resistance, a low resistance placed in parallel with the meter movement converts the meter to a DC ammeter function.

Basic, generic D'Arsonval movements are available with arbitrary 0 to 10 and with many other scales. These meters can be wired into equipment to fill the functions of voltmeters or ammeters as mentioned previously. In those applications where this has been done (using external multiplier or shunt resistors), meters having the same full-scale deflection and similar coil resistance can be substituted. See Figure 5–36.

PHYSICAL MOUNTING — Meters of different physical appearance but same internal structure as the original may be pressed into service if the new mounting requirements can be met.

Speakers
Speakers are made with two principal ratings, the impedance of the speaker and its maximum power handling rating. The impedance of the speaker represents the amount of loading the speaker will present to the driving source. The replacement speaker should be the same as the original. Using a different impedance speaker will result in a small loss of audio volume. Loss of volume might be noticeable on a stereo system where only one of the two speakers has been replaced. In applications that are operating at low volume levels and where matching to another speaker is not required, almost any common speaker will substitute for another.

The larger the physical size of the speaker, the better it will respond to low frequencies. On the other hand, a small speaker is more efficient in coupling high frequencies into the air. Combinations of large and small speakers are used for these reasons in high-fidelity audio installations.

Mounting of a substitute speaker may require drilling new mounting holes. Be careful not to stress or put a twist on the speaker frame, as this can easily cause misalignment of the speaker cone, causing rubbing of the voice coil on the frame or magnet. Tighten the mounting screws only as snug as necessary.

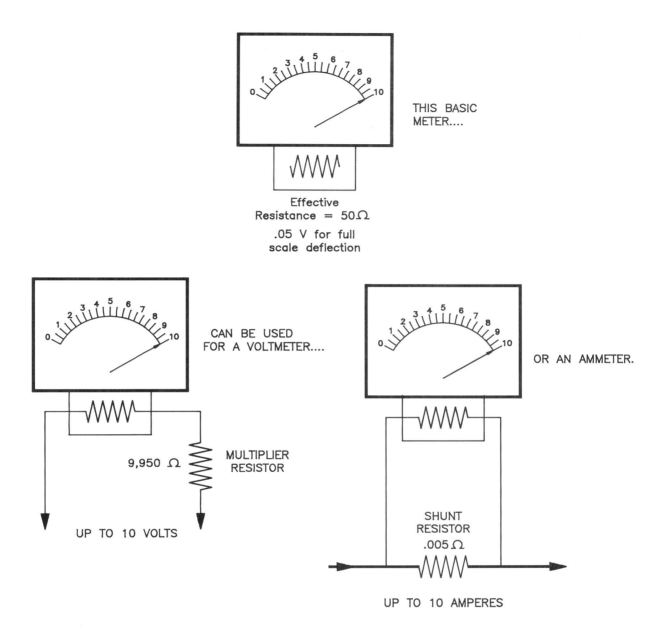

Figure 5-36 Sample circuits using the "standard" 50-ohm coil, 1 milliampere full-scale D'Arsonval movement for either a voltmeter or an ammeter

Microphones — Microphones in common use today are the dynamic, ceramic, and crystal microphones. A few carbon microphones are still in use for some special applications such as communications in aircraft.

The dynamic microphone is a high-quality instrument which is operated very much like a speaker, but in reverse. Sound moves a diaphragm that moves a coil in a magnetic field. A tiny voltage is produced by the coil that is coupled, usually, into a small transformer that converts the small voltage with relatively high current into a sound waveform with higher voltage and a lower current. This transformer is called an impedance matching transformer. It typically has a 600 ohm output that is then fed into an amplifier. These microphones should be replaced only with other dynamic microphones having the same specifications. See Figure 5-37.

Crystal and ceramic microphones can sometimes be interchanged without circuit modifications. Compensation for the different microphone output levels can be made by adjusting the gain of the amplifier. See Figure 3-58.

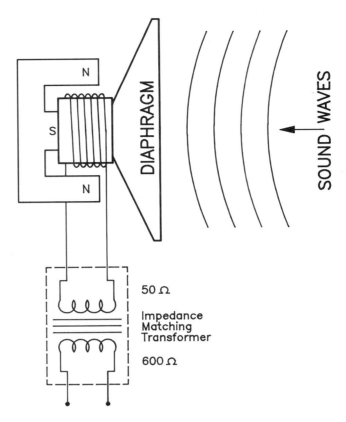

Figure 5-37 Internal wiring of the dynamic microphone makes it operate like a speaker in reverse, with an impedance matching transformer somewhere in the circuit

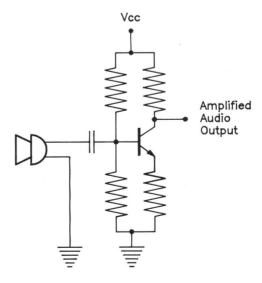

Figure 5-38 Amplifier circuit for a typical crystal or ceramic microphone

The carbon microphone is basically a variable resistor, the value of which is varied in accordance with the sound waves striking its diaphragm. This microphone requires an external source of current through the resistor element. Variations of the current flow result from sound striking the diaphragm. See Figure 5-39.

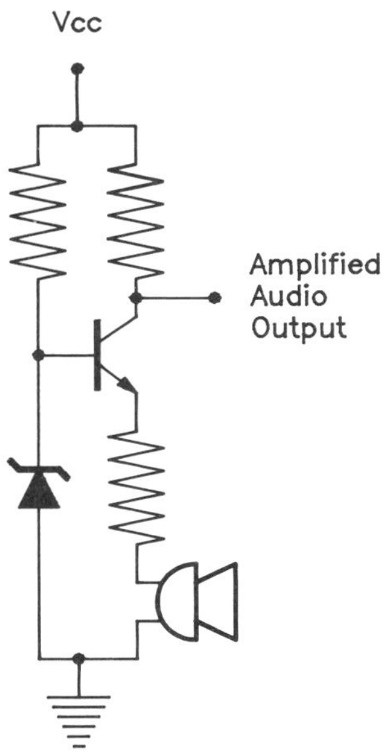

Figure 5-39 Typical circuit for a carbon microphone

One of the common problems with a carbon microphone is the packing of the granules of carbon that make up the resistance element. These can sometimes be jarred loose (this is an unusually rugged instrument) by a gentle rap against something hard. Be reasonable, but a rap such as this can sometimes vastly improve a carbon microphone that is reported to sound "distorted" or "ragged."

Any of these microphones can have a switch installed. This switch might be used separately from the microphone to switch on a transmitter or public address system. Other applications use a microphone-mounted switch to momentarily short out the microphone if the speaker needs to cough, sniffle, or talk to someone off-mike. In an emergency, an external switch could be used to perform this function if one is not provided on the replacement microphone.

5.11 Install the New Component

Mechanical installation — Be sure that the replacement component is put into the circuit properly. Although some parts like resistors and small capacitors may be inserted either way, electrolytic capacitors, transistors, and diodes must be properly oriented. Even a resistor might look better if inserted with the color bands oriented the same as the other parts.

If the original component was attached to the chassis or a heat sink, it probably also had a heat sink compound applied that must be renewed between the component and the mounting surface. Refer back to Figure 5-12.

The original component may have been electrically isolated from the heat sink by a mica washer and special bushings or washers to prevent the mounting bolts from shorting the case to the heat sink. If such was the case, installation of the new part should use the same or new insulating components if available. Before connecting any wires, check with an ohmmeter to be sure that there is no leakage or a short from the component mounting to the heat sink. See Figure 5-40.

Making a soldered connection — The physical requirements of a soldered connection will vary with the industry in which the technician works. Aeronautical applications will probably require a firm mechanical connection of several wraps of wire before solder is applied. The television repair shop will require only that the connection is secure after soldering. Either way, the connection must be kept still during and shortly after the soldering process.

248 □ Replacing Defective Components

Figure 5-40 Checking to be sure the new part is insulated from the heat sink

Sometimes hemostats or other clamping devices can be used to hold things together until the solder hardens.

After the connection is heated to the proper temperature to melt the solder, *apply the solder to the connection, not to the iron.* If the solder is applied to the iron first, the rosin will burn up before getting to the connection where the rosin is needed for cleaning. The best place to apply the solder is generally to the opposite side of the connection from the iron. The solder will tend to flow toward the heat, thus covering the entire connection.

A soldered connection can be gently blown upon after removal of heat to aid in quickly cooling the connection. This is recommended in the case of soldering semiconductor components, including integrated circuits.

A soldered connection that tends to leave a spike of solder chasing the iron as it is removed is an indication of too little soldering flux. The cure for this is simple. Apply a tiny amount of rosin-core solder to the heated connection and then remove the iron a split second before the solder is removed. A good connection will have smooth fillets and the solder will be shiny.

Removing components with the soldering iron was covered in detail in paragraph 5.2.1.1.

5.12 IS THIS A CRITICAL, ANALOG CIRCUIT? — Some analog circuits must be realigned or readjusted if certain components are replaced. Precision analog instruments are typical of equipment that require this care. Circuits that have precise gain or bandpass requirements, for example, may be sensitive to replacement parts. The instruction book for the equipment in question will give directions on what must be done if critical components are replaced.

5.13 *Realign, Readjust as Necessary* — Calibration procedures must be accomplished after the replacement of critical components in some precision analog equipment. An example of this kind of circuit is a precision DC amplifier used in an oscilloscope. Replacement of one of the amplifier transistors would probably change the overall gain of the circuit. By a simple adjustment of a calibration gain control, the gain of the circuit can be brought back into specification.

The directions for realigning or adjusting a circuit must come from the documentation for the equipment in question. There should be detailed, step-by-step instructions as to the inputs, outputs, and the criteria of adjustment as calibration progresses. Some applications require precise input signals, stable in frequency or amplitude. Without these precision input standards, the instrument may have to be sent to an outside facility for proper calibration.

Realignment procedures must be followed in the order stated in the instruction manual. Deviation from the precise order of adjustments can result in totally unacceptable results of the entire procedure.

5.14 Reassemble Everything

Now that the components are replaced and the equipment works as it should, put everything back into place. Remember to secure all mounting bolts. Incidentally, when replacing a panel having a number of mounting bolts, start all of the bolts first before tightening any of them. Putting one or two in and tightening them down will usually result in having to go back and loosen them before the others will start properly in their holes.

Be careful not to pinch wiring under bolt heads. This is especially easy to do when re-mounting circuit boards. Wiring underneath, out of sight, can often be caught between the circuit board and the mounting bolt.

5.15 Run All Operational Tests

No matter how certain the technician may be about the repair just made, there could be other problems within the equipment. Every repair should be checked carefully, beginning with a verification as to whether or not the reported problem is now gone. Then the equipment should be operated through its entire set of operating parameters to be sure that all is well with it. This means checking the operation of all controls, and, as far as practical, exercise of all of the circuitry within the equipment.

The final checkout may also use generic test instruments. Generic test equipment includes wattmeters, spectrum analyzers, and signal generators. Some electronic systems have dedicated system test procedures and specialized test equipment that is required for system checkout.

5.16 ANY PROBLEMS REMAIN?

Perhaps a new problem has arisen, requiring further repair. Sometimes the technician inadvertently introduces a new problem that must be corrected. Discovering a problem may mean that all of the troubleshooting must begin from scratch, right from Phase 1.

A nontechnical note belongs here. When all of the troubleshooting is done, it is a good professional touch to give the equipment cabinet and front panel a bath. Using a mild cleaner such as "409," put a bit on a clean cloth and wipe the equipment down. You may be surprised at all the dirt and gunk that comes off apparently clean equipment! Although the customer may not notice it, psychologically this is a good move. It makes the customer feel good about the equipment repair at first sight.

5.17 Do Required Paperwork

Once all the repairs have been completed and all other discovered problems are gone, the equipment can be considered completely repaired and returned to service. Now is the time for completing all of the sometimes unwelcome paperwork. There is a good reason for paperwork, however.

Warranty repairs can be paid to service shops by the manufacturer only if the proper information is provided to the manufacturer. Principal items needed are date and time, a problem brief, the action taken, parts used, hours of labor, and the serial number of the equipment. Such a record can be useful if the same customer comes in saying that the equipment he brought in last week is not working again. An inconspicuous check of the serial number might indicate that this considerably-less-than-honest person has brought in a *different unit* than last time and now expects you to repair it free of charge.

Be sure to update the shop's record of parts used so that they can be re-ordered. Failure to do this will eventually cause delays while waiting for parts, or special trips must be made to get them from the local distributor—if the part is in stock there.

Many employers also expect the technician to keep additional records, such as the job number for the equipment. This job number is an accounting device to enable the employer to charge the customer for payment. Without proper records, the employer cannot claim your time and does not get paid for your work.

Equipment history cards are sometimes required. These records are important if the equipment is maintained by different people throughout the lifetime of the equipment. The histories bring later technicians "up to speed" quickly with regard to a specific piece of equipment.

It is a good idea to keep your own "black book" of work. Jot down those things just mentioned, along with other useful information such as mileage to and from a job, the time you left home on an emergency service call and when you returned, plus any meals and/or gas you used for which you will expect reimbursement. Other good items to keep a record of, besides those mentioned above, include the contact person on a job and the telephone number. These little record books can become very valuable. Suppose you must backtrack and resubmit a claim for reimbursement when the original you sent in a month ago gets lost somewhere in the red tape abyss.

If the equipment will be shipped, pack it carefully and be sure to include all of the required paperwork inside the box, such as the shipping invoice and billing documents.

☐ Phase 5 Summary

This phase has discussed some of the pitfalls and given some valuable tips on replacing components and getting the equipment back into condition for delivery to the customer. The technician was made aware of the need to look for failures other than the one initially found. Soldering and desoldering of components were covered because of their importance.

☐ Review Questions For Phase 5

1. After turning off power in preparation for removing or replacing a component, what should be done next?
2. What is the best kind of soldering iron to use on printed circuit boards?
3. The tip of your temperature-regulated iron is oxidized and very dirty. How should you attempt to clear it?
4. Name the two kinds of solder and the one that should never be used on electronic circuits.
5. An option to keep in mind after desoldering an IC and before installing a new one is _____.
6. What should be done just before removing a component from a circuit?
7. What is the indication on an ohmmeter of a good, medium-value capacitor?
8. What is the unit of measurement when testing diode junctions on the diode scale of a digital multimeter?
9. You are not sure if a reading of 3 ohms is correct for a given inductor after removal from a circuit. How could you know if this were the correct reading?
10. What are the three resistance checks desirable for a suspected defective transformer?
11. A certain switch in your junk box measures 3 ohms from the common to the normally closed connection. Is this a good switch?
12. How many junctions can be detected in an SCR, using a DMM on the diode scale?
13. A JFET shows a diode junction between gate and source. Is this normal?
14. A JFET shows a diode junction between gate and drain. Is this normal?

15. What is the simplest test of an IGFET, other than finding there is no short present?
16. Can a digital multimeter on the diode function test a zener diode?
17. What is the most conclusive test for a suspected defective IC?
18. How should a battery be tested to see if it is good?
19. What is the basic pattern for a resistor as tested on a solid state tester?
20. What is the basic pattern for a capacitor as tested on a solid state tester?
21. What is the basic solid state pattern for a diode junction?
22. How should a bipolar transistor be tested using a solid state tester?
23. What instrument provides the ability to trigger and thereby test an SCR or triac out of circuit?
24. Can a solid state tester check zener diodes in the avalanche mode?
25. Can ICs be effectively tested with a solid state tester?
26. You have removed a component, only to find that it tests normally. Should you replace that component immediately, then proceed to make further tests?
27. You have found, verified, and replaced a defective component. Is it time now to apply power to test the circuit without further analysis?
28. Name three of the six factors that must be considered when replacing a resistor with another, one that is not exactly the same as the original.
29. Name three of the five factors that must be considered when replacing a capacitor with another, one that is not exactly the same as the original.
30. You must replace a transformer that originally had a 1 ampere, 24 VAC winding. All you have is a transformer with the same primary, but it has four 6 VAC windings of 2, 2.5, 4 and 5 ampere current capabilities. Can this transformer be used?
31. What is the likely result of connecting a 400 Hz transformer to a 60 Hz line?
32. You must replace a SPST switch that operates a 2 ampere circuit. The only switch you have available is a DPST switch, with contacts rated at 1 ampere. Can you use it?
33. How are spring-loaded switch contacts usually marked?
34. What type of rotary switch should be used to switch the multiplier resistors of a voltmeter, a make-before-break or break-before-make?
35. What is the purpose of a semiconductor replacement guide?
36. Placing batteries in series affects both voltage and current capabilities. What are they?
37. What benefit is derived from operating batteries in parallel?
38. How many 2 V cells would be required to supply a 4 ampere, 4 V load if each cell is rated at 2 amperes?
39. Name the two most common types of rechargeable batteries.
40. Where should heat be applied when making a soldered connection?
41. What causes the smoke of a soldering operation?
42. You have replaced a component in a precision attenuator. What would you expect to have to do before certifying that the equipment is ready for use?
43. You have found a problem, replaced a component, then reinstalled the cabinet. What must then be done to be certain the equipment is ready for the customer?
44. You have used three components from the stockroom. What is the logical expectation to ensure there will be no future shortages of these parts for next time?

PHASE 6
DEAD CIRCUIT TROUBLESHOOTING

☐ PHASE 6 OVERVIEW

There are two reasons why the technician might find this phase of particular help: when he or she wants to try for a quick fix without signal tracing to find the problem, and when, in spite of lack of documentation or schematics, the board must be repaired anyway.

Two instruments are useful for troubleshooting without power, the ohmmeter and the Solid State Tester. Both of these instruments inject a small signal into the circuit under test and measure the response returned by the circuitry. The ohmmeter provides a DC signal, and the SST provides an AC signal. The SST provides two dimensions of troubleshooting information compared with the two possible points of the ohmmeter. This makes the SST the instrument of choice.

ALWAYS TURN OFF ALL POWER AND DISCHARGE LARGE FILTER CAPACITORS BEFORE USING AN OHMMETER OR SOLID STATE TESTER!

6.0 IS A SOLID STATE TESTER AVAILABLE? — The Huntron Tracker is one model of a solid state tester, a unique instrument for testing circuits without applying power. See Figure 6-1.

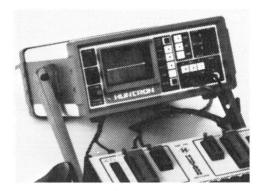

Figure 6-1 The Huntron Tracker is a solid state tester that provides a great deal more information about circuit operation than that offered by an ohmmeter (Photo courtesy of Huntron Instruments Inc.)

An overview of the SST and its use is given in paragraph 4.9.5. The use of the SST is discussed in detail for out-of-circuit troubleshooting in paragraph 5.5. If the reader is not familiar with the basic patterns of typical individual components, they should be reviewed before going farther, as the basic waveforms will now be combined. Paragraph 6.0.1 covers typical patterns that may be encountered when testing components while they are still installed in a circuit. This means that the waveforms shown will be combinations of the basic out-of-circuit waveforms covered in paragraph 5.5.

6.0.1 Using the SST for In-Circuit Tests — It would be unrealistic to give more than a few examples of combinations of waveforms because of the variety of components and their responses to differing voltages and frequencies.

The SST will show combinations of devices as combinations of their individual waveforms. The diode and the resistor is a good combination to consider first. An ohmmeter might well show this circuit to be only a resistor because of the relatively low value of resistance. The SST is able to separate them. See Figure 6-2.

Another example is the capacitor and diode combination. See Figure 6-3.

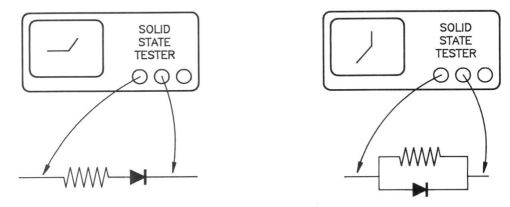

Figure 6-2 SST patterns for series and parallel resistor/diode combinations

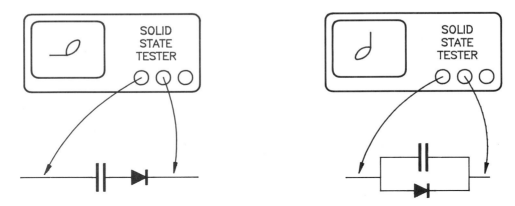

Figure 6-3 The capacitor and diode combination in parallel and series, and resulting solid state tester patterns

The testing of integrated circuits may produce more complex patterns. These patterns are the result of reading junctions within the chip. A waveform taken on a chip installed in a circuit is usually the result of reading back into one chip output and ahead into at least one input. See Figure 6-4.

6.0.2 Use Pattern Recognition or Documented Waveform Methods

PATTERN RECOGNITION — Once the basic patterns of various components as seen on an SST are familiar, the SST can be used in the "pattern recognition" mode. If the technician knows what components are across the points probed with the instrument, the resulting waveform can be predicted and any differences investigated further.

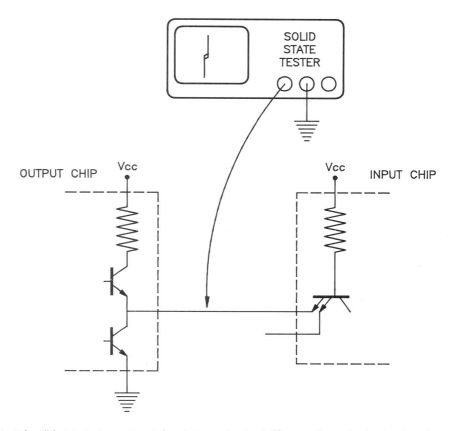

Figure 6-4 A solid state tester pattern taken between in-circuit ICs usually reads circuitry in at least two chips, one output and one or more inputs, all in parallel with each other

DOCUMENTED WAVEFORMS — An alternative to the "pattern recognition" method is to refer to an actual record of how test points on a good board are supposed to appear on the SST display. This method of using an SST is much like comparing actual circuit waveforms seen on an oscilloscope to recorded oscilloscope displays for a normal circuit.

6.0.3 IS THERE AN EDGE CONNECTOR? — Many failures of electronic components are the result of over stressing them electrically, often by static discharge. The edge connector of the card is likely where most of the inputs and outputs will be stressed. See Figure 6-5.

Connectors and wiring to the board can also bring in voltage spikes and overloads. Be sure to check these points on a circuit, too.

6.0.4 Test the Edge Connector — Testing the edge connector is simple enough. Just find either the ground bus or the Vcc bus (these are used because they are connected to most of the components on the board) and connect either of the two SST probes there. Using the other probe, test each of the connections one at a time, looking for abnormal patterns. At the least, a technician can watch for "fractured knees" or curved lines. Any indications of shorts should also be investigated to see if they are normal.

6.0.4.1 Test Power and Major Components In-Circuit — The largest components are often the first components to give trouble. Large transistors, for instance, are made that way to throw off power in the form of heat. Heat breeds failures, thus the large semiconductors should be among the first components to be tested. Large capacitors are usually electrolytics, which are prone to failure by shorting because of the small spacing within them and the relatively high voltages they routinely handle. Look for opens and shorts

Figure 6-5 The edge connector of a card is a likely place to receive electrical stress and detect a component failure

on the transistors and primarily for shorts on the capacitors. There is a good chance that testing these components will show at least one of them to be bad.

6.0.4.1.1 ARE YOU USING THE COMPARISON METHOD NOW? —
The comparison method is a third way to use the SST. This method depends on observing the change in the SST display while the SST automatically switches inputs from two boards, one good and the other bad.

6.0.4.1.2 IS A GOOD IDENTICAL BOARD AVAILABLE? — The SST
is very good at comparing two boards. This method can be used only if a good *identical* board is available. The good board must also have all switches or jumpers at exactly the same settings as the bad board before the comparison will be valid.

Keep in mind that ICs of the same marked type (e.g., 7400) may have different waveforms if manufactured by different companies. This is because some may be provided with input protection diodes or other small differences. The SST will show these differences, but both circuits may be normal at this point. Look for radical differences between the boards or small differences between identical chips from the same manufacturer. These will probably lead to the problem.

6.0.4.1.2.1 Use Comparison Method — The comparison method amounts to having
the SST electronically switch inputs, alternating from one "hot" test lead to another "hot" lead, placing the resulting patterns of each on the screen. If the two waveforms are identical, the pattern seems to sit still. If different, the display toggles back and forth in time with the test lead switching.

Setting up for the comparison method is easy. The technician connects the common lead of the SST to the ground bus of both boards. The two "hot" test leads are then connected to the two remaining input jacks and the instrument set to automatically switch between them. See Figure 6-6.

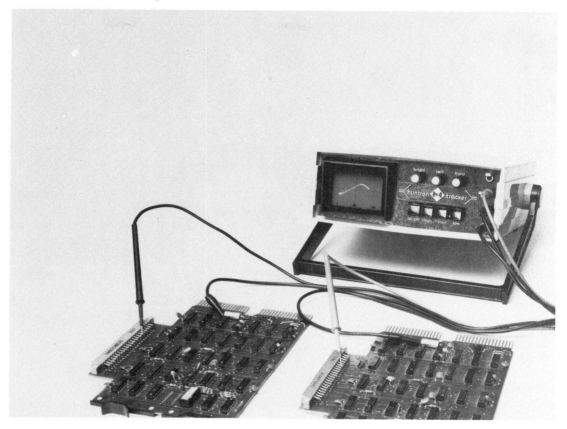

Figure 6-6 Using the SSTs' comparison function

6.0.4.1.2.2 Use a Huntron Switcher if Available — If there are many ICs involved, an accessory to the SST is available to help. The Huntron company has put together an efficient little device to allow the comparison of the many pins to be checked by comparison on good and suspected ICs quickly and easily. See Figure 6-7.

Using the switcher is simple. Connect it to the normal inputs of the SST. Connect the ribbon cables, which are already equipped with the proper IC clips, onto one IC on the good board, and the identical chip on the suspected board. Cycle through the pins of the ICs by pressing the buttons provided. The waveforms should remain essentially stable as the switcher alternates the display from one board to the other. A large difference between waveforms can be either the problem sought or possibly the result of looking at chips made by different manufacturers.

6.0.4.1.3 Evaluate Time Required vs. Probability of Repair — To arrive at this point, there is probably no information available, no way to signal trace through the card because of this, and, as far as we can tell, all of the major components check out as good. If the major components check OK and there is no identical card to compare against, the technician will probably have to spend a great deal of time, with only a small chance of success, in tracing the problem further.

Consider whether it is worth perhaps another hour or more of troubleshooting time with a small chance of finding the problem, or if the board simply should be discarded. The cost of the board and availability of a new one must be weighed against the cost of having the technician pursue the problem. Sometimes these "dog" boards can be put aside for a time when the technician has no other productive work for a few hours.

258 ☐ Dead Circuit Troubleshooting

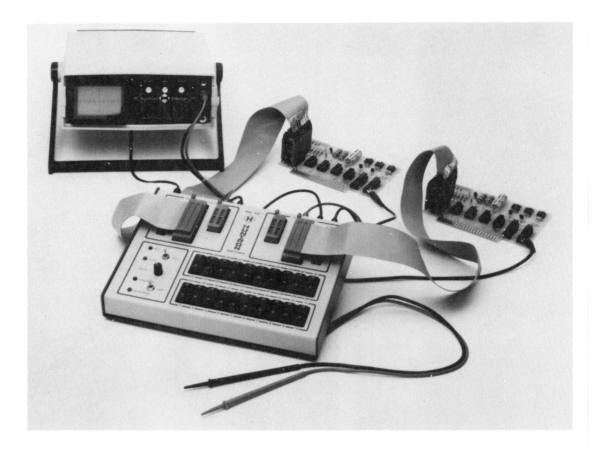

Figure 6-7 The Huntron Switcher was made to enable a technician to compare a multitude of IC pin patterns, good and suspected, in a short period of time with relative ease (Photo courtesy of Huntron Instruments Inc.)

6.0.4.1.4 Check Every Component on the Board — Checking every component on a board may be a bit tedious. It may be the only course of action available, however, if the board cannot be signal traced for any reason, such as lack of documentation or proper power supplies.

Remember that the SST will not show very high-resistance values, very small-value capacitors, or avalanche voltages beyond the voltage sweep of the SST range selected. If all of the waveforms for a given board seem to check out, suspect one of the components that the SST cannot "see" and test it further with an ohmmeter, or, if really necessary, replace it on the chance that it may be bad.

6.0.5 Follow PC Board Trace Back to Defective Component —
If you find a bad waveform on the edge connector, follow the circuit trace back to the offending IC. There is a good chance that this is the one that caused the problem on the board. Beware of two or more inputs being tied together, on a single trace, as one bad input can cause all the paralleled inputs to look bad. When there are several ICs tied to the same bus, the offending IC can be isolated by removal, one at a time, testing after each removal.

Using this technique, many failures can be found in under a minute, even without knowing what the board is used for or how it works!

6.1 IS AN OHMMETER AVAILABLE? — If neither an ohmmeter or an SST is available, it is futile to try to troubleshoot any complex card. If there are a few socketed components and replacements are available, a shotgun approach may work. Replace what you can with ease, and try the card. If it works, fine. If it doesn't, however, it should be set aside until the proper instruments and/or documentation is available. Another option is to simply discard the board.

6.2 Use Ohmmeter for Testing — Review paragraph 4.9.4, which gives basic information on using the ohmmeter for testing components while they are still in the circuit.

6.3 Test Power and Major Components In-Circuit — Use the ohmmeter to test the most likely components to fail, usually those that are largest. Transistors, SCRs, triacs and electrolytic capacitors should all be tested for shorts and opens. It may be necessary to disconnect one or more leads to verify a suspected component if the ohmmeter is reading through paralleling components.

☐ Phase 6 Summary

Dead circuit troubleshooting is successful in a high percentage of cases if the technician has access to an SST and knows how to use it properly and interpret the patterns displayed. The chances of success are considerably reduced if the technician must troubleshoot with only the ohmmeter. Note that there is little need for a schematic during these troubleshooting techniques. Because of this, the SST can often be used with a high rate of success by relatively untrained persons. Familiarity with the patterns shown, or the availability of proper waveforms, can enable the technician to find most problems quickly.

☐ Review Questions For Phase 6

1. What are the two instruments used to test circuits without power applied to the circuits?
2. What caution must always be observed when using an ohmmeter or solid state tester?
3. You are troubleshooting a plug-in printed circuit board. When using a solid state tester, what should be tested first?
4. You are required to troubleshoot a box full of identical PC boards. What method would be the most efficient, using dead-circuit techniques?
5. What components are most likely to cause a problem on a PC board?
6. What instrument makes it possible to efficiently troubleshoot circuits without knowledge of the circuit, its use, with no documentation, and without the application of power?

PHASE 7
TROUBLESHOOTING THE POWER SUPPLY

☐ **PHASE 7 OVERVIEW**

Power supplies are more or less standardized in the circuits used. Once a particular power supply is well understood, all similar types are easily recognized and understood. Since power supplies share common circuitry, the next few flowcharts are considerably more component- and circuit-specific than the preceding charts. Since a specific circuit will vary in detail from generic supplies, the technician must be aware of and make allowances for minor differences.

7.0 NOTES ON SAFETY FOR YOU AND THE SUPPLY — Safety recommendations were thoroughly covered in paragraph 2.0. The new technician is well-advised to review that information carefully and abide by it at all times when working on the potentially dangerous voltage within power supplies. The switching supply, in particular, often generates about 365 VDC in the input circuits. Troubleshooting should be done in an area with safety rubber matting on the floor. Another person qualified to administer CPR should always be present in case of electrical shock.

Troubleshooting in high-voltage equipment should be done using only one hand. For this purpose, any voltage over about 50 VAC or DC should suffice to qualify as "high voltage" for a normal person. Persons with health problems must use lower voltage thresholds of caution. The unused hand should be placed behind the back, preferably in a pocket. This habit helps prevent the possibility of current flow across the chest, the most lethal path.

Only test leads in excellent condition with sharp points should be used for voltage measurements. This avoids problems that can be caused by a slipping probe.

The servicing technician must remember that one side of an oscilloscope is connected to ground. Use of an oscilloscope can cause circuit damage if the grounded oscilloscope "common" is touched to the "hot" side of the incoming power line. Review Figure 4–73.

The flowcharts for these power supplies would be much more complicated if every instance of the need for power off/power on were cited. It will be assumed that the servicing technician has enough respect for electrical power to turn off power before touching any exposed metal surfaces of the power supply, and that any large filter capacitors will be discharged. The use of a dead man stick in rendering circuitry safe to service was covered in paragraph 2.0.

Naturally enough, the power must be reapplied when voltage measurements must be taken. It is up to the technician's good sense to keep safely coordinated in using the power switch while testing power supplies.

7.0.1 CARE, FEEDING, AND TROUBLESHOOTING OF BATTERIES

Testing batteries — While it is easy to make a voltage check of a battery of any kind to determine whether it is good or not, a simple voltage test is not a valid test. *A battery must be placed under load while testing its terminal voltage* to give a meaningful result.

Primary cells — Primary cells are those that are bought fully charged and discarded when the charge is insufficient to perform the job. Because of their internal chemical composition, they cannot be recharged to regain their original chemical state. They are actually consumed during the normal course of their life.

The two most common types of primary cells are both familiar in their use for flashlights and small toys—the Carbon-Zinc type and the Alkaline.

Since these cells are not recharged after use, care and maintenance amounts to keeping the contact areas of the cells clean and the battery clips or contacts firmly against the cells.

Since all cells are chemical in nature, and since lower temperatures slow chemical reactions, it follows that long-term storage of cells is best done by keeping them cool. Freezing will ruin them, but cooling in an ordinary refrigerator will extend their life a great deal. This is a good idea for flashlights kept for emergency service. Store them in the bottom of the refrigerator in a zip-lock plastic bag and they should be ready for use for a long, long time.

The carbon-zinc cell consumes the outer container of zinc during the discharge of the cell. Only a cardboard cover prevents immediate leakage of the corrosive paste. Holes in the container can allow the inner corrosive paste to leak through the cardboard and out onto equipment. This is why it is a good idea to remove even fresh cells from anything valuable if it is not to be used for an appreciable length of time. Discharged carbon-zinc cells, on the other hand, should be removed *immediately* from equipment to prevent damage. See Figure 7–1.

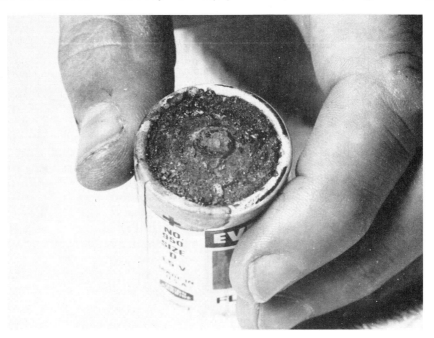

Figure 7–1 Result of leaving a discharged cell in a flashlight. The flashlight was ruined, too!

Alkaline cells are not as prone to causing corrosion problems as carbon-zinc cells. Alkaline cells can be stored for a long time at lowered temperatures. Discharged cells should still be removed from equipment as soon as possible, however.

Troubleshooting either carbon-zinc or alkaline cells is simple. If the cells are not delivering 1.5 volts or 1.4 volts respectively *while under normal load,* the cells are either defective or the load to which they are connected is overloading them. Badly overloaded cells may heat up, particularly the alkaline cell with its lower internal resistance and resulting higher capacity for heavy current flow. An example of this procedure might be troubleshooting the four cells used to power a small portable radio. If powered by carbon-zinc cells, there should be more than about 5.5 volts for the four fully loaded, used cells. Alkaline cells, on the other hand, should measure about the same even though their terminal voltage is slightly lower.

Individual cells within a battery of cells can be tested by comparing the voltage drops across individual cells while under normal load. If one cell is particularly bad, there may even be an apparent reversal of voltage across the weak cell. This is because the cell is no longer a cell at all, but has become a series resistor. See Figure 7-2.

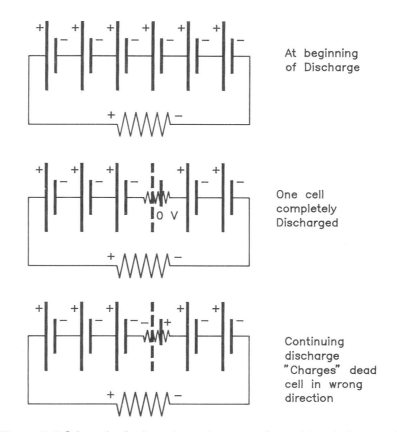

Figure 7-2 Schematic of voltage drops when one cell goes into polarity reversal

Secondary cells — There are two principal types of secondary, or rechargeable, cells in common use today. The lead-acid battery so familiar as the electrical reservoir of our automobiles is the first type; variations on the nickel-cadmium cell are the second.

THE LEAD-ACID CELL — Before delving into the details of the lead-acid cell, the technician should beware of three dangers in regard to them:

- ☐ Generation of explosive gas (hydrogen), particularly when near the end of a charging cycle
- ☐ Presence of dangerously corrosive sulphuric acid within the cells
- ☐ Danger of burns from metal accidentally shorting the terminals of the battery

The lead-acid battery is relatively cheap to make and because of the great many applications in which it is used, fairly inexpensive to buy. The technician should be familiar with the maintenance and troubleshooting of these batteries.

Maintenance of a lead-acid battery includes occasionally adding distilled water to replenish that lost by normal evaporation and the increased evaporation caused by overcharging. Some "maintenance-free" batteries claim that no water need be added for the life of the battery. Water consumption can be reduced by using special metals in the manufacture of the battery, but eventually, aided by high temperatures and overcharging, the cells will probably need water—or the battery may fail because of its lack. The proper level for the electrolyte of a lead-acid cell is well above the tops of the plates. A level below the tops of the plates allows them to dry out and permanently reduces the capacity of the cell.

Allowing a lead-acid cell to remain discharged for long periods of time causes a buildup of a hard surface on the plates. This condition is called sulphation. A small amount of sulphation can be removed by proper, long-term charging. If allowed to progress too far, the cell will be ruined and will not accumulate a charge. The cell will act more like a resistor than a chargeable cell. The high points of maintenance of a lead-acid cell are to keep sufficient liquid in the cells and to keep the battery charged. Long-term storage, such as keeping a motorcycle battery over winter when the vehicle is not used, is best done by keeping a constant voltage on the battery of about 14.0 volts. This will cause a small charging current flowing all the time. This small current is sometimes called a "trickle charge."

Problems in lead-acid cells can cause them to cease to accept a charge. The true test of such a battery is to first be sure that there is sufficient charge in the battery for a load test. The state of charge of a lead-acid cell can be determined by using a hydrometer.

The hydrometer measures the relative "thickness" or density of the acid solution in the cell, comparing it to water. A cell in a fully charged state should have about 1.260 times the specific gravity of water. The hydrometer reading, together with any temperature correction necessary, will measure this figure. If the battery cells are not up to par, charge the cell with a moderate rate of charge for several hours. Keep an eye on the hydrometer readings to note when the cell reaches full charge.

A "quick charge" is not good for lead-acid batteries. Such a charge overheats the cells and can warp the plates, causing internally shorted cells. Once a cell is shorted, the battery can only be scrapped. Shorted cells will result in the battery producing less than normal voltage. Current capability can be sharply reduced because of loss of an active cell.

Once all the cells indicate a normal charge with the hydrometer, the battery should be tested *under a heavy load* to see if it can maintain a reasonable voltage. A 12-volt car battery, for instance, should show greater than about 9 volts when cranking the engine, which is a very heavy load on the battery.

The gel-cell — The gel-cell is a lead-acid cell that has a pasty electrolyte rather than a free-running liquid. This makes this cell particularly nice for applications where the battery might be momentarily turned upside down or where the production of hydrogen gas may be dangerous. The gel-cells are sealed from the outside atmosphere. They should be charged only within the ratings for the specific battery in question. Charging at too high a rate can rupture the cell, ruining it.

The nickel-cadmium cell — The nickel-cadmium (ni-cad) cell is most familiar when it resembles the cells used in flashlights. These cells find use in many electronic applications, providing rechargeable power for portable equipment ranging from handheld transceivers to portable test instruments. The first cell to consider is the sealed ni-cad cell.

Whereas the lead-acid, carbon-zinc, and alkaline batteries gradually lower in output voltage as the state of charge decreases under load, the ni-cad cell holds output voltage relatively constant over the discharge curve. When the cell voltage drops below about 1.0 volt, the cell is completely discharged. See Figure 7–3.

This cell is available in many different sizes. Popular sizes include the "D" or large flashlight size and the smaller "AA" cell, about as big around as a fat pen, but half the length. These cells are meant to be charged at a specified charge rate depending upon the ampere-hour rating of the cell. Large "D" cells may have a rating of 4 Ah, or 4 ampere-hours. This rating means the cell will provide about 4 amperes for an hour, while maintaining nominal voltage, about 1.1 to 1.2 volts, with normal discharge currents. The same cell would have an endurance of about 10 hours at 400 milliamperes. *The standard charge rate for ni-cad cells is 14 hours at 1/10 the ampere-hour rating.* The 4 Ah cell, for instance, should be charged at 400 mA for the 14 hours. Since charging for 10 hours would make a 100 percent charge, the 14 hours of charging results in about a 40 percent overcharge to allow for the inefficiency of the charge/discharge cycle of the cell.

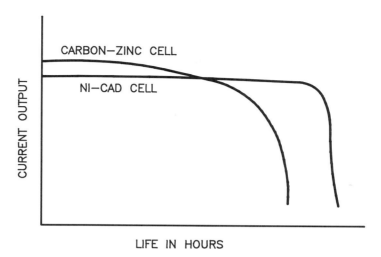

Figure 7-3 Discharge curve comparison between a carbon-zinc and a ni-cad cell

The state of charge of a ni-cad is difficult to determine. It is best to discharge the cell until it reaches 1.0 volt (nearly dead) and then charge for the 14 hours at the proper rate. It can then be assumed that the cell is at 100 percent charge.

Ni-cads do not hold charge well over long periods. They should be trickle-charged with about 1/10 the normal charge current to maintain cell readiness. Weeks of overcharging with the standard rate of charge will shorten the life of the cell.

A unique problem occurs with ni-cad batteries. Several cells in series, charged and discharged together over many cycles, can result in some cells becoming unbalanced in capacity with the remaining cells. One or more of the cells may discharge quickly, thus making the battery pack deliver less than its total capability. The best solution is to discharge each cell *individually* to zero volt, then charge the entire string normally with a full 14-hour charge. Full performance from the battery may require that this procedure be repeated several times. Individual discharging of the cells within a battery is shown in Figure 7-4.

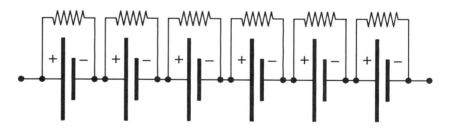

Figure 7-4 Individually discharging the cells of a ni-cad battery prevents reverse-charging of weaker cells

Reverse charging of some of the cells within a ni-cad battery is caused by some cells discharging to zero voltage before the remaining cells. The remaining cells force a current through the discharged cell in the reverse direction of a proper charge. Continued current flow through a ni-cad after the cell reaches zero volts can cause the cell to short internally. Once a cell has developed an internal short, the cell is no longer useful. Although sometimes a short can be "cleared" by forcing a large surge of charging current flow through the cell, the cell can no longer be considered dependable. It should then be used only in applications where reduced reliability will not do harm.

Normal charge and discharge currents produce gases within the ni-cad cell which are reabsorbed before causing problems. Heavy current can make these gases accumulate faster than they can be absorbed again. This increases the internal pressure within the cell to dangerous levels. Each sealed cell is provided with a diaphragm which will, if necessary, puncture and vent the gases. Once a ni-cad has vented because of internal pressure it should be discarded. See Figure 7-5.

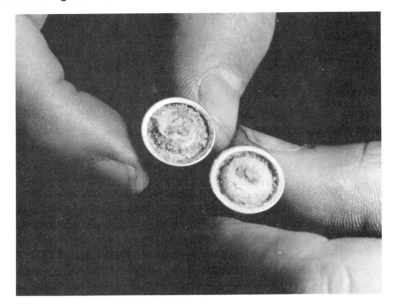

Figure 7-5 A ni-cad cell that has vented, showing corrosion due to the leakage of the electrolyte

Ni-cad cells are sometimes used in quick-charge circuits that force a much greater than standard charge through the cells. Full charge is determined, not by longevity of the charge rate, but by the amount of heat suddenly produced when the cell reaches full charge. This heat is detected by a small thermal switch mounted in physical contact with the cells. The heat opens the thermal switch, causing the charge rate to reduce to trickle levels for charge maintenance. Needless to say, cells in this kind of service cannot be expected to last through as many charge and discharge cycles as those treated with proper finesse.

7.0.2 REVIEW OF THREE TRANSFORMER POWER-SUPPLY TYPES — A power supply is generally understood by technicians to mean a module of electronic equipment that takes one form of electrical energy and, using a transformer, produces at least one DC output voltage. While this may not be a technically accurate definition, it is the picture that comes to a technician's mind when the term is used.

The power supply almost always uses a transformer. There is a good reason for this, because the transformer gives the desirable isolation from input source to output and because a normal decrease in voltage at the output is accompanied by an increase in the current available over that drawn from the original source. The efficiency of the transformer is also an attractive advantage over a simple resistive voltage dropping arrangement.

Since a transformer is now accepted in our definition of a power supply, we will consider now how supplies might vary from a basic design.

Input circuit variations — Input voltage is often selectable between 120 VAC and 240 VAC. A transformer can have a single winding for 120 VAC, or it may have two such windings, intended to be connected in series (in phase) to accept the 240 VAC input voltage. Such an arrangement may or may not require that the two windings be connected in parallel for full-power output operation on 120 VAC. See Figure 7-6.

A transformer will work quite well with a squarewave input waveform, too. Switching power supplies use this principle. See Phase 7B for explanation of this circuit. Partial sinewaves are also eligible for driving transformers. See Phase 7C, the phasing supplies, for more detail on these circuits.

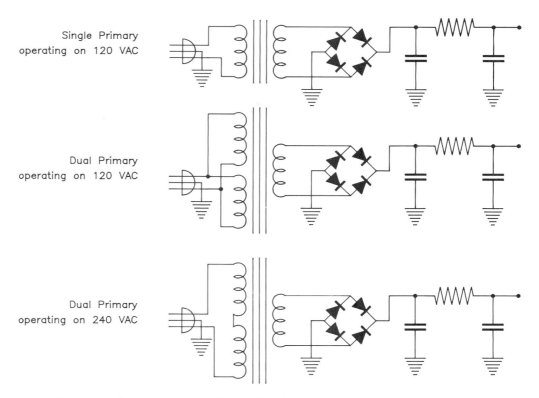

Figure 7-6 Common methods of connecting the power transformer to the AC power lines

Output circuit variations — The half-wave output is simple and inexpensive to produce. It uses a minimum of parts. See Figure 7–7.

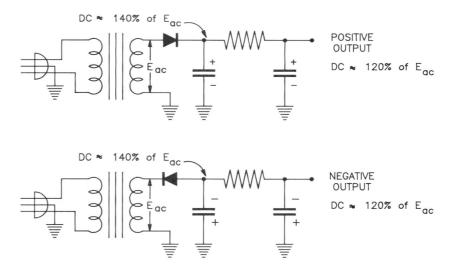

Figure 7-7 The half-wave rectifier is relatively simple. Reversing the diode and the filter capacitor polarity produces a negative output supply

The bridge rectifier must be thoroughly understood by the technician, since it is perhaps the most common kind of rectifier in electronic use today. It provides full-wave output and uses the entire secondary for each half of the power cycle.

Some of the variations of a bridge circuit appear in Figure 7–8. Each of these circuits should be understood thoroughly because of the different voltages that they produce at their outputs.

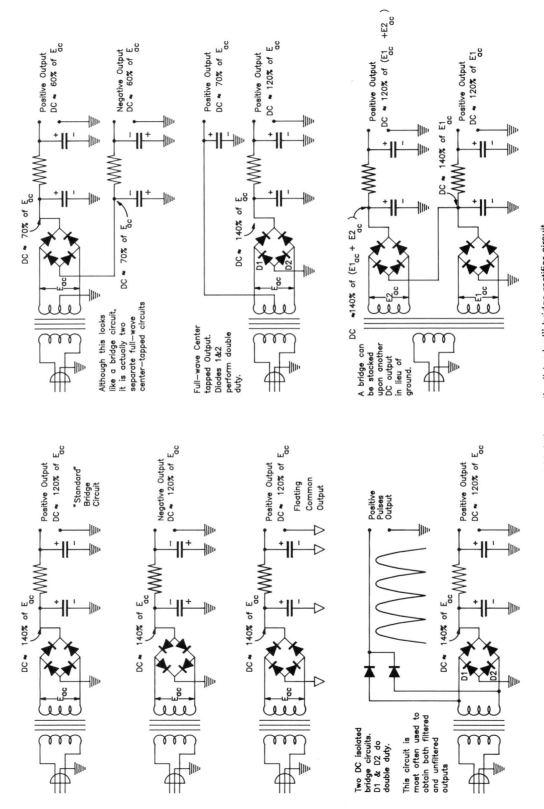

Figure 7-8 Variations on the "standard" bridge rectifier circuit

The principal detail to watch for in analyzing power supply output circuits is the location of the common or ground connection.

The tuned secondary winding is another variation used with power transformers. A special winding and a capacitor of proper value are selected to put the transformer into approximate resonance with the incoming line frequency. This winding will have a heavy current flowing, and will tend to dampen any transient voltage spikes and partial cycle power failures, due to the "flywheel" effect of the tuned circuit. See Figure 7–9.

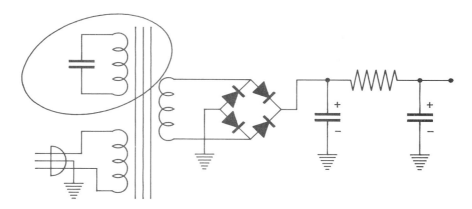

Figure 7–9 The tuned winding sometimes used in power supplies helps in keeping the power input free of free of short interruptions and voltage spikes.

How to recognize the three-power supply types by hardware analysis
— The transformer-rectifier power supply is characterized by the use of relatively few components. These are the usual fuse and switch in the primary, the relatively large transformer, and a rectifier and large electrolytic capacitors. The rectifiers and capacitors are sometimes located separately from the transformer, on a PC board. See Figure 7–10.

Figure 7–10 A common transformer-rectifier power supply. Note the relatively large power transformer

The switching supply, on the other hand, has many more components than the previous type of power supply. There is a high-voltage rectifier and its filter capacitors, a large switching transistor or two that are probably mounted on heat sinks, a relatively small power transformer *or two,* and another one or more rectifiers and filters on the output. These power supplies often are placed on a single board that has a distinct separation of the primary and secondary PC board traces. See Figure 7–11.

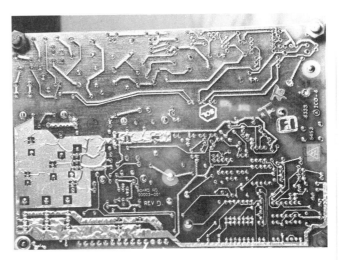

Figure 7–11 A switching power supply. Note the two transformers on the left view (a) and the separation of the input and output circuitry on the right view (b).

The last supply, the phasing supply, will usually have a unijunction transistor that drives the triac or SCRs that, in turn, controls the primary power to the transformer. The phasing supply is best identified by the presence of the SCRs or triac and the unijunction transistor. These supplies are often used in heavy-duty battery chargers.

The presence of an SCR may only be an indication that there is over-voltage protection in the power supply output circuit, rather than identifying the power supply type. See Figure 7–12.

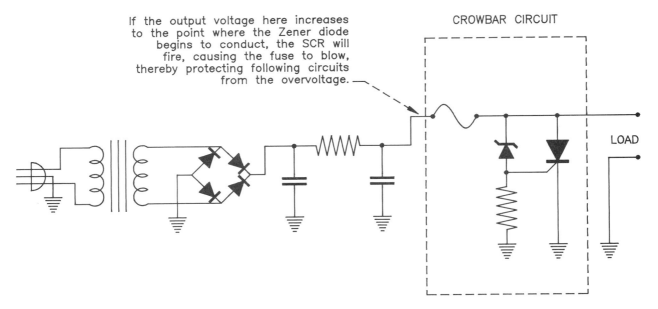

Figure 7–12 The crowbar circuit is another use for the SCR in power supply circuits

7.0.3 GENERAL NOTES ON POWER-SUPPLY TROUBLESHOOTING

Shorts and opens — A review of these terms as they relate to power supplies is in order. Remember that a short circuit causes abnormally high current flow and often causes a fuse to blow. An open circuit is a break in the normal current flow, resulting in no load voltage because of that break.

What is ground, chassis, neutral, and common? — Figure 7–13 shows the schematic symbols for these three common electronic terms.

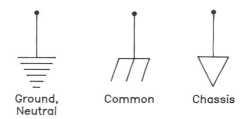

Figure 7–13 Electronic symbols for ground, chassis, and common

Ground is a term that is often used quite differently from its original context of meaning *earth* ground. Chassis connection and voltage common points that are connected back through wiring to earth ground have also picked up the incorrect but popular term "ground." The third wire, colored green in three-wire electrical cords, is by definition a true ground connection, being connected to a good earth connection near the power distribution panel somewhere in the building, where the electrical service comes into it.

Neutral refers to the wire that comes from the center tap of a power line transformer, connected to the earth ground near the transformer or at the distribution panel. The neutral wire is different from the ground wire, however, in that the neutral wire is expected to carry current flow normally. Thus, the neutral wire, because of its resistance, may not be at true earth ground potential when carrying heavy load currents. See Figure 7–14.

Chassis refers to the metal frame portions of the equipment in question. The chassis of equipment should be connected to a good earth ground to prevent possible electrical shock. This is accomplished in high-quality equipment by connecting the third or green wire of three wire cords directly to the chassis. In this case, the chassis is actually a ground connection at the same time. Equipment using a two-wire power cord cannot offer this protection, and the chassis is sometimes used as a common connection that could be dangerous to service.

Circuit common is that point from which most voltage measurements should be taken during troubleshooting. Circuit common is sometimes incorrectly shown on schematic diagrams as a ground symbol, however. The circuit common symbol is used mostly in those applications where the circuit common is *specifically isolated* from ground. In this case, the use of the ground symbol would be incorrect, as would be symbols of chassis and neutral. Refer back to Figure 3–55.

Voltage reading from what reference? — Knowing where to take meaningful voltage readings requires a bit of thought. DC voltages are generally taken from the circuit ground or common connection. AC readings, on the other hand, may or may not be directly referenced to these common points. AC readings must often be taken without reference to an earth-grounded connection. The secondary windings of power transformers are a good example of an AC reading that must be taken without reference to the circuit common.

Signal tracing in power supplies — There are two basic types of problems that result in no output voltages from power supplies: short circuits and opens. Short circuits usually blow the fuse. Open circuits generally result in no power output from the supply, but the fuse does not blow.

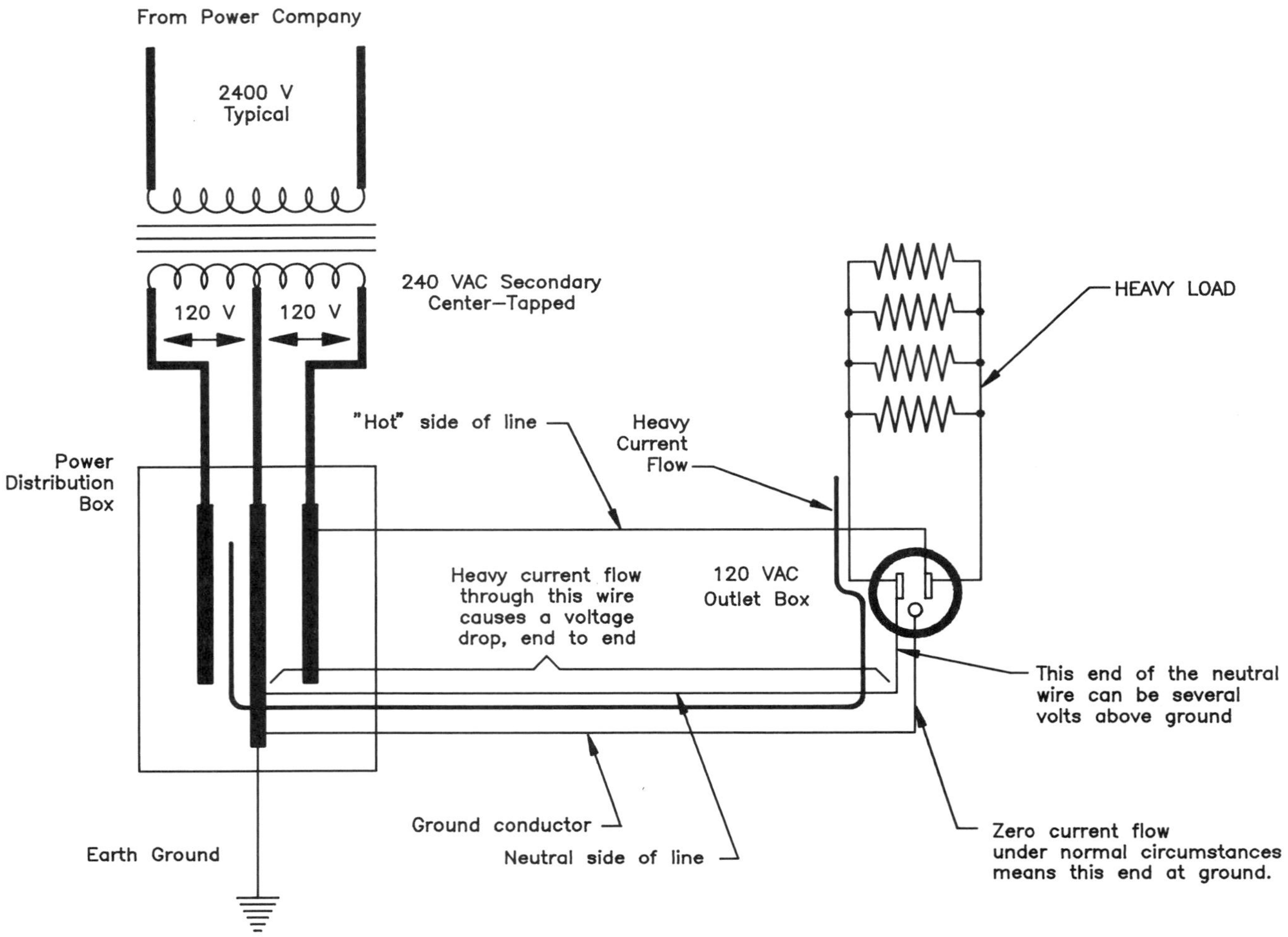

Figure 7-14 Why current flow through the neutral wire puts it at a different potential from true ground

Open circuits are found by "signal tracing" of a sort, using a multimeter on the DC or AC voltage functions to locate the cause of a problem. Short circuits, on the other hand, are found by opening the circuit at various points in the circuit and seeing if there is still an overload present. Each of the Phase 7 flow-charts begin by asking an important question: whether or not the power supply blows fuses. Depending on the answer to this question, the troubleshooting takes one of two very different paths.

Partial voltage outputs — The problem of partial voltage output from a power supply is usually caused by an open in one of the bridge rectifiers or a filter capacitor. A less likely cause could be half of a tapped transformer secondary developing an open circuit. These problems are relatively easy to troubleshoot. Voltage readings on the rectifier diodes will give clues to a bad diode. An open diode in a half-wave rectifier circuit will have no DC but plenty of AC across its leads. An open diode in a bridge will have half the secondary AC and slightly less than the DC of the good diodes. See Figure 7-15.

Lack of any AC voltage output across only one side of a center-tapped winding indicates an open in that side of the transformer winding.

The last probable cause of a partial output voltage is an electrolytic that has developed an internal open circuit. Sometimes the lead wires corrode in two, inside the component. Such a failure would be evident by a reduced DC output voltage and identical AC and DC readings across each of the rectifiers. Verification of this problem is shown by an excessive AC voltage

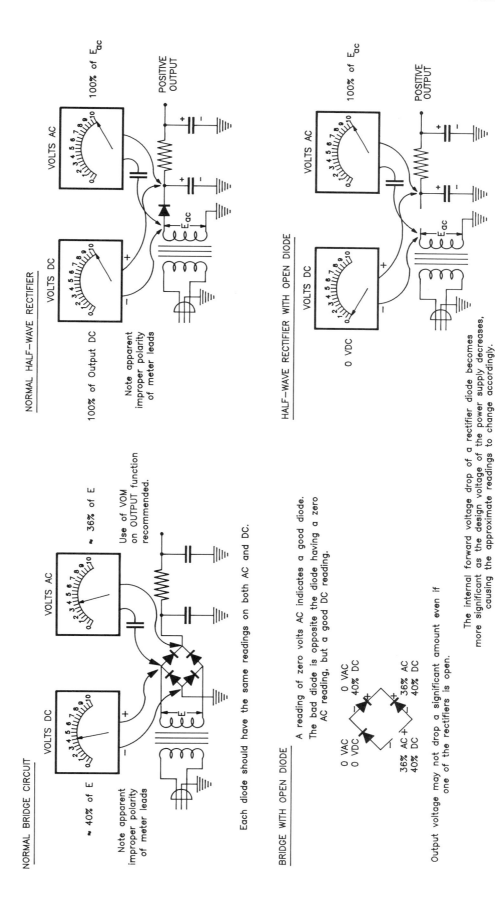

Figure 7-15 AC and DC voltage readings for bridge and half-wave rectifiers, with good circuits and circuits with an open rectifier

directly across the filter capacitors. Test for this condition by paralleling a like capacitor across the old. Appearance of normal DC under these conditions verifies an open cap.

An indication of how much current a supply is delivering can be easily calculated if there is a resistor in use as part of the filter circuit. Using the known value of the resistor, divide this resistance into the voltage measured across it. The resulting number is the current in amperes being delivered to the load. This figure may be compared to the current specification for the power supply to evaluate if the power supply is carrying a normal load. See Figure 7-16.

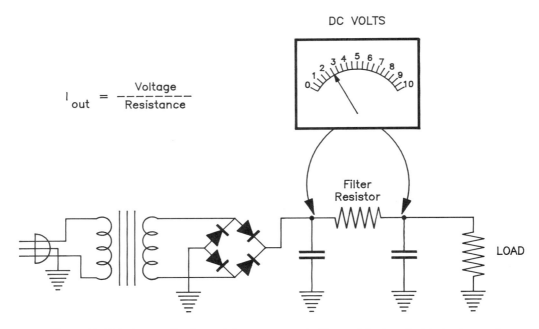

Figure 7-16 Using a voltmeter and the power supply filter resistor to calculate load current

Although this voltage measurement is seldom given on schematics, it is an important reading to have available. The service technician should get a normal filter resistor voltage drop reading on a good unit and record it on the schematic.

☐ Phase 7 Summary

Phase 7 has provided general background information that pertains to all power supplies in general. Variations on basic circuitry were given to help in the recognition of circuit variations often found in actual equipment. Partial voltage output from any type of supply is caused by a few, easily tested failures that were shown and discussed. The details of troubleshooting any of the three types of power supplies and four types of regulators are covered in the following phases.

PHASE 7A
TROUBLESHOOTING THE TRANSFORMER/RECTIFIER POWER SUPPLY

☐ PHASE 7A OVERVIEW

Phase 7A discusses in detail troubleshooting what is perhaps the most common power supply in use today, the transformer/rectifier power supply. This power supply is capable of heavy currents at low- or high-voltage outputs, depending upon the turns ratio of the transformer. It is commonly operated on 120 VAC in small equipment and 240 VAC or more in high-powered applications.

7.1 IS THE POWER SUPPLY BLOWING FUSES?

— The answer to this question will determine if the troubleshooting will progress using the AC and DC voltage scales of a multimeter to locate where the circuit is open, or whether the voltmeter will be useless and it will be necessary to break the circuitry by desoldering or otherwise opening the circuit until the defect is identified. The presence of interunit plugs on a particular piece of equipment can modify the purely half-split method by making it more convenient to break the circuit at slightly different points than one would normally choose. In either case, the half-split method of narrowing down the problem will be used, a method that is clearly evident by study of the flowcharts.

7.1.0.1 Select Option

— If the power supply is blowing fuses, there is too much current flowing. Before going much farther, though, it is a good idea to confirm that the fuse that blew was of the proper current rating and of the proper type for the circuit. It would be embarrassing to spend several hours looking for a problem that wasn't there, only to find that the originally installed fuse was too small for the circuit. Don't interchange slow-blowing and fast-blowing types of fuses. Installation of a fast-blowing fuse in a circuit that should have a slow-blow fuse will usually result in the fuse immediately opening when power is applied, even if the current rating is correct.

A fuse that blows violently, leaving a deposit of vaporized metal on the inside of the glass, indicates a very heavy current and a "dead" short, one having near zero resistance. This is characteristic of many power supply failures. If the fuse blows gently, merely opening the internal wire, it may be worth replacing the fuse and putting the power supply under normal load to see if the problem is gone. Occasionally a fuse will blow without necessarily indicating a problem in the circuitry.

When it is determined that the power supply will not energize without blowing the fuse, one of the three following methods should be used to apply power to the circuit for the purpose of isolating the exact location of the problem.

7.1.0.2 USE OF A VARIABLE TRANSFORMER

— The use of a variable transformer allows the application of carefully controlled, low-input voltages to the defective supply. It should be used with an ammeter to monitor the current being drawn by the supply. See Figure 7A–1.

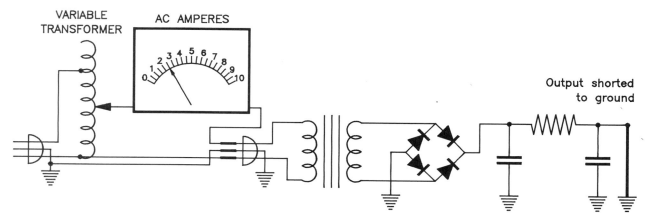

Figure 7A–1 Using a variable transformer and an ammeter to keep current flow to a safe value

The variable transformer should be brought up slowly in output voltage—until the AC ammeter indicates the normal fuse rating current flow. In other words, if the fuse was rated at 1 ampere, the variable transformer should be adjusted for a current of about this value into the defective supply. This is the highest safe current flow to allow while looking for the short circuit.

7.1.0.3 Use a Soft Fuse
— This is the preferred method of troubleshooting shorted power supplies. It is inexpensive and conclusive in its indications. See Figure 7A–2.

The size of the incandescent lamp to use is one that will give near normal current flow through the power supply when fully lighted. For instance, a power supply fused at 1 ampere and operating at 120 VAC requires a lamp that is rated at about 100 W at 120 V. The current flow through the 100 W at 120 VAC is nearly 1 ampere, the normal current flow through the power supply.

The resistance of the filament of an incandescent lamp is non-linear with respect to its brilliance. At low voltages, the filament is low in resistance. Thus, if the power supply is near normal working current, the lamp will not glow very much, and will develop a fairly low resistance. This shifts most of the available voltage to the load, a normal supply input. A shorted supply, on the other hand, will place nearly all of the available voltage across the lamp, which then raises its resistance to limit current to a safe value.

The lamp is also a built-in indicator of current flow. The brighter the lamp, the more the power supply input looks like a short circuit. Even with a "dead short" input, the power supply input circuit is limited to the current drawn by the lamp on the full-line voltage.

Soft fuses can be made by using an incandescent lamp, an ordinary socket, line cord, and a burned-out fuse to hold the wires at the proper spacing so they may be clipped into a fuse holder.

7.1.0.4. Feed It Fuses
— This is the least desirable method of verifying if the short has been isolated. At least one more overload must take place at full-line voltage for this method to work. This can further stress already strained components. Without a variable transformer and AC ammeter or a soft fuse, it may be the only course of action.

When feeding fuses to a power supply to find a short, the normal half-split method should be abandoned. With power off, a new fuse of the proper type and rating is installed and circuitry is disconnected one point at a time, *beginning from the input end* of the supply. Break the circuit at point A on Chart 7A. Reapply power. If the fuse blows, the problem must be prior to that break. If the fuse does not blow, *reconnect* point A, break it at point C, then again apply

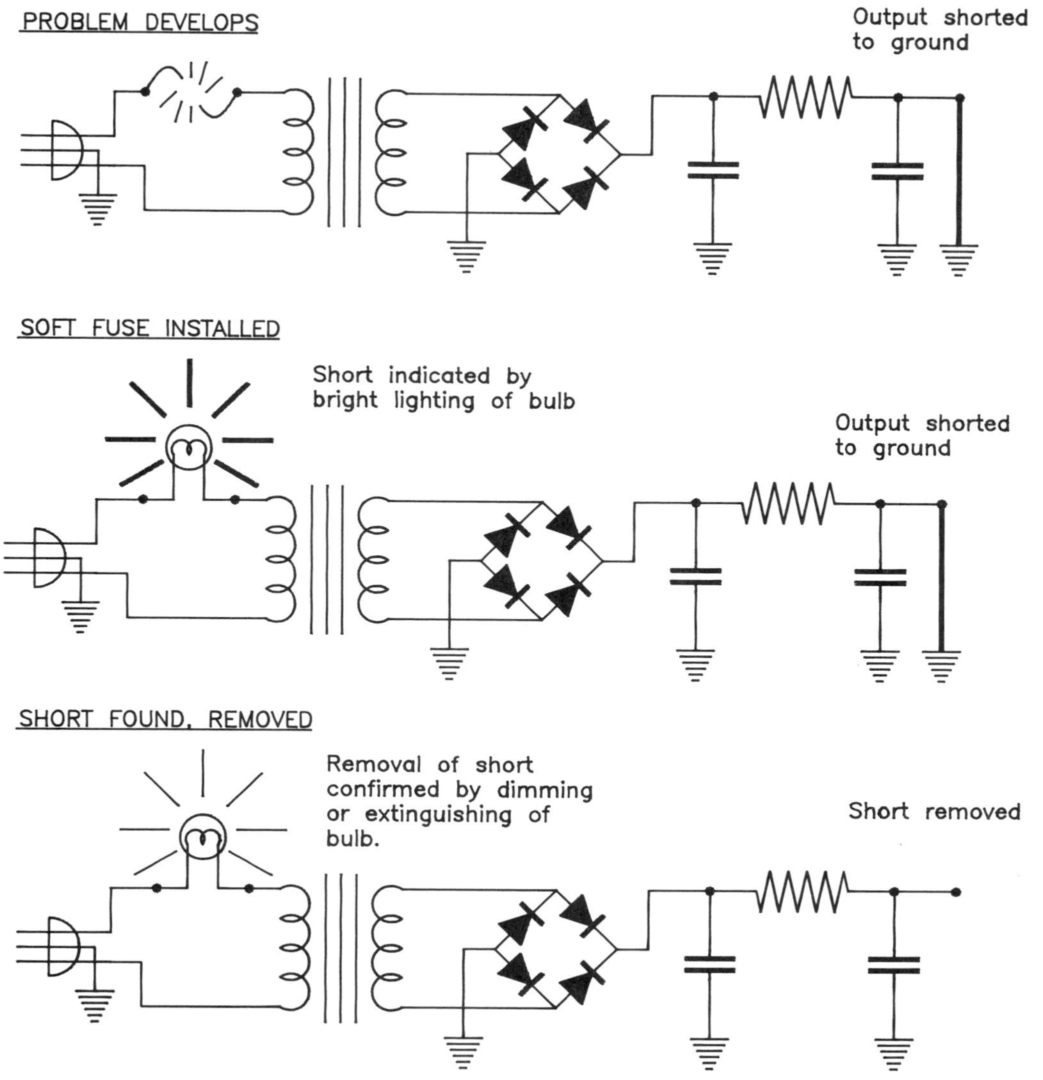

Figure 7A-2 The principle of the soft fuse in schematic form

power. If the fuse now blows, the problem is between points A and C, the transformer itself. If not, continue in this manner progressively to point F until the fuse does blow. When it does, the problem is between this last and the previous break in the circuit. Be sure to reconnect previous breaks as you proceed.

7.1.0.5 Disconnect All After Filter at Point E
Point E is between the filter and the regulator, if any, or the load. Breaking the circuit at the output of the capacitor filter and the input to the regulator (or the load if no regulator is used) effectively splits the circuit into two halves, that of the basic rectifier/filter and that of the regulator and load. This is an example of the half-split method of troubleshooting. Dividing the circuit in half is a fast way of going by the most efficient route to the bad component.

7.1.0.6 Apply Power
Power should be reapplied at this time. Depending upon the kind of option selected for the application of power, the presence of the overload should be monitored. If the short is still there, the variable transformer and AC ammeter will still indicate the strong current flow as before. If the current has dropped substantially or to zero, the short

has been disconnected. The soft fuse, if used, may still show a full brilliant glow if the supply is still shorted. If the soft fuse glows dimly or not at all, the short has been removed from the supply.

7.1.0.6.1 Reconnect Point E, Disconnect Point F — Point F is between the regulator, if any, and the load. If the overload is now gone, there must be a short within the regulator or the load itself. This new break in the circuit will determine whether the regulator or the load is at fault.

7.1.0.6.2 Load or Output Filter Capacitor Is Shorted — When the overload indication showed that the regulator was not at fault, the problem has been localized to the load or a filter capacitor across the output of the regulator. This filter capacitor may be located on the power supply subassembly or it may be somewhere on the load circuit board. If it is obvious, disconnect it and try this test again. You may find that this was the problem, and reconnecting the load no longer causes the overload. Replace the filter capacitor if it was shorted, of course. If this is not successful, the next paragraph continues to help in finding the problem.

7.1.0.6.3 Identify the Shorted Component — Once the load on a power supply is determined to be overloading the supply, the technician should be aware of the different possibilities that might cause an overload. See Figure 7A-3.

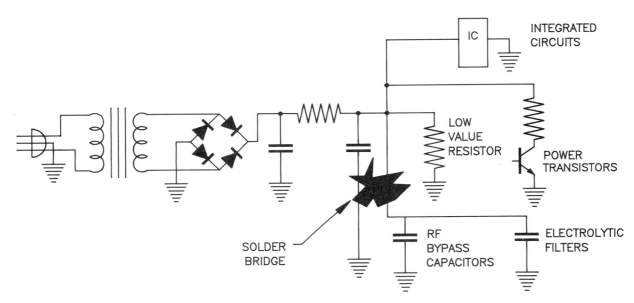

Figure 7A-3 The power supply often provides current to many parallel paths, any one of which might cause an overload

A clue to the source of a short can be obtained by taking a simple resistance reading from the supply bus of the bad board to its ground. The power supply and filter should be disconnected for this test if possible. This table will give clues as to the three probable causes of the overload problem.

RESISTANCE READING	PROBABLE CAUSE
Less than 2 ohms	Solder bridge, shorted bypass capacitor
Greater than 2 ohms, less than about 100 ohms	Shorted IC or other semiconductor
Greater than about 100 ohms	"Pseudo-zener" load

The semishort — When a semiconductor fails by shorting, it does not go to a complete "dead" short of zero ohms. Common values of resistance (which will also depend on the current the ohmmeter uses) may vary from 2 ohms to perhaps 20 ohms in actual circuits. This fact can help the technician narrow down the possibilities of a shorted component on a circuit board. This fact was discovered and verified by the Huntron Instrument Company during their research into failures while developing their instruments.

The "pseudo-zener" load — Once in a great while a technician may find a load that has a very reasonable resistance when measured with an ohmmeter, yet it causes an overload of the power supply when connected to it. These symptoms can be caused by a semiconductor that has developed an internal failure that makes it *seem* like a zener diode. Only when a substantial voltage is applied will the load suddenly begin to draw excessive current. This problem can be verified by monitoring the amount of DC current drawn while the DC input voltage to the board is slowly increased. There will be a sudden and abnormal increase of current before the normal supply voltage is reached. Of the five methods available for finding a shorted component, only the current tracer and logic pulser option of troubleshooting *will not* find the pseudo-zener problem. See Figure 7A–4.

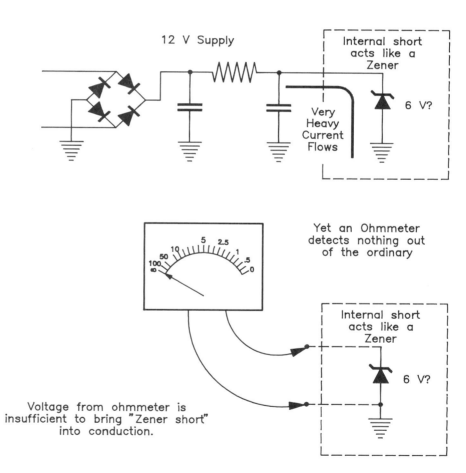

Figure 7A–4 Schematic representation of a load that has a pseudo-zener type of overload

7.1.0.6.4 DOES THE LOAD HAVE A SEMISHORT?
— A semishort is a severe load on a power supply, but one that is not a direct or "dead" short. If the problem is shown to be a semishort or a "pseudo-zener" load, the smoke-out method, explained in paragraph 7.1.0.6.9, is probably the best method to choose.

7.1.0.6.5 Cut-Trace-and-Try Method
— This method involves cutting the PC board traces at selected points using the half-split method. It is the most destructive method to use in finding a short circuit.

Using this method will make it necessary to physically trace the supply line trace on the board. This may take considerable time, and there may be many branches.

There are five different structures for Vcc and power lines: the parallel bus, the tree, the loop, the plane, and the random line structures.

The only time the kind of bus structure becomes significant during troubleshooting is when using the cut-trace-and-try method for finding a short circuit.

Vcc bus structures — The parallel bus structure looks like the pattern shown in the upper right of Figure 7A–5.

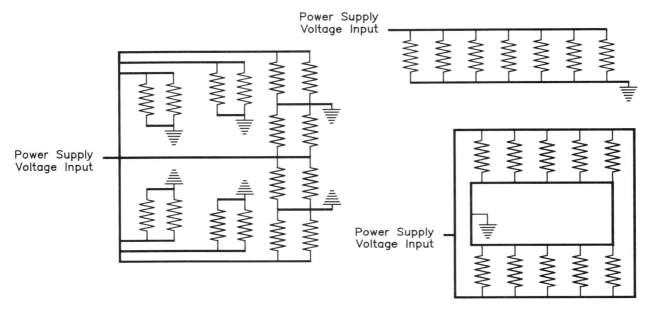

Figure 7A–5 Three representations of Vcc bus types: parallel, tree, and loop

Note that the parallel bus may wrap back and forth on a board, but the layout is neat. This structure is often used on PC boards with many ICs installed.

The cut-trace-and-try method with the parallel bus structure follows the half-split method. Cut the supply line (Vcc) half-way through the length of the bus for the first test. Apply power and test for the presence of the overload. Reconnect the break and proceed toward the power input point or away from it according to the result of the test on the first cut. Keep cutting, testing, reconnecting, and cutting again until the short is located.

The tree-structured power bus looks something like the lower left of Figure 7A–5.

The tree structure does not lend itself well to the half-split method. Major branches must be successively cut, proceeding to the short. Cut the first major branches to isolate the main "trunk" that is causing the problem. Repair the cut and proceed to isolate the bad "branch," repair the cut, and proceed to the bad "twig."

The loop is the last Vcc structure shown in Figure 7A–5, in the lower right.

The cut-trace-and-try method of finding a short on a PC board is effective with the loop structure *only* if the loop is first broken for troubleshooting purposes. Any paralleling path *not* broken will result in very misleading symptoms.

As long as the loop is left intact, no single break will result in isolation of the short from the power input points on the board.

The power plane bus structure — The power plane structure is a separate layer within a PC board, usually one of the layers in a multiple layer board. Because the layer is inaccessible, the freezer method or the removal of likely components are the only two methods available to find a short circuit between Vcc and ground planes.

The random power bus structure — The random bus structure is one that has no particular pattern. The Vcc supply line meanders all over the board and can form loops or even swap sides of the board now and then. The cut-trace-and-try method can be used, but the path of the Vcc line can be complex and therefore confusing. Beware of a looped supply line on such a board for this reason. The symptoms of such a loop could be very confusing if not detected beforehand. See the explanation of the looped bus structure covered earlier in paragraph 7.1.0.6.5.

7.1.0.6.6 Current Tracer and Logic Pulser

— The current tracer is particularly useful in finding short circuits on PC boards without damage to the board or removal of too many parts. This is probably the most valuable use for the current tracer.

The procedure is to supply a pulsed signal with a logic pulser directly into the shorted Vcc and ground connections of the PC board. Using the current tracer, follow the path of heavy current flow. Refer back to Figure 4–13.

When the current tracer is used in this manner, it should be constantly rotated back and forth for maximum response to the current flow. If the current turns a corner, it will be necessary to also turn the current tracer the same amount to have maximum sensitivity.

The current tracer can follow the current up to *and through* a defective component. Look for an IC, bypass capacitor, or a solder bridge as the most likely components to be causing the problem. Although using the current tracer requires a bit of skill to operate properly, it only takes a few minutes to get the "feel" of this instrument.

Current is traced by first adjusting the tracer for a brilliance just *slightly less than maximum,* at a point very near the pulser input. Once this level is set, the brilliance shown will vary according to the density of the current flowing directly beneath the sensing coil at its tip. Because of this, a trace that has a considerable narrowing will show an apparent increase of current flow, whereas one that widens out will indicate less current flow.

While tracing the current, beware of the current tracer showing a slightly dimmer response as the circuit is traced. This is usually an indication of having left the main current flow and taken a side branch. Double back until the instrument shows the same brilliance as that shown at the beginning, allowing for wider or narrower traces.

Another thing to watch for is current flow that goes through the board and continues on the opposite side. Multi-layer boards also may confuse troubleshooting by a similar trick, having current flow within the board. Trace the fault current using the side of the board that keeps the trace closest to the current tracer tip. This will help narrow in on the proper trace, even though the current tracer is quite capable of tracing current through the thickness of a board.

The logic pulser/current tracer method of finding a short is most effective in troubleshooting the parallel bus, tree, and random bus structures for the Vcc and ground paths.

The logic pulser and current tracer method cannot find a "pseudo-zener" short because insufficient voltage is produced by the pulser.

7.1.0.6.7 The Freezer Method

— Place a shorted board in a very cold freezer for perhaps an hour. Upon removal from such conditions, the board will quickly accumulate a light covering of frost. This frost is the indicator we need. Quickly, put the board into a mockup or connect a power supply so that a heavy current will flow through the short in the board. *The board will show the heavy current flow path as the first place that melts the frost.* Follow both the Vcc and ground traces and they will point to the shorted component or solder bridge.

This method should work well with all but the looped and plane bus structures, as explained under the cut-trace-and-try method.

7.1.0.6.8 Removing Likely Components — A careful look at the schematic will probably show that some of the components in the circuit *cannot* cause a major current flow, even if they short completely. Using this fact and carefully studying the schematic will show the most likely components that could cause the problem. Refer back to Figure 4–65.

If there are only a few components on the board, it may be that a lucky guess will be quicker than attempting to pin down the cause of the short through more exact methods. For instance, if there is an SCR or electrolytic capacitor on the board and little else, it makes sense to just disconnect one or the other and test them with the ohmmeter. With a bit of luck, you can sometimes go directly to such a failure. This approach will work with any power bus structure.

7.1.0.6.9 "Smoke Out" Method — A resistance of more than about 2 ohms makes available the option of applying a carefully controlled current of proper polarity to the board and feeling for a component to overheat.* *Be careful that the power supply used will not apply too much voltage if the defective component should suddenly open!* The hot component should be the one that is causing the overload. Verify this by disconnecting it and reapplying power before replacing the component.

This method works well with any power bus structure.

7.1.0.7 Disconnect Secondaries at Point C — Point C is at all the secondaries of the power transformer. Be sure to reconnect the circuit at Point E. Disconnecting the secondaries will eliminate all of the circuitry to the right, including the rectifiers and filter capacitors. Disconnect all of the windings on multiple secondary transformers to eliminate all of the loads, rectifiers, and the circuitry following them.

7.1.0.7.1 Filter Capacitor or Inductor Shorted to Ground — Once the problem has been identified as either the filter capacitor or inductor, the end is in sight. Although the capacitor is probably more likely to short internally than to have the inductor short to its case, it might be that dismounting the inductor would be easier than disconnecting the capacitor. If the inductor can be taken off the chassis and supported by some sort of insulation, it does not need to be disconnected. If the short is gone when the inductor is dismounted, then the inductor is shorted to its case and should be replaced. In some emergency cases it might be permissible to allow the circuit to operate with the inductor insulated in this manner until a replacement inductor can be obtained. When this is done, however, the inductor becomes a safety hazard. One might assume that the inductor case is grounded, as it usually is, and could be shocked severely if touched. Be sure to tag the inductor plainly that its case is "hot" to avoid an accident.

7.1.0.7.2 Determine Which Component Is Defective with Ohmmeter — The ohmmeter should be able to show conclusively which of the two suspected components, the inductor or the capacitor, is at fault. This may require disconnecting one end of the component from the circuit to avoid paralleling components.

7.1.0.8 Disconnect Primary at Point A — Point A is the input to the primary windings. This step may require that several of the wires going into the transformer be disconnected. If the transformer in question has more than two wires connected to the main power input, *be sure to mark how each of the wires is connected* because they must be properly phased with respect to each other. Connecting it back together later in a different manner could cause a short because of the improperly phased windings. The windings would be fighting each other instead of helping each other.

Disconnect all of the wires to the primary windings, preferably at both ends of the windings. If the transformer is shorted to the chassis or transformer core, leaving one end of the transformer connected to the power line could still cause an overload.

---

*Overheating semiconductors or resistors are indications of a defective, shorted component. An overheated resistor is probably not defective, but indicates another component in series with the resistor is shorted.

7.1.0.8.1 Transformer Is Shorted
— Since most transformers are relatively expensive components, it is a good idea to run one additional test of a transformer, when it is completely out of the equipment and lying on the bench. Apply primary power to the transformer and look for the same indications of overload that were present while it was installed. Of course, one must use the fuse, soft fuse, or variable transformer and AC meter method to prevent sustained excessive current flow.

If by some chance there was a fault in troubleshooting and the transformer is apparently OK now, be sure to check for a short from *any* of the windings to the core or frame of the transformer. Removal of the transformer would have broken such a current path to the chassis, making the problem apparently disappear. Remember also that an internal short could very easily make the transformer dangerous to touch while power is on.

7.1.0.9 Short Is Between Power Input and Transformer Primary

There are probably only a few possible components that can cause such symptoms. Any transient suppression components such as MOVs (metal oxide varistors), capacitors, or zener diodes are by necessity connected across the line input. If any of these short, the fuse blows, even if the power supply itself is not overloaded. Finding them is easy. Just disconnect one end and test each component with the ohmmeter for a shorted component.

Another component that might create these symptoms is a shorted EMI (electromagnetic interference) filter. Inside these components are capacitors connected from one side of the line to the other. See Figure 7A–6.

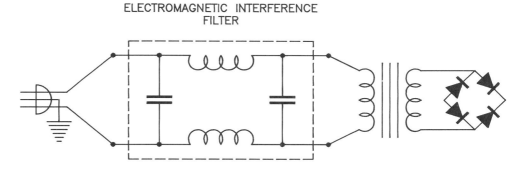

Figure 7A–6 Schematic of the inside of an electromagnetic interference (EMI) filter

When a capacitor shorts in an EMI filter, the line input is shorted. Remove the input leads where they enter the filter to see if the short is still present.

7.1.0.10 Check PC Board for Short with Ohmmeter
— If likely components are not causing the short, the only other possibility is that the PC traces have a solder-splash or wire from one side of the line input to the other. Use an ohmmeter and, if necessary and applicable, resort to the cut-trace-and-try method, covered in paragraph 7.1.0.6.5.

7.1.1 Measure AC at ALL Secondaries, Point C
— Point C is at the output of all the secondaries of the transformer. Remember that the AC at this point is probably *not* referenced to ground. Measure from one end of the secondary to the opposite end. The outcome of this test indicates whether or not the open circuit is within the transformer, before it, or after Point C.

Be sure to check all of the secondaries on the transformer. It is possible that only one of several output windings is defective. Each one must be checked for normal voltage.

The lack of voltage at any one of several secondaries means that one secondary winding is open.

7.1.1.1 Check ALL Power Supply Output Voltages Between Filter and Regulator, Point E
— Be sure that all of the power supplies are producing DC for the inputs to the regulators. One of several supplies may not be producing voltage at all. This could cause the circuitry to follow to malfunction even though the remaining sections of the power supply are working properly.

7.1.1.1.1 Check ALL Regulator Outputs, Point F
— Point F is the output of the power supply after the filter or, if provided, the regulator. A good input to a regulator but no output from it is a definite indication that the trouble lies in the regulator or somewhere after. Remember that the input to most regulators must be at least 15 percent or more above the expected output voltage of the regulator, as most regulators operate on the principle of simply reducing a higher input voltage to a specified output voltage.

7.1.1.1.2 Trouble Is Not in the Power Supply
— It is quite possible that the wire carrying power from the regulator to the load (the normal circuitry) is open. In any case, if the regulator voltage output(s) are normal, there is no problem within the power supply. The original conclusion that the power supply was bad must be reexamined.

7.1.1.2 Check all Rectifiers for Opens
— An open bridge rectifier must have at least two open rectifiers to have no voltage at all on the output. Partial voltage output can be caused by one or two of the bridge rectifiers being open. The best way to test rectifier diodes for open circuits is to use a solid state tester while the diodes are in-circuit (use the lowest range and frequency) or to remove at least one end of each diode and test with an ohmmeter using the diode function.

7.1.1.2.1 Replace All Rectifiers in Circuit
— The discovery of one open diode suggests that the failure may have been due to a short, which caused the diode to open. Since the failure of a single diode by shorting may cause the matching (opposite) diode to fail as well, it is a good practice to replace all of the diodes involved. Before reapplying power, be sure to check for possible causes of the diode failure. Check the output filter capacitors and the following circuitry for a short. This is most easily done by checking the circuit with an ohmmeter, across the filter capacitors, before connecting any of the new diodes.

7.1.1.3 Check Filter Resistor/Inductor for Open Circuit
— With power applied to the circuit, check the voltage across the filter inductor or resistor. Only a small amount of voltage should be lost here, generally less than 10 percent of the total normal output voltage of the supply. If the entire voltage output of the supply is across this component, it is open. Such a failure is often the result of a short to ground in the following circuitry, somewhere after the filter resistor or inductor.

7.1.1.4 Circuit Is Open Between Transformer Secondary and Filter Input
— Since it is now verified that the rectifiers and the filter inductor or resistor are good, there must be an open in the wiring or circuit board trace somewhere between the AC output of the secondary and Point D, the input to the filter. The open circuit is traceable using the AC scales of the voltmeter to the rectifier inputs, and with the DC scales from the rectifier outputs to the filter input. Trace from where the voltage is normal just to the point where it is lost, and the open circuit should be right at this point. Remember when troubleshooting the bridge rectifier circuit that the AC voltage is not referenced directly to ground, but the DC coming from it *is* referenced to ground.

7.1.2 Check AC Input at Point A
— This is an important check point. It is usually easily found in the equipment and quickly determines if line voltage is actually reaching the transformer primary. Failures of fuses, switches, interlocks, power cords, and wiring will all be detected with this step. Be sure to measure across the input winding with an appropriately high AC scale and not with reference to ground or the chassis.

7.1.2.1 Transformer Has an Open Winding. Replace — Line voltage
input without voltage output from a transformer indicates an open winding. The transformer must be internally repaired or replaced. Sometimes a transformer can be repaired. If an attempt would be acceptable, removal of some of the insulating tape on open-core transformers will sometimes reveal the open connection. The open most often occurs where the lead-in wire is spliced to the small, single strand copper wire of the winding. With care, an open at this point can be cleaned, wrapped again, and resoldered. Doing a small repair job like this can sometimes save the expense of a new transformer.

7.1.3 There Is an Open in the Primary Wiring. Check Interlocks, Fuses, Switches — Primary wiring for some equipment must make a
long run before it finally reaches the transformer. Interlocks are sometimes overlooked by a servicing technician, an embarrassing thing to happen if someone happens to be watching! Power and interlock switches do fail, sometimes without an obvious change in the feel or sound of the switch.

7.1.4 Check with AC Meter for Opens from Line Input — The
loss of AC input voltage is relatively easy to trace. Determine that AC voltage is actually coming into the equipment (the cord itself could be at fault) and leave one of the AC meter leads connected at this point. Using the other lead, trace the AC voltage along the opposite side of the line from the point where the reading was normal, progressing toward the transformer. The lack of a voltage anywhere on this line indicates that the open is between this point and the last point where the AC voltage was normal. If voltage is found all the way to one side of the transformer, then the open circuit is in the side of the line where the first probe was left. Leave the first probe at the transformer primary, and trace with the second probe until the open circuit is found. See Figure 7A-7.

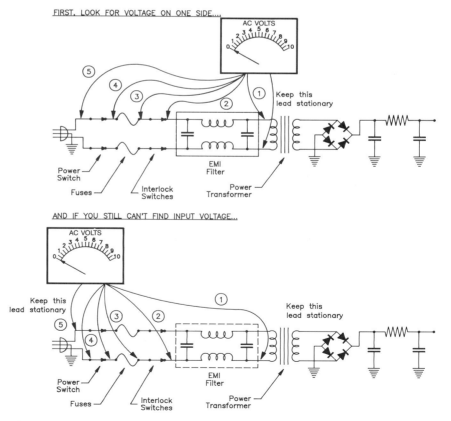

Figure 7A-7 How an open might be found between a good AC line input and the power transformer, moving only one VOM lead at a time

It is also possible that the circuit has a "hidden" fuse, in an out-of-the-way place, or that a tiny fuse called a pico fuse may have been used. See Figure 7A–8.

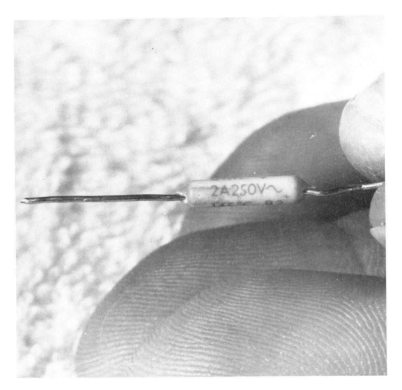

Figure 7A–8 The pico fuse is cleverly disguised and looks like a small resistor

☐ Phase 7A Summary

The phase and Flowchart 7A should be well understood by the technician. The concepts of progressively splitting a circuit in half until the problem is isolated is basic to this level of troubleshooting. General troubleshooting tips must be interpreted to fit any actual circuit, using some common sense and reasoning. Once the bad component is isolated, replace it by referring to Phase 5.

PHASE 7B
TROUBLESHOOTING THE SWITCHING POWER SUPPLY

☐ PHASE 7B OVERVIEW

The switching power supply is becoming commonplace in electronic equipment. It provides the power for most personal computers, for instance. With its interesting, unique characteristics, it must be thoroughly understood by a service technician. It can be dangerous and confusing to troubleshoot.

7.2 NOTES ON HOW THE SWITCHING POWER SUPPLY OPERATES —

At the simplest level, the switching supply uses a DC power input that is broken up into a squarewave by one or two transistors. This squarewave is acceptable to a transformer. The transformer performs the familiar function of isolation from the incoming supply and a step-up or step-down of the input voltage. See Figure 7B–1.

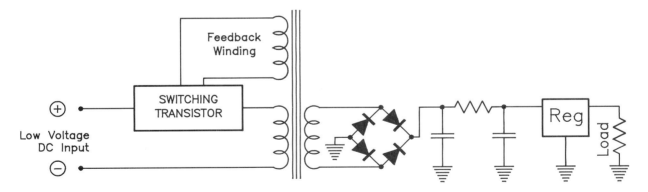

Figure 7B–1 Block diagram of DC-to-DC conversion, which uses an additional feedback winding to produce powerful oscillations in the transformer

There are several variations on the basic switching power supply. The first to consider will be the supply used in personal computers. The signal passed through the Class A optical isolator is proportional to the output voltage of the supply. If there is too little voltage at the output, the signal passed to the input circuitry increases the duty cycle of power applied to the transformer primary, thus raising the output voltage to where it belongs. See Figure 7B–2.

Electronic Troubleshooting ☐ 287

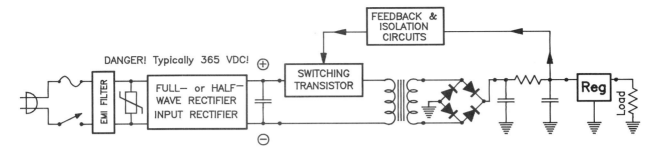

Figure 7B-2 Block diagram of personal computer power supply, typical of switching power supplies

Another variation is a power supply that uses a transformer for coupling feedback. The power supply output voltage is compared to a reference voltage. Circuitry converts differences between these two voltages to a waveform whose average pulse width varies in such a manner as to compensate for those differences. If the output voltage should sag a bit, the pulse width waveform broadens, increasing the duty cycle. This is coupled to the switching power transistor, which delivers more energy to the transformer, thus counteracting the sag with the application of more energy. This action results in raising the output voltage to normal. See Figure 7B-2.

Yet another variation on the switching power supply is the supply that does not regulate at all. This simpler supply takes feedback from an extra winding on the power transformer and uses it to produce a power oscillator. Energy is tapped off a winding at the output side and rectified to produce DC. This output DC is completely isolated from the input DC power to the oscillator. This kind of power supply was once popular to convert 12 VDC of a vehicle battery to the higher DC voltage requirements of typically +300 VDC or more needed to run vacuum tube equipment from a 12 V power source. See Figure 7B-1.

Any of these circuits could have more than one secondary circuit, each producing a different voltage. Personal computers, for instance, often have three windings, providing separate supplies of +5 VDC, +12 VDC, and -12 VDC. The two 12-V supplies produce power needed to run disk drives and the RS-232C serial communications ports.

A switching supply should never be operated without a load! Excessively high voltage transients (spikes) would be produced, probably shorting the switching transistor. A normal load helps to suppress these voltage transients, keeping them within reasonable voltage values. Operation without a load may result in the destruction of the switching transistor, the most likely component to be damaged by the high transients that would result. Insulation within the power transformer could also be punctured, shorting it.

One additional point needs to be made in regard to these power supplies. Since the frequency of operation is now variable, no longer tied to the 60 Hz of the power line operated supply, the frequency of operation can be made much higher with no additional cost. In fact, the cost drops when using higher operating frequencies. At higher frequencies, less core is needed in the transformers for a given power rating. This makes these supplies able to produce the same currents as 60 Hz supplies while using smaller transformers. The operating frequencies of switching supplies are usually beyond the upper limit of hearing of 20 kHz to avoid the irritation of a continuous squealing sound, which would result from the laminations of the transformer moving slightly.

7.2.1 IS THE POWER SUPPLY BLOWING FUSES? —

If the power supply is blowing fuses, the problem is probably a short somewhere *before* the transformer. Switching power supplies usually cease to oscillate and draw minimal current if they are overloaded anywhere in the secondary circuit or beyond. The short causing the fuse to blow is usually the switching transistor, since it is subject to some very high switching transients, or the input rectifiers or filter capacitors. If the transistor shorts, the DC current flow through the transformer is excessive, and blows the fuse. Only AC can be used by a transformer.

Troubleshooting a switching supply that blows fuses makes it necessary to choose one of the three methods of applying power as explained in detail in Phase 7A, paragraphs 7.1.0.2, 7.1.0.3, or 7.1.0.4. Review those paragraphs and choose one of those methods.

7.2.1.2 Remove Switching Transistor

Be sure to remove all power and short the input filter capacitors before touching the switching transistor! Voltages to 365 VDC are commonly used to feed these transistors when operating from power line voltage.

If the switching transistor should short, the DC input to either the regulated or unregulated switching supplies is placed directly across the transformer. The transformer will seem a direct short to the DC, thereby blowing the fuse.

Since the switching transistor is probably the most likely cause of an overload in a switching power supply, remove it first in eliminating where the short might be located.

7.2.1.3 Apply Power

Using one of the three methods mentioned in paragraph 7.1.0.1, apply power to the circuit. If the overload is gone, the variable transformer and AC ammeter method will show almost no current flow at all, since the power supply is essentially an open without the switching transistor active in the circuit. The soft fuse method will show a bright light if the overload is still there, and no light at all if the transistor was shorted. Feeding of fuses will result in the fuse remaining good if the transistor was shorted, blowing violently if the transistor was not the problem.

Of course, if the transistor is determined to be shorted, it should be replaced by referring to Phase 5, "Replacing Defective Components." On the other hand, if the short still exists with the transistor out of the circuit, re-install it. Remember to use heat conducting compound if necessary to provide proper cooling to the heat sink, if used.

7.2.1.4 Test Filter Capacitors across Point B for Shorts

The next most likely cause of a shorted input are shorted filter capacitor(s) between the input rectifier and the switching transistor. These electrolytics are identifiable by their unusually high voltage ratings, considering today's common solid-state voltages of 5 V and 12 V. Electrolytics in this service have voltage ratings of about 250 V. If one of these shorts, it will be detectable with an ohmmeter. *Be absolutely sure that the capacitors have been discharged before handling them or testing them with an ohmmeter!* It may be necessary to disconnect a capacitor and test it out of the circuit to get a conclusive reading.

If an input capacitor is shorted, it would probably be a good idea to replace or at least check the input rectifiers. A shorted capacitor would severely overload the diodes and could very likely cause their failure too. Replacing the diodes might avoid a future failure even if they test as good.

7.2.1.5 Test Input Rectifiers Out of Circuit for Shorts

The rectifiers in the input circuit can develop short circuits. *The capacitors in the input circuit must be thoroughly discharged* before handling them or testing the diodes with an ohmmeter.

If one or more of the input rectifiers are shorted, it is a good idea to replace all of the rectifiers, as they have probably all been overloaded.

A versatile input circuit — The input circuit of some switching power supplies is easily converted by changing a single wire to allow operation on either 120 VAC or 240 VAC. See Figure 7B-3.

Careful circuit tracing of this figure will show that when operating on 120 VAC, this is a voltage doubling power supply, alternately charging one or the other of the two capacitors connected in series. Two of the diodes in the rectifier configuration are not used on 120 VAC input. The result is a voltage across the two capacitors of about 365 VDC. On 240 VAC, the circuit acts as a simple bridge rectifier. This arrangement will also produce the same 365 VDC at the output.

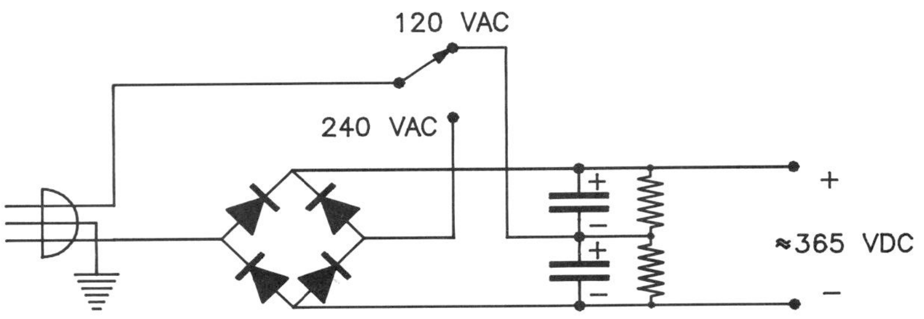

Figure 7B-3 The switching supply used in personal computers uses this input circuit because of its ability to operate on either 120 VAC or 240 VAC

7.2.1.6 Test EMI Filter and MOV for Shorts

— Since the overload has been localized to somewhere prior to the input rectifiers and filter, all that remains is the wiring and those components that connect from one side of the input line to the other. This would include any transient or electromagnetic interference components. Take one end of the MOV or zener transient suppression component loose and test it with an ohmmeter for a short. Shorted components must, of course, be replaced. The EMI filter is tested most effectively by temporarily disconnecting it and bypassing the connections from input to output and then testing to see if the short is now gone. If it is, the EMI filter was causing the problem and will have to be replaced.

The EMI filter is commonly used on switching supplies because the normal operation of the supply produces quite a bit of electromagnetic interference by the very nature of its operation. Transient voltage and current spikes could, if not suppressed, be conducted via the power line to other equipment using the power mains. The EMI filter should be replaced rather than operating the equipment without it because of the chance of causing problems in other, unrelated equipment that might be sensitive to the transients. Computers are a good example of equipment that will not tolerate transients well. A schematic of a typical EMI filter is given in paragraph 7.1.0.9.

7.2.1.7 Inspect Board for Solder Bridges, Burned Spots

— If the components of the switching power supply are not causing the short circuit, the wiring or the PC board traces must be at fault. Careful inspection may disclose the problem. As a last resort, it may be necessary to find the short by using the cut-trace-and-try method, shown in paragraph 7.1.0.6.5 of Phase 7A.

7.2.2 Check for Proper DC Across Input Capacitors, Point B

Do not use an oscilloscope on the input circuits of a switching power supply! These circuits are not at all isolated from the power line. Using a grounded instrument such as the oscilloscope can cause great damage to the circuit.

Be very careful when measuring the DC across these capacitors. Up to 365 VDC is typically present. Figure 7B-3 shows the typical circuitry used in switching power supplies that operate from 120 VAC sources. The DC filtering capacitors may be operated in series as shown, which allows the circuit to produce the same output voltage on 120 or 240 VAC input power. Each capacitor should have half the total DC output voltage across its terminals. Since the power supply in question at this point is not blowing fuses, begin troubleshooting by using the half-split method.

7.2.2.1 Circuits to Left of Point B Are Normal

— Point B is the input to the switching transistor. There is no need to troubleshoot any farther to the left if normal DC is present at point B. Normal voltage at this point means that the problem lies to the right of point B on the block diagram.

7.2.2.2 Test for Open Switching Transistor

BE VERY SURE TO HAVE ALL POWER TURNED OFF AND THE INPUT CAPACITORS COMPLETELY DISCHARGED BEFORE USING AN OHMMETER OR SOLID STATE TESTER.

Paragraphs 4.94 and 4.95 may be of help at this point. These paragraphs deal with in-circuit testing of components with both the ohmmeter and the solid state tester. An open transistor will probably not have affected any of the other components in the circuit and can simply be replaced.

7.2.2.3 Short in Output Circuits or Defective Feedback Circuit Is Indicated

— It is characteristic of most switching power supplies that, if there is an overload on the supply, they simply stop oscillating and idle with minimal current flow. When the overload is removed, the supply recovers and operates normally. This characteristic can be misleading, since the fuse does not blow, yet the problem is a short. Note that the short must be *after* the transformer, however. A short prior to the transformer will cause the fuse to blow as expected.

If there is a problem in the feedback circuit, this can also cause symptoms of little or no output when the input power circuit is normal.

It has been determined at this point that there is either an overload on the power supply or there is something wrong in the feedback circuit. As a last possibility, the power transformer could be defective. First, the overload question will be resolved.

7.2.2.4 Check Output Circuits for Shorts

— Remove power from the circuit and, using an ohmmeter, determine if the power supply sees an overload condition. Measure the resistance from point E, the input to the regulator, to ground. Anything less than a few ohms is considered a short. If a short is evident, determine which component is responsible for the overload by half-splitting the output circuitry of the power supply transformer (all the circuitry after the transformer), tracing the short to ever-narrowing sections of the circuitry. As an example, if the output filter capacitor was shorted, this would be evident by a short appearing when measuring across point E, but the short would not be as evident when testing the output of the regulator. The regulator will give some isolation and make finding the problem easier. The wiring can be desoldered, the PC board can be cut, or a series component such as a regulator IC or the filter resistor can be desoldered to break the circuit, thus isolating portions of it.

7.2.2.5 Trace Operation of Feedback Circuit

— The lack of a short or overload on the power supply output suggests that a malfunction in the feedback circuit is causing the problem.

There are many variations on the feedback circuits that are used, so no specific troubleshooting procedures can be given other than those suggested in paragraph 3.15.1. *Do not use an oscilloscope on the input circuits of a switching power supply!* These circuits are not at all isolated from the power line. Using a grounded instrument such as the oscilloscope can cause great damage to the circuit. Review Figure 4–73.

Basically, the feedback circuit should sample the DC voltage output of the power supply and compare this voltage to a standard reference voltage. The difference between the two should control the width of the pulse fed to the switching transistor. An output voltage that is less than normal should, through the feedback circuit, provide a wider pulse to counteract the low output.

The feedback circuit must also provide some means of isolating the voltage levels of the input from the output. This means that different grounds are used on either side of the power switching transformer. The signal flow from output back to input requires a means of voltage isolation in the signal path. Two common means are the optical isolator and a small pulse transformer. The pulse transformer will be smaller in size than that used for the power switching.

The optical isolator can be operated either as a pulse coupler of varying pulse width, or it can be part of an analog feedback scheme with the pulse width generating circuitry mostly on the primary side of the switching transformer.

The unregulated switching supply has no complex feedback circuit, only a winding on the power transformer. About the only thing that can go wrong with this winding is that it might open. This possibility is easily checked with an ohmmeter.

7.2.2.6 All But Transformer Tests OK

— The transformer is about the only component not directly tested to this point. Replacing a transformer can involve a great deal of work so it has been left to last. If all of the preceding tests reveal no problem, either an open or a short within the power transformer, including shorts to the case or core, could produce these symptoms. It is time to change the transformer. Phase 5 should be of help from here on. Note that an unregulated switching supply must be phased properly to operate. Write down the details of the original part and the color coding, if applicable. The replacement part must be installed exactly as the original.

7.2.3 Problem Is an Open Before Point B

BE ABSOLUTELY SURE THAT THE HIGH-VOLTAGE INPUT CAPACITORS HAVE BEEN DISCHARGED BEFORE HANDLING THEM OR TESTING THEM WITH AN OHMMETER!

If the switching transistor is not getting proper input voltage and the supply is not blowing fuses, the problem must be an open circuit prior to the input transistor. The AC-operated power supply has a rectifier and filter stage between the switching transistor and the line input, the full- or half-wave rectifier. Paragraph 7.2.1.5 explains this circuit in some detail. Look for open diodes or bad connections in the rectifier area.

7.2.3.1 Check Input Rectifiers for Opens

— Paragraphs 4.94 and 4.95 will be of help in using the ohmmeter or solid state tester to find an open diode in the input circuitry. If operating on 120 VAC, the presence of about 175 volts across only one of the output capacitors is an indication that one of the diodes is open. If operating on 240 VAC, half-normal voltage across the two capacitors in series indicates the same problem.

7.2.4 An Open in the Circuit Board Between Point A and Point B Is Indicated

— An open circuit feeding the input rectifiers can be an unexpected interlock left open, a defective interlock, or possibly a pico fuse somewhere on the board. See paragraph 7.1.4 for more details on what to seek.

☐ Phase 7B Summary

This phase assists the technician in avoiding further damage to the switching power supply during troubleshooting. The unique symptoms of a shorted output on a switching supply are discussed and troubleshooting methods based upon them are described. The importance of not using a grounded test instrument, such as an oscilloscope, on the primary side and always having a load on such a supply has been discussed. The switching power supply will continue to be of importance in the years ahead, and should be a familiar circuit to the servicing technician.

PHASE 7C

TROUBLESHOOTING THE PHASING POWER SUPPLY

☐ PHASE 7C OVERVIEW

The phasing power supply is the third major type of power supply in common use. Its theory of operation and maintenance tips should be familiar to the service technician. This is an efficient power supply, wasting little energy in heat. A typical common use for the phasing power supply is in maintaining the charge on batteries used as the standby power source in no-break power systems.

7.3 NOTES ON HOW THE PHASING POWER SUPPLY OPERATES —

The phasing power supply varies its output voltage by varying the point on the incoming AC waveform when an SCR or triac is fired. Once they are turned on, these semiconductors will continue to conduct until the line voltage reaches the zero-crossover point, at which time they become an open circuit. By varying the time into each cycle at which they fire, the amount of energy coupled through them can be varied. See Figure 7C-1.

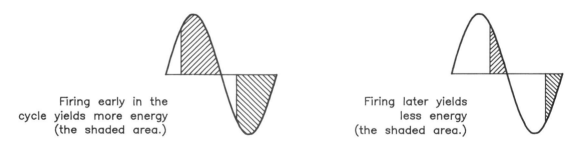

Figure 7C-1 How varying the time of firing during each AC cycle can vary the total energy coupled through the SCR/triac of the phasing power supply

All that is needed to operate the SCR/triac is a circuit that will pulse at the proper timing into each AC input cycle. The unijunction transistor (UJT) makes such a circuit very simple.

When operated as a simple unsynchronized oscillator, the UJT produces a string of pulses at the output. See Figure 7C-2.

When the UJT is powered by a rectified but *unfiltered* waveform, it can be caused to fire during that waveform at varying times, under the influence of a DC signal input. See Figure 7C-3.

This circuit can now be used to make a regulated power supply of great current capability and high efficiency. The DC output of the supply is sampled and fed to the UJT to influence

294 □ Troubleshooting The Phasing Power Supply

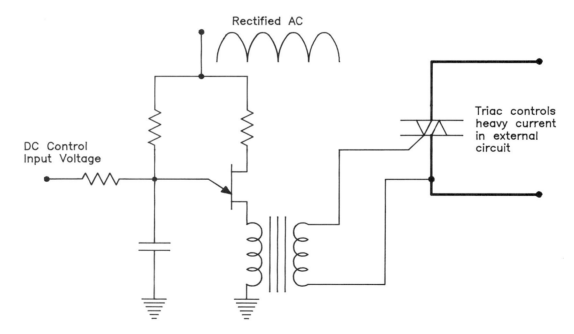

Figure 7C-2 The UJT in a DC circuit produces a train of sharp pulses

Figure 7C-3 When the UJT circuit is powered by an unfiltered ripple voltage, the time of the first firing on each pulse of the ripple is greatly influenced by a DC applied to the emitter circuit

the firing time. The output pulses are then coupled into the SCR/triac and used to control the main power input. See Figure 7C-4.

Since the input to the phasing power supply is already an AC acceptable to the power transformer, the *shorting of the SCR/triac will not cause the fuse to blow*. The output voltage will increase above normal voltages and will stay there without control. An open in the SCR/triac will, of course, result in zero output voltage from the supply.

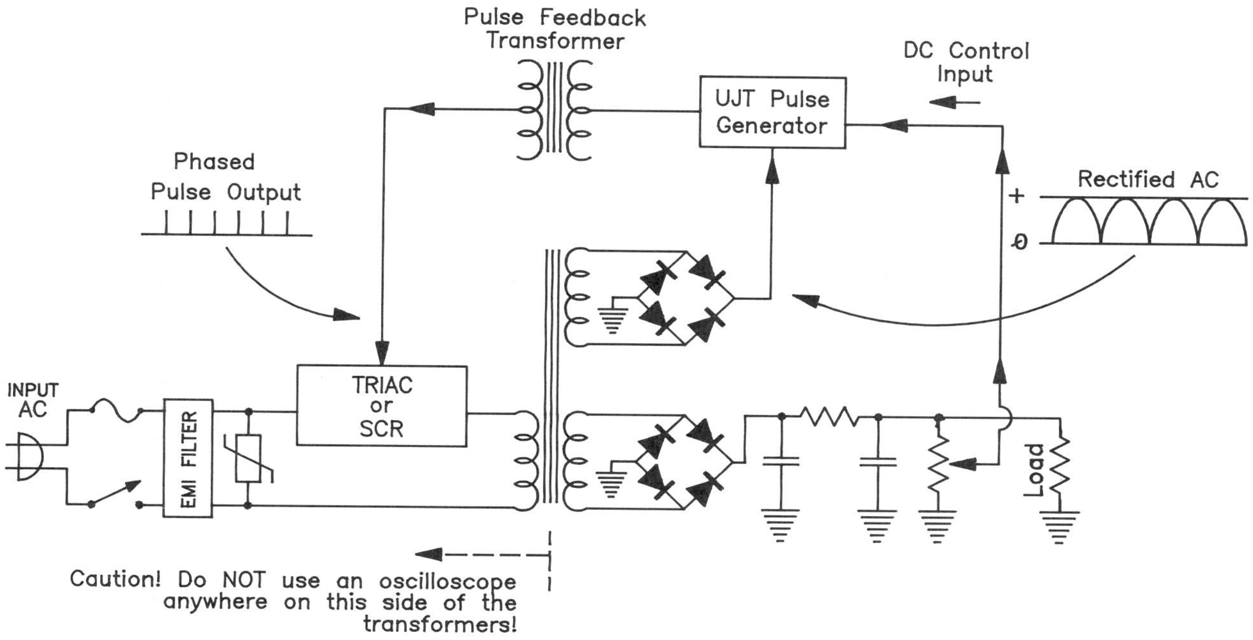

Figure 7C-4 Block diagram of the Phasing Power Supply

7.3.1 IS THE POWER SUPPLY BLOWING FUSES? — If the power supply is blowing fuses, the short can be in the load on the power supply or in the power supply itself.

Use one of the three methods of paragraphs 7.1.0.2, 7.1.0.3, or 7.1.0.4 to prevent damage during troubleshooting. Disconnect all the secondaries of the power transformer, point C. If the application of power still shows a short, either the transformer or circuits prior to it are shorted. To eliminate the transformer further, be sure to disconnect the winding supplying unfiltered AC to the UJT circuit. If the short persists, troubleshoot as for a similar problem in Chart 7A. Go to point S on that chart and continue troubleshooting.

On the other hand, if the power supply does not blow fuses when disconnected at point C, this is an indication that there is a short after the transformer. Again, this problem was covered for a similar case. Refer to Chart 7A, point R and continue troubleshooting on that chart.

If the power supply does not blow fuses at all, the next thing to consider is the output DC voltage.

7.3.2 DOES CIRCUIT RESPOND TO CONTROL? — If the circuit responds normally to any manual controls, as are often provided, the circuit is probably working normally. Verify the initial report of problem.

A controllable output that is out of tolerance may be due to a defective or improperly connected triac. Some triacs act like an SCR, firing only on half of the cycles, if the input gate is pulsing with the wrong polarity. Check the polarity of that pulse or simply reverse the pulse input wires to see if the problem is gone. This problem would only occur if another person has worked on the unit before and improperly wired the unit.

If the power supply is at maximum output without responding to changes of the manual control, the SCR/triac is probably shorted. Turn off power and test the semiconductor. If it was shorted, replace it.

If the power supply has zero output, this is a classic symptom of an open SCR/triac. Test this component for an open, and replace if appropriate.

7.3.3 Trace Problem in DC/UJT Circuit

— In the event that the SCR/triac is not shorted, yet there is no control of a higher-than-normal output, the UJT circuit is probably not getting the proper DC changes that it should. Check the input DC circuit to the UJT emitter.

If the SCR/triac is not open, yet the power supply has no output voltage, the UJT circuit is indicated as the problem. There are probably no pulses being produced to fire the SCR/triac. Tracing problems through the UJT circuit should be done with care. *Do not connect an oscilloscope to the output (SCR/triac) side of the pulse feedback transformer!* This circuit will be damaged if it contacts the grounded input of the oscilloscope. Do all the necessary troubleshooting on the UJT side of the pulse feedback transformer.

☐ Phase 7C Summary

With a few differences, the phasing power supply is somewhat similar to a common transformer/rectifier power supply. Only part of the incoming AC waveform is switched into the power transformer, however. With only the added circuitry of the UJT and the SCR/triac switch, troubleshooting this power supply is in many ways similar to that of the transformer/rectifier power supply.

PHASE 7D
TROUBLESHOOTING VOLTAGE REGULATORS

☐ PHASE 7D OVERVIEW

The different types of regulators and a brief explanation of how they work is included in this phase. Unique failure symptoms and quick fixes, when appropriate, are suggested. Voltage regulators are used in most electronic equipment and are prone to failures. A technician must be familiar with them because a good percentage of troubleshooting will involve these important circuits.

7.4 NOTES ON REGULATOR TYPES — There are four types of voltage regulators:

1. The zener or shunt regulator
2. The linear IC regulator
3. The discrete semiconductor regulator
4. The switching IC regulator

Each of these has characteristics that should be understood in order to troubleshoot them effectively in actual circuits.

A regulator has an input voltage which, by various means, is reduced to a specified value for the output load on the regulator.

7.4.1 Apply Input Power — It may be necessary to feed input voltage to the power supply via one of the three methods discussed in paragraph 7.1.0.1. If an overload is indicated by any of these methods, go to paragraph 7.4.3.

7.4.2 INPUT VOLTAGE OK? — Because voltage regulators work by "getting rid" of some of the available voltage in order to produce a firm, lower output voltage, the input must of necessity be higher in voltage than the output. A regulator cannot be expected to operate properly if there isn't sufficient voltage available for it to use. At least about 15 percent higher voltage input than regulated output will be required for proper operation.

7.4.2.1 Disconnect Load and Filter Capacitor from Regulator

Some regulators, including linear IC regulators, have internal overload protection. If the output of the regulator is shorted, the IC goes into a current limiting mode, producing only a small output. The overload will not usually cause the fuse to blow, yet the overload is there. The purpose of this test is to see if the IC is in this current limiting state.

Disconnect the loads and any filter capacitors from the output of the regulator. There may be more than one filter capacitor, and these may or may not be physically located close to the regulator, but may be on the load PC board or circuit. It is our intent at this point to see if the regulator is at fault, and leaving anything on the output of the regulator that could be shorted defeats this step.

7.4.2.2 OUTPUT VOLTAGE OF REGULATOR OK NOW? — If the regulator is shorted, an overload will still be evident on the power supply. Troubleshooting of an overload problem continues with paragraph 7.4.2.6.

There may also be a problem within the regulator that prevents it from having the normal output voltage, even without a load connected. Lack of proper output voltage with a normal input voltage, without a load, is a sure indication of a problem within the regulator.

If the output voltage of the regulator is normal when the load is disconnected, there could still be a problem. See the next paragraph.

7.4.2.3 Substitute Resistor for Load — The load or the output filter capacitors may be at fault if the regulator is operating with normal output voltage without a load on it. To determine for sure where the problem lies, substitute an appropriate resistor or another good PC board of the same type as a normal load on the regulator. An appropriate resistor is one that will draw the same current as the normal load on the regulator.

With a known good load, the regulator must be able to sustain the regulated voltage at the input to the load.

7.4.2.4 OUTPUT VOLTAGE OF REGULATOR OK NOW? — If the regulator output voltage is normal with a resistor or known good load, then the problem lies in the load itself. Chart 7A, point P is the starting point for the procedures that help to find the problem in the load circuits.

7.4.2.5 A Current Limiting Problem Is Indicated — If the regulator cannot sustain the proper voltage into a proper load, a resistor, or known good PC board, then the regulator has a problem in supplying the necessary output current. The regulator is "seeing" an overload current problem where one does not exist. In other words, the current overload circuitry is too sensitive. Paragraph 7.4.2.6 continues the troubleshooting procedures.

7.4.2.6 IS THIS A SHUNT ZENER OR AN IC REGULATOR? — If this is a simple one-component regulator, just replace it. Replacing a component is covered in Phase 5. If the circuit consists of discrete components, it will be necessary to go deeper into the circuit to find the exact component responsible for the failure.

7.4.2.6.1 Discrete Semiconductor Regulator — A typical discrete regulator is shown in the block diagram of Figure 7D–1.

The discrete regulator may or may not have circuitry to monitor the current drawn from the regulator. Without this added circuitry, there is no electronic way to guard against current overloads on the supply. Without current sensing circuitry, excessive currents should be guarded against by using a fuse in the regulator circuit.

The discrete regulator samples the output voltage and compares this to a stable reference voltage. The difference is amplified and applied as a signal to a power transistor in series with the output current flow. By varying the amount of conduction of this power transistor, the transistor is made to look like a variable resistor supplying voltage to the output circuit. By varying this "resistor," the output voltage can be precisely controlled.

7.4.2.6.2 Check Pass Transistor — The pass transistor of a discrete regulator takes the brunt of the current attack. In so doing, it produces heat, and heat is a major factor in semiconductor failures. The principal symptom of a shorted pass transistor is an excessive output voltage over which there is no control. An open pass transistor results in zero output voltage. Either of these symptoms should lead the technician directly to the pass transistor as a first check. Phase 5 should be consulted for additional tips when replacing the transistor if it is found defective.

7.4.2.6.3 Check Voltage Reference Component — The voltage regulator operates by comparing a variable output voltage with a fixed reference voltage. Differences are amplified and applied as a correcting signal to change the variable output voltage in the

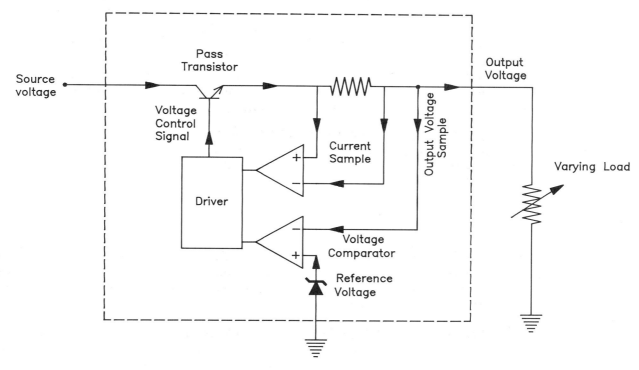

Figure 7D-1 Block diagram of a discrete voltage regulator with current sensing

proper direction to bring it back to the correct voltage. The voltage reference component may be a special precision IC or it may be a simple zener regulator. See Figure 7D-2.

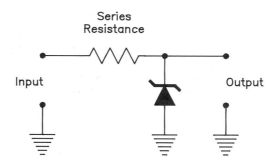

Figure 7D-2 The zener is often the reference voltage for a high current discrete regulator

If the reference component is a zener diode, and if the zener voltage reference is not correct, troubleshoot it as a separate circuit. If the zener voltage is too high, replace the zener, as it is open. If the zener output is too low, even zero, either the zener is shorted (test it with an ohmmeter) or the circuitry it supplies has a short. See Figure 7D-3.

7.4.2.6.4 Problem Is in Error Amplifier Stages —
If the reference source voltage is normal and the pass transistor is neither shorted nor open, then there must be a problem with the circuitry that compares the regulator output voltage with the reference. Another possibility is that the current monitoring circuit thinks that there is too much current flow—usually caused by an open shunt resistor—or there is a failure within the current monitoring circuit. Each of these possibilities must be traced using the DC signal tracing methods of paragraph 3.15.1.

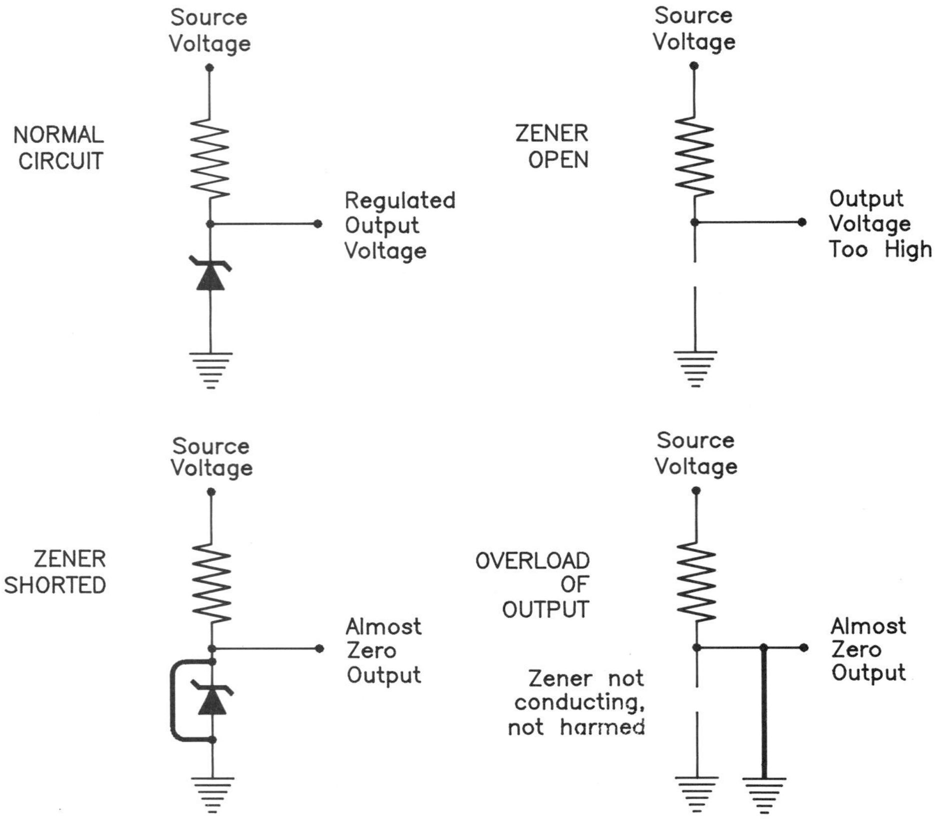

Figure 7D-3 Problems of the zener voltage regulator

7.4.2.7 Switching Regulator — The switching regulator is usually an IC with a few external components as shown in Figure 7D-4.

The switching regulator is an improvement over the linear regulator in that it wastes very little power. The linear regulator must literally burn up excess voltage at the current demanded of the load. This is wasted power. The switching regulator is either fully on or fully off, and therefore is not wasting power. The duty cycle of the on and off times determines the average output voltage of the regulator.

7.4.2.7.1 Disconnect Load Including Filter Capacitor — Either the load or the filter capacitor on the output of the switching regulator could short, thus causing an overload on the regulator chip. Disconnecting them both eliminates these possibilities, but another capacitor must be substituted before the circuit is operated. The output circuit must be filtered to a smooth DC for the internal voltage circuitry of the chip to regulate properly.

If the output voltage of the switching regulator is normal with only a substitute capacitor, the overload is probably in the output circuit, the load, or the original filter capacitor. Test the filter capacitor for a short. If not shorted, reconnect only the capacitor and try the circuit again. If the circuit works with the original output filter capacitor, the load must be at fault. Further help in locating a short in the load circuit is found beginning at point P of chart 7A.

7.4.2.7.2 Check Inductor for Open — An open inductor will cause a symptom of zero volts output. Turn off all power and discharge the filter capacitor on the output of the regulator. Test the inductor with an ohmmeter for an open circuit. It may be necessary to disconnect one of the leads of the inductor for this test. If the inductor is not open, it could be shorted and therefore would be difficult to test. A shorted inductor would cause the output

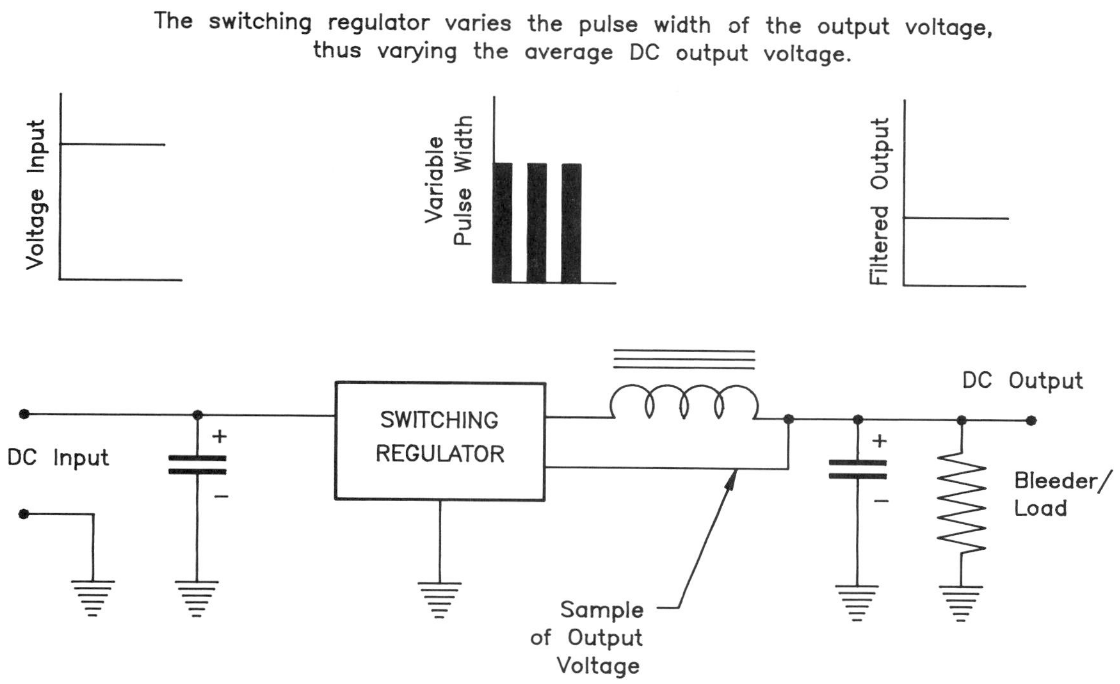

The switching regulator varies the pulse width of the output voltage, thus varying the average DC output voltage.

Figure 7D-4 The switching regulator is similar to the linear regulator, but includes an inductor and capacitor, plus a voltage feedback pin

voltage across the capacitor to have voltage "spikes," as visible with an oscilloscope. These spikes will interfere with the regulator and cause improper, "dirty" output voltage. In this case, replace the inductor to see if it was the problem.

7.4.2.8 Output Voltage Is Too High
— If the output voltage from the switching regulator is too high, there is an excellent chance that the regulator pass transistor inside the IC has shorted and no longer controls the output.

7.4.2.9 Regulator IC Is Defective
— If the regulator is determined to be at fault, it should be replaced by referring to Phase 5, "Replacing Defective Components."

7.4.3 Disconnect Regulator from Power Supply
— If there is insufficient input voltage, the regulator should be disconnected to see if the power supply is being overloaded and for that reason cannot supply the proper input voltage to the regulator.

7.4.4 Problem Is in the Power Supply
— If the power supply cannot provide the proper 15 percent or more overvoltage required to drive the regulator, the problem must be prior to the regulator. See the appropriate chart, 7A, 7B, or 7C, according to the kind of power supply supplying the regulator.

7.4.3.1 Regulator or Load Has a Short
— If the power supply voltage comes up to a normal level when the regulator is disconnected, then there must be a problem after this breakpoint. In order to find the overload, the technician must perform another test.

7.4.3.2 Reconnect the Power Supply Input
— Connect the power supply to the regulator again. This will give the power necessary to see if the problem is in the load, or if it is in the regulator.

☐ Phase 7D Summary

The ability to quickly locate and repair malfunction in a regulated power supply is required of all electronic technicians who are expected to troubleshoot to the component level. The basic principles of the four major kinds of regulators must be understood, along with the proper troubleshooting procedures to use in finding problems and repairing them in a minimum of time.

☐ Review Questions for Phase 7 through 7D

1. Why is it important to use sharp test probes when troubleshooting?
2. Why is it recommended to work with only one hand in high voltage circuits?
3. You have purchased several ni-cad cells with a capacity rating of 4 Ah. What is the standard charging rate?
4. What are the three general types of transformer power supply?
5. What is the principal detail to look for when analyzing power supply output circuits?
6. Discuss the difference between ground, neutral, common, and chassis.
7. Power supply failures come in two basic types. How can you tell the difference?
8. What are the three methods by which the technician can effectively troubleshoot the shorted power supply?
9. You find that you must troubleshoot a power supply using the feed-it fuses method. How would you approach this problem, with a minimum number of blown fuses?
10. What is the difference between a short, a semi-short, and a pseudo-zener short?
11. What must the technician be aware of if the decision is made to use the cut-trace-and-try method of finding a short?
12. There are a great many ICs and bypass capacitors on a board on which you must find a semi-short. All the ICs are soldered in. What method would be best to use in finding the short?
13. Could a shorted transistor having a 4.7 kΩ collector resistor cause a Vcc to ground short on the board?
14. You suspect that one of the windings of a transformer is shorted. What test would conclusively indicate the transformer condition?
15. If one of the rectifiers in a bridge circuit is opened, what should the technician look for before applying power to a newly installed diode?
16. Discuss the troubleshooting method of Figure 7A–7.
17. What caution should be remembered when troubleshooting defective switching power supplies?
18. A switching power supply blows fuses. Where is the probable problem?
19. What is the most likely component to be causing a switching power supply to blow fuses?
20. If a switching power supply has no power output, what must the technician keep in mind about this symptom?
21. What caution must be kept in mind when using the oscilloscope to troubleshoot a power supply?
22. You decide that the switching transistor of a switching power supply is probably shorted. You turn off the circuit power and prepare to make an in-circuit check with a solid state tester. What *must* be done first?
23. What are the symptoms of a shorted SCR in a phasing power supply?
24. The input voltage to a regulator must be at least _____% higher than the output voltage in order for the circuit to operate.
25. An unadjustable three-terminal IC regulator is producing too much output voltage. What is the most likely cause?
26. What is the main advantage of the switching regulator over a linear regulator?

PHASE 8
TROUBLESHOOTING SOFTWARE-DRIVEN CIRCUITS

☐ PHASE 8 OVERVIEW

This phase will discuss the special troubleshooting methods necessary when dealing with microprocessor circuits. The microprocessor is the principal integrated circuit, the "brain," in computers. It is also finding applications in controlling functions, such as in printers, plotters, industrial controllers, robots, copy machines, toys, automobiles, modems, and copy machines, to name only a few. The well-rounded technician must be competent in the repair of these circuits, especially as their importance and applications grow.

This phase will assist the technician in finding the way through typical hardware, with an occasional reference to operating a personal computer. A digital technician should be familiar with personal computers, as it is probable that a great deal of work and even testing functions will be done with them in the future.

8.0 NOTES ON SOFTWARE, HARDWARE, AND THREE-STATE CHIPS

Software — The software dictates which chip does what function, and when it is done in relationship to other operations. Each step of the program takes care of a tiny part of the overall operation. A loose analogy to this might be an orchestra. The printed sheet music could be thought of as the master program, the exact detail of what happens, and when it happens. Similarly, each chip in the equipment has a part to play, some small and some large. The entire operation is under the control of the conductor, or the microprocessor in our analogy. The microprocessor exercises tight control, however, for if the micro dies, the entire operation comes to a screeching halt within a few microseconds. None of the other chips does anything substantial on its own without the direction of this master control chip.

The microprocessor blindly follows the instructions of the program. It is capable of no action on its own initiative. If the program were to tell the microprocessor to stop dead in its tracks and do nothing, it does exactly that. The program usually tells the microprocessor to literally go in circles, doing the same operation over and over until something more important comes along for it to do. Thus, *the software determines what the hardware will do.* If there is a defect in the software for any reason, the microprocessor will not operate properly, even if all of the hardware is in perfect working order. *The software is as important as the circuit hardware that it runs.*

Hardware — The microprocessor circuitry, or the hardware as it is most often called, is similar to other digital circuits with one important difference. Information is passed back and forth on not one, but many lines, all of which are active at the same time. Information is commonly passed on 8, 16, or more lines simultaneously. These lines are called the *data* lines of the equipment. Once data is passed to its destination, the microprocessor may pass another bundle of information along the same route, or it may call on completely different chips to transmit and receive data on the data lines. Figure 8–1 contains a block diagram of a simple microprocessor controller that performs many of the fundamental functions of a computer.

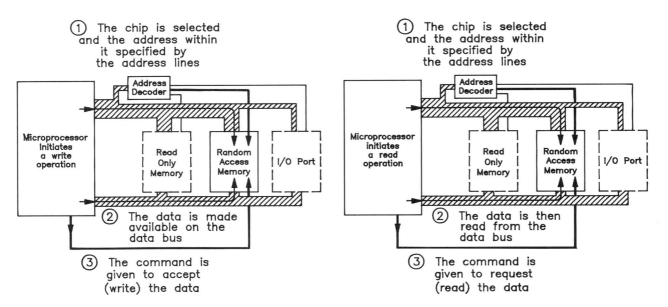

Figure 8–1 How several chips are tied to a common set of data lines, but only one of them can transmit information on the lines at any given time

Note the data lines of the circuit. These lines can be used by any of the chips connected to them to transmit data, and they are used to pass information to chips that at the proper moment are switched to the receive function. The timing required to read the data at the receiving chip is under the direct control of the microprocessor through the *address* lines. These lines select the proper chip to operate at any given moment. The *read and write* control lines, which tell the chips whether to transmit data or to accept it as a receiver, are also involved in a typical microprocessor operation.

A typical microprocessor operation might be to obtain a byte of information from the permanent storage in the Read Only Memory (ROM) chip and to then write that information into a temporary location in Random Access Memory (RAM). The block diagram of Figure 8–1 can be used to follow this chain of operations. The sequence of operations is as follows:

1. The ROM chip is selected by the upper bits of the address. At the same time, the lower bits of the address determine exactly which location will be accessed within the ROM. Accessing the ROM implies that this will always be a read operation, since the ROM cannot be written to.
2. The READ (RD) control line goes active. This tells the ROM chip to transmit the data located at the specified address on the data bus. Time is allowed for this to be accomplished, at the end of which the microprocessor reads the information from the 8 data bus lines.
3. The microprocessor now deselects the ROM, thereby disabling it and placing it into a "floating" or three-state status. The ROM is now "transparent," and acts as though it is no longer connected to the data bus.
4. The data received from the ROM is then placed on the data bus by the microprocessor.
5. The microprocessor now selects the appropriate address within the RAM chip by a similar means as that shown for the ROM chip. The RAM chip is selected by the upper bits, and the specific location by the lower address bits.

6. This is a write operation, so the WR line goes active, which causes the RAM chip to accept the byte of data on the data lines. The RAM chip will place that data into the address specified on the address lines.

Three-state devices — The microprocessor orchestrates the operation of software-driven circuits and, in the process, must control the input and output functions of other integrated circuit chips. Digital information flashes back and forth at high rates of speed. This operation requires that first one chip and then another takes over as the source of signals on the data lines that are common to many chips. Since only one chip may send a signal on the line at any given moment, there must be a means of disengaging all other chips that are not sending signals at that particular instant of time. This disengaging of a chip makes it become "invisible" or "transparent" to other chips on the data bus. This state is sometimes called the three-state or floating state.

8.1 Reboot the System
— If a microprocessor somehow gets an improper instruction for any reason, it cannot recover on its own. Software can be damaged, and all it takes is a single improper bit to send a microprocessor into silicon insanity. The cure for a "crash" such as this is to stop everything and to tell the microprocessor to start over again, right from the beginning of the program. This process is called rebooting. Some computers require a removal of all power and reapplication of it to begin all over. Others have a special pushbutton that accomplishes the same purpose without turning off the power. This is called the reboot or cold reboot switch. Personal computers have a special key combination, as an example of a rebooting operation, called the warm boot. It consists of pressing the Control, Alternate, and delete keys together. This is not a really complete reboot, however, and may not always work. Turning off the power and turning it back on again is the ultimate correction for a severe microprocessor crash.

A computer can be made to crash, or loose its place in the list of instructions, by seemingly innocent things. Electrical equipment that draws heavy current, a hairdryer for instance, can cause a computer on the same circuit to crash because of the sudden surge of current to the dryer. Equipment sharing a common power line that causes voltage spikes (or transients as they are called) is another source of computer problems. Some computers are more susceptible to these voltage spike problems than others because of less effective filtering in the power supply. Sometimes even touching the keyboard of a computer after walking across a synthetic carpet can cause a small static spark that can cause the computer to crash.

If a reboot of the computer is necessary and successful, so much the better. This ends the problem. Frequent crashes, however, should be tracked to their cause. See Phase 9 for details of troubleshooting these problems.

Rebooting several times after a failure might be a good idea to be confident that the system is indeed up and running normally.

8.2 CAN YOU CHANGE TO A PROVEN PROGRAM?
— Microprocessors operate on software. This software can be stored in several ways, some of them very susceptible to damage. Some of the more common ways that software is stored are in special chips called ROMs (read-only memory), PROMs (programmable read-only memory), EPROMs (erasable programmable read-only memory), on floppy diskettes, or on hard disks. Programs stored within microchips are sometimes called firmware.

Changing software can accomplish two troubleshooting purposes: 1) it can isolate the original program as the cause of the problem if an identical replacement program works, or 2) special software can be used to test and/or exercise specific parts of the hardware. This kind of hardware troubleshooting software is called *diagnostic software.*

A proven program is one that, preferably, has been operated on this specific system in the past with no problems. New software cannot be considered as proven software, because it frequently must be customized before it will operate with a particular system. This process of making the modifications necessary to run the new software is called configuring the software. Unconfigured software is not proven software. Software that has run on identical systems could be expected to qualify as a proven program, but the other system on which it ran successfully must indeed be identical in all hardware aspects. Any differences in the two systems could be reason for the software not to work at all, thus making it useless as a proven program. An example is a word processor program. If another system has a different printer, the hardware of the two systems is not identical and the program used on one may not work properly on the other.

Sometimes it is not convenient to change to a proven program. If the program is contained within a chip, then it is not reasonable to begin changing chips. These chips do not give problems that often. Media such as floppy diskettes and hard disks are more easily damaged or otherwise changed so that they will not function properly. This is where changing to a proven program is especially applicable.

Consider an example on a small personal computer. If the system ran fine with a particular floppy disk program yesterday or even a few moments ago and does not now perform as it should, there is a very good chance that the floppy diskette has developed a problem. Diskettes are particularly prone to failure when mishandled, as when placed on a pair of scissors (magnetized, of course), or when a paper clip puts a small crimp in the outer edge of the diskette. Either way, the diskette may be damaged. Remember, it takes only one tiny, misguided bit of information to make a microprocessor crash. See Figure 8–2.

Figure 8–2 The fragile floppy diskette, perpetrator of many a computer crash

An alternative to using different software is to use the suspect software in a different system of the same type. If the software works in other equipment, there is probably something wrong with the first system's hardware.

8.2.1 DID THE TROUBLESOME PROGRAM EVER WORK? —

Don't expect new software to work properly the first time. If a new program does not work as it should, don't blame the hardware. The software should be changed to match the hardware, a process called configuring the software. The instructions that came with the software should give full details on how the configuration should be carried out. A second way to get information on how to install software is to look for installation or "read me" files that may be part of the software package.

8.2.1.1 Make New Copy of Program from Backup Copy

The original program diskette — Original program diskettes are not configured for the specific system on which they will be operated, and can be thought of as "virgin" programs. *This original diskette should only be read from and copies made from it.* The original program diskette should never be written to or changed in any way. See the recommendations of Figure 8–3.

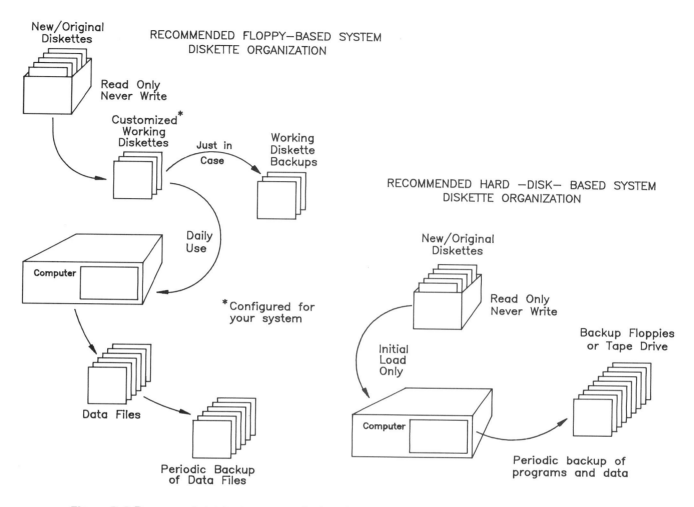

Figure 8-3 Recommended data storage organization of floppy- and hard-disk based personal computer systems

The program working diskette — The working diskette is one that has been copied from the original program diskette and then changed to meet the computer system needs. It can be further customized for more efficient use of the computer through such embellishments as the building of a special file that sets up the system and the software each time it is turned on. Personal computers call this an AUTOEXEC.BAT file. Additional smaller programs such as CONFIG.SYS can be added to accomplish special tasks along the way to a fully customized operating system.

The program backup diskette — Once a working diskette has been set up and is running normally, it is a good idea to make a copy of this diskette just in case the working diskette is damaged. This diskette is called a program backup diskette. After this copy is made, it must be put away in a safe place. Personal computers use the DISKCOPY program for copying entire disks. If any changes are made to the working diskette's program itself, a new backup should be made to record those changes. An example is when a different printer is installed and the program working diskette must be reconfigured to operate it.

Making a new program working diskette — If the program working diskette has failed, the simplest thing to do is to make a new copy from the program backup diskette. It is best to use a different diskette, in case the suspect diskette has been physically damaged, to make this copy. The DISKCOPY program is again used to make a duplicate working diskette.

Replacing a suspected diskette with one made from the backup copy may result in the system again running properly. The problem has thus been confirmed as having been the suspected diskette.

8.2.2 DOES SOFTWARE MATCH THE HARDWARE?

— New software must be basically compatible with the hardware on which it is to be run. For example, software written for an ITT personal computer will not run on an Apple computer. The microprocessors used in each are entirely different and the binary code is of necessity entirely different, too.

8.2.3 IS THE SOFTWARE CONFIGURED PROPERLY?

— Software purchased for a specific system must be changed, modified slightly to operate with a multitude of variables. For example, printers vary from system to system, as do the kinds of monitors and other accessories. Software writers make provisions for these differences, allowing for changes to be made to the software accordingly. It may fall to the technician to set up new software to run with a specific system. The instruction book for the new software should detail all of the steps necessary. Personal computer programs often provide a special SETUP or INSTALL program to run, which in essence gets into the main program and modifies it for operation with a specific system.

Another hint for the technician installing software is to look for any README file that may be included with the new software. The microcomputer TYPE command is used to put the contents of such a file on the screen where it may be read.

8.2.4 Program Is at Fault

— If the program under consideration is compatible with the microprocessor used, is configured to work with the specific hardware of the system at hand and still does not work properly, and if a different program works well with the hardware, it may be necessary to get into the program itself to find the problem.

8.2.5 IS THE SOURCE CODE AVAILABLE?

— If the software program is written in a high-level language like COBOL, C, or PASCAL, it will be of little use to the technician in tracing a problem because of the complexity of the codes involved in finally breaking down operations into machine code. Troubleshooting such software is the domain of the software engineer or programmer. Such incompatibility problems should be referred to the software dealer or the software author for resolution.

A competent technician should be familiar with the concepts of machine and assembly language programming, however. Machine code is the basic "ones" and "zeros" of binary digits, the level at which the technician should be effective in tracing problems. Assembly language provides a means of manipulating the ones and zeros effectively. A digital technician should be able, given sufficient time, to write short programs for a specific microprocessor that will test memory, input and output ports, and various other function of a microprocessor system.

While field technicians will not normally have access to source code, the technician working with engineers on new microprocessor-driven hardware products may very likely work directly with assembly language programs. See Figure 8–4.

8.2.6 Determine Where the Software Fails

— Determining where the software fails to operate a particular assemblage of hardware usually entails using one of two methods: either using the logic analyzer or putting special stopping points, called break points, in the software. The break points may allow the technician to verify the digital signals present on any of the chips while the microprocessor is halted. At the least, the use of a halt instruction as a breakpoint provides evidence of having at least reached a specific point in the program.

8.2.6.1 Use the Logic Analyzer

— The logic analyzer will provide information on the operation of a digital circuit in very detailed form. It can be likened to the freeze-frame capability of a moving picture projector. Once the record is obtained, the display and analysis

```
;
PORT1   EQU     PORT3A          ;D8-D0
PORT2   EQU     PORT3B          ;D3-D0 only
;
        ORG     TPA     ;Robot RAM beginning address
;
        LD      A,10010010B     ;Initialize #3 8255 chip for Port A
        LD      (CONT3),A       ; input, Port C Lower Input (environ-
                                ; ment switch inputs).
;
        LD      A,00000000B     ;Memory must be set to all 0's in the
        LD      HL,4000H        ; Learning Memory area of RAM.
ERASE:  LD      (HL),A          ;This routine clears memory from 0400H
        INC     HL              ; to 0500H.
        LD      A,H
        CP      50H
        JP      Z,CLEAN
        LD      A,00000000B
        JP      ERASE
;
CLEAN:  LD      HL,INT4         ;Set up INT4 to respond to an
        LD      BC,LEARN        ;environment problem.
        LD      (HL),C
        INC     HL
        LD      (HL),B
;
;
        EI
        CALL    ON
;
; Without an interrupt, caused by something in the environment
; going "wrong", the robot endlessly selects random motion for a
; random time.
;
OVER:   LD      A,R
        RLA
;
```

Figure 8-4 An example of assembly language source code. This small program was written for a Z80 microprocessor

of the digital information can be done at leisure in several different ways. An explanation of the logic analyzer was covered in paragraph 3.10.17.1.

Using a logic analyzer will require that the technician have a working knowledge of the analyzer—many of its special functions are not normally needed—and a thorough knowledge of the circuit on which tests are to be made.

Although a full explanation of the logic analyzer will not be given here, enough can be given to introduce the instrument and give its most frequent use—obtaining digital data on a suspected circuit that can be compared with data from a good circuit.

CAPTURING DATA WITH THE LOGIC ANALYZER — The logic analyzer must be set for at least two items before it can be of use in tracking a problem through the hardware and software of the circuit:

1. External clock
2. Triggering word

The clock signal determines when samples of the circuit are taken. Normally, samples are desirable at every tick of the clock that operates the system under test. Thus, the technician most often uses "external clock" option on the logic analyzer setup menu. One of the wires at the test probes or pods are connected to the clock of that system.

The triggering word determines when the analyzer will stop recording digital information. Where an oscilloscope trigger *begins* the oscilloscope trace, the logic analyzer trigger word *stops* a process of continuous recording of digital information in a high-speed loop of RAM memory.

After these two selections are made, the analyzer is readied for data collection and the circuit in question started. When the trigger word occurs, the data collection usually is set to continue for a short time, then stop. Data collection is now complete. The digital data is now stored in the RAM of the logic analyzer. It can now be saved on disk, displayed on the screen, or compared with similar, previously recorded data from a good circuit.

DISPLAYING DATA WITH THE LOGIC ANALYZER — The timing display of the logic analyzer provides a graphic display that can be compared with the logic levels expected in the circuit under examination. Differences can be investigated and the problem found from the information provided. The timing display can sometimes be printed out to provide a "hard copy" of troubleshooting information. Other data formats are available such as binary and hexadecimal tabular forms.

8.2.6.2 Using Break Points
— If no logic analyzer is available, the break-point option might be used. This option will require some means of writing a customized program and directing the microprocessor to execute the new program. An example would be to use the assembler and debug programs of a personal computer to write or modify existing PC programs and to install break points to troubleshoot the execution of the program.

Another example would involve the equipment necessary to program an EPROM chip and then insert the newly programmed chip into the microprocessor circuit. See Figure 8–5.

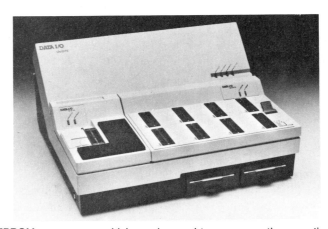

Figure 8–5 An EPROM programmer, which can be used to reprogram the operation of a microprocessor (Photo courtesy of DATAI/O, Redmond, WA)

It is sometimes advantageous for the technician to write short diagnostic routines to test particular portions of the circuitry. Insertion of a HALT command at appropriate points in a given program is particularly useful. This is a powerful option, because it is simple and effective. The microprocessor thus leaves the circuitry in a known state. The halt statement can be successively moved through a program, providing a sort of stop action that can be effective as a troubleshooting tool.

Of course, using break points will require a knowledge of machine and assembly language. While learning these programming necessities may not be particularly thrilling, especially at the beginning, programming at the ones-and-zeroes level of microprocessors is both complex and satisfying. A person cannot be a thoroughly trained digital technician without a good working knowledge of pin-for-pin troubleshooting of microprocessors.

8.2.7 Revise Software as Necessary
— The revision of software will require the use of a computer. The original source code revisions will require the technician to recall the original source code file, type in the new sections of program, and delete the old. See Figure 8-4.

The new source code will then have to be converted into a form suitable for programming into an EPROM. The conversion of the typed words of assembly language into hexadecimal codes is accomplished by an assembler program. One of the new files produced by the assembler program is called an object or Hex file. See Figure 8-6.

```
:03000000C36900D1
:10006600C341033E8332031021005831005801007A
:10007600382B36AA7CB820F97DB920F52100582BFB
:100086003EAABEC2420478BC20F579BD20F121000B
 10009600582B36557CB820F97DB920F52100582B10
 100CA6003E55BEC2420478BC20F579BD20F13E140F
:1000B6003211393E00321239060221CF3811B10110
:1000C6001A77231310FA3E38ED47ED5E210210CB66
:1000D6005EC408053E54CDB6013E01CD85043E49B9
:1000E6003201303E1432013021FF390E003A011040
:1000F600E604CA8F01CD6801163ABA20F0CD730125
:10010600470D7301CD7301CD7301FE00C27B01CDD6
:1001160073012377109CD2901ED44B9C242040ECB
:10012600018C8CD6801FE30DA4204FE47D2420408
:10013600CB07CB073818CB07CB07E6F057CD6801BE
:10014600CB07CB073812CB0FCB0FE60F82C9CB07F5
:10015600CB07E6F0C69018E4CB0FCB0FE60FC60927
:1001660018EA3A0130E60228F93A0030C9CD2901E9
:10017600F5814FF1C9FB3E02CD85043E4FCDB60158
:100186002102010CB46C4DD047601002011003A217D
:10019600058EDB0FBC3003A060221CF3811B40176
:1001A6001A77231310FAFBC3003AC39E01C3003A21
:1001B60000F5C5D5E5FE20CA2E02FE41DA3602FE5E
:1001C6005BD23602D6411711660526006F195623F3
:1001D6005E0E007ACDE6017BCDE601E1D1C1F1C923
:1001E60047E680CD170278E640CD170278E620CDA7
:1001F600170278E610CD170278E608CD170278E6E2
:1002060004CD170278E602CD170278E601CD170273
:10021600C92804CBC1180ACB41C8CD3904CDA603E1
```

Figure 8-6 Example of a Hex file, produced by an assembler program from the original source code

The Hex file is then sent to a PROM programmer where it is converted into binary bytes. The EPROM can be erased and reprogrammed if it is first inserted into an EPROM eraser and flooded with ultraviolet radiation for a few minutes. See Figure 8-7.

8.3 Substitute Equipment
— The substitution of known good equipment for suspected equipment is the first step in troubleshooting a problem on a digital, microprocessor-based system. Using this technique, it is a relatively simple matter to isolate the bad unit. For troubleshooting purposes, the cables used between equipments should also be replaced with other good cables to isolate them as possible problems, too.

For example, if the printer of a computer system is giving problems, by all means change to another similar printer and see if the problem is still there. If it is not, try using a new

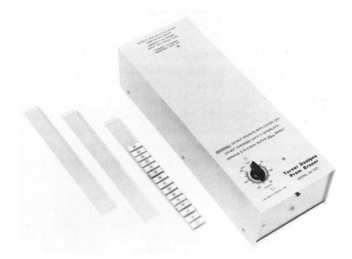

Figure 8–7 An EPROM eraser unit (Photo courtesy of Turner Designs)

cable (assuming it is removable at both ends). You may find that the problem was the cable, not the printer at all.

8.4 Check ALL Power Supply Outputs — Once the bad equipment is identified, the next step is to be sure that all of the required power supply outputs are normal. This is usually a quick check, and can be done effectively right at a familiar chip. Most 74XXX ("X" meaning "don't care what is here") chips are TTL, probably the most popular digital chip today. Try checking for Vcc from the highest numbered pin of one of these chips to the pin diagonally opposite. Most of the time, this test will show Vcc, which should be 5.0 volts, +/− 0.25 volt. CMOS circuits can run on any voltage from 3 to 18 volts, so further information on the equipment will be necessary to identify and measure the voltage that is proper for that equipment.

Obviously, if the power supply is not producing the proper voltages, this problem must be addressed first. There must be a problem associated with that part of the equipment. Refer to Phase 7, which deals with power supplies.

8.5 ANY SWITCH, HARDWARE CHANGES MADE? — Many PC Boards and peripheral equipment operated by microprocessors use small switches to determine initial operating parameters. If these switches have been changed, or could have been changed, then it is a good idea to double-check the new settings. If the switches are external, accessible to the "general public," then this is certainly a good place to check early in the troubleshooting session. If changes have somehow been made, proper settings should be set into the switches and the system tested *from a cold start* for normal operation. On the other hand, if the switches are internal to the equipment, there is probably little chance that these switches have been changed.

An important point must be made about configuration switches. These switches are generally read by the equipment only once, during the initial setting up of the equipment at power-up time. If the switches are to be changed, the equipment must then be shut off and re-energized for the new settings to become effective.

Problems can develop when making changes to existing equipment, too. Installing new cards into a personal computer, for instance, may result in the computer refusing to work. The switches just mentioned should be checked to be sure that the computer is configured properly for the hardware change being made.

Whenever changes are made to a computer and the system switches are then changed, it is a good idea to *write down the switch settings before any changes are made.* In so doing, the technician provides a "way out" if the proposed changes do not work properly. The old settings and hardware can be restored as a starting point for another or different attempt at the hardware change if necessary.

Some computers and peripheral equipment have software-settable switches. These software switches are set with a special program, often called SETUP. The settings of these switches are saved in non-volatile (permanent) memory so that the settings can be used next time without manually resetting them.

8.6 ARE DIAGNOSTIC OR SELF-CHECK PROGRAMS AVAILABLE?

— Diagnostic programs are software test instruments. They can be part of the firmware of a printer, for instance, or they may be stored on a computer's floppy disk, ready for use when needed. Such programs provide a quick, often thorough check of the microprocessor-driven circuitry.

The self-check program provided for many printers can be initiated by a special combination of keys, a combination usually activated as the printer is turned on. This test will check the hardware of the printer and most of the electronics. About all that it doesn't check is the circuitry involved in the passing of data from the input connector to the internal circuits of the printer. If such a printer passes the self-check, there is no need to check power supplies that power the internal circuits. They must be working. Only those circuits involving the reception of data from the computer need to be further tested. See Figure 8–8.

Figure 8–8 Printout resulting from a printer's diagnostic self-check

Diagnostic programs are also available for a great many other applications besides printers. Such programs are available for testing the hardware of personal computers, for instance. These programs can test many of the functions of the monitor, keyboard, system board, and the input/output ports of the computer. Using the self-check or diagnostic programs can save a great deal of time for the servicing technician. See Figure 8–9.

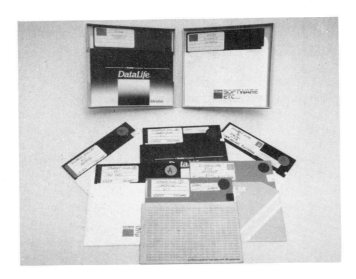

Figure 8-9 system diagnostic diskettes

Specially recorded software is available to effectively test the hardware of a floppy disk drive, too. Deliberate off-track recording of data is one method used by the XIDEX Corporation on their INVESTIGATOR and DIGITAL DIAGNOSTIC diskettes. These diskettes provide a handy method of troubleshooting the hardware of these diskette drives.

The INVESTIGATOR diskette, shown in Figure 8–10, is an unusually easy-to-use program that tests all the important parameters of a drive. It is easily used by any reasonably competent computer user. Periodic checks of the diskette drives with this program can provide a warning of impending mechanical failures due to gradual degradation of the hardware of the drive.

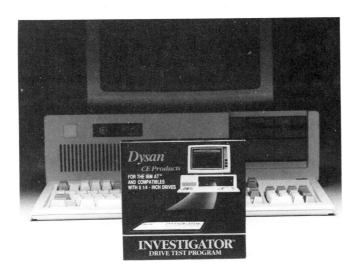

Figure 8-10 Xidex's Investigator diskette, a user-friendly check on the condition of disk drive hardware (Photo courtesy of Xidex Corporation)

The INTERROGATOR diskette is another product from the same company that is a technician-oriented product used with a special digital diagnostic diskette. These provide in-depth analysis of the drive hardware. It is a specially recorded diskette with off-track and deliberately "twisted" data recorded upon it. The ability of a given drive to read the data symmetrically within specified limits provides a fast, precise measurement of performance. This information can then be used to analyze, adjust, or reject a drive. See Figure 8–11.

Figure 8–11 The Interrogator diskette is used with a digital diagnostic diskette as a precision troubleshooting tool for the computer technician (Photo courtesy of Xidex Corporation)

Large computer systems can be analyzed with diagnostic software while the computer runs other programs, on a time-sharing basis. The technician gives the proper commands from a terminal that can be located anywhere—across the room or across the country. The results of running the diagnostic software localize the problem within the computer system and help determine whether or not a technician must be dispatched for a field trip.

8.7 WILL DIAGNOSTIC OR SELF-CHECK PROGRAMS RUN AT ALL?

— If the diagnostic or self-check programs do not run at all, begin troubleshooting the equipment to the card level next. See Phase 2 for help from here.

On the other hand, the diagnostic program may point to a defective card, in which case the technician can replace or repair the bad card. Troubleshooting to the component level, once the bad card is identified, is covered beginning at point M of flowchart Phase 3B.

8.8 KNOWN GOOD CARDS AVAILABLE?

— If diagnostic or self-checks are not available or the problem is not otherwise localized to the defective card, it may be that extra, good cards are available for substituting. Isolating the problem by card substitution is a fast, efficient way of narrowing down the problem.

8.8.1 Substitute a Card

— *Be sure to turn off power before removing or inserting PC boards into equipment!* This step helps prevent damage to the boards by the application of power in the wrong order to various circuits. It also avoids some of the transient voltage problems that can occur. It is a good idea to touch the chassis of the equipment with one hand before inserting the card to discharge any stray static charges that may have accumulated.

The new card substituted into the equipment must be identical to the old one. This must include the switch settings and jumpers on the new card, if any.

8.8.2 Problem Is in Motherboard, Backplane, or Mechanical Portion of Equipment
— Once all of the cards within a piece of equipment have been replaced, only a few things remain that can be the cause of a problem. Whatever it may be called, the motherboard, backplane, or system board, the sockets into which the circuit boards plug could be the problem. See Figure 8–12.

Figure 8–12 A backplane

Look very carefully at each of the small springs that contact the card edge. Perhaps one of them has been bent out of position and is now either not making contact or is shorting to another contact. See Figure 8–13.

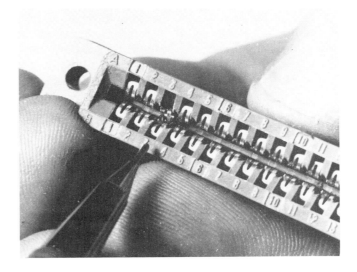

Figure 8–13 A damaged connector like this one is a common cause of problems

Careless or hurried insertion of circuit cards during troubleshooting can cause problems like this. Be careful when inserting cards, making sure they are lined up correctly before applying a reasonable pressure to seat them.

When all of the circuit boards are eliminated, the only thing remaining is the wiring within the equipment and whatever that wiring may connect to. Each of these wires must be inspected for breaks. If necessary, the items connected by the wires will have to be tested or temporarily replaced.

An example of this situation might be encountered when working on a particularly difficult problem with a printer. When all of the circuit cards are eliminated, all that remains to cause a problem are the power wiring, an interlock perhaps, and the printhead.

8.9 ARE CARD DOCUMENTS AVAILABLE?

— Without good cards to substitute and without documentation, it is almost useless to try to troubleshoot digital equipment further. While it is possible that the problem might be found using the methods of Phase 5, "Troubleshooting Dead Circuits," the sheer number of chips involved may make this futile.

If neither substitute cards or documentation are available, the equipment should probably be put aside until either or both become available.

Documentation for the card in question must include all of the items of paragraph 3.6. Microprocessor-driven circuits are usually quite complex, and the theory of operation is especially important.

8.9.1 NOTES ON TROUBLESHOOTING THREE-STATE DEVICES

— Because of the necessity of sharing common lines for data flowing from many chips, in turn, to others, each of the chips that transmit information on the lines must be capable of "letting go" of the line, allowing others to assume the transmitting role. This is why the three-state, or "floating" state, is necessary.

The output lines of a chip designed for inclusion in a microprocessor circuit must have this capability of disengaging the output lines. When the chip is "told" to transmit information on the common data lines, the signal that does it *enables* the output stages. When the chip is enabled, it can transmit information. Common pin designations to transmit data are output enable (OE), or chip select (CS). The chip select pin must also have a READ pin enabled to get data from that chip. Writing to a selected chip makes that chip a receiver. See Figure 8–1.

8.9.2 8-BIT MICROPROCESSOR?

— Some of the most common 8-bit microprocessors are the Z-80, 8080, 8085, 6800, and 6502. These are called 8-bit microprocessors because the data flow to and from them is 8 lines wide. Later processors use 16 or more lines for data. If the circuit uses one of these 8-bit processors, the use of another inexpensive instrument becomes optional. Such an instrument is not yet available for processors having more than 8-bit wide data busses.

8.9.2.1 STATIC STIMULUS TESTER AVAILABLE?

— The static stimulus tester is a set of manual switches that allow the technician to directly manipulate the hardware of an 8-bit microprocessor circuit. The original microprocessor is removed from the circuit and the tester cable inserted into its socket. See Figure 8–14.

8.9.2.2 NOTES ON USING THE STATIC STIMULUS TESTER

— The documentation that is provided with the tester is easy to follow in performing manual operations of memory or input/output read or write. The process of setting up the proper address, generating the read or write command, and the reading of the information on the data lines is done at human speeds, not microprocessor speeds. Any combination of signal conditions can be held indefinitely for troubleshooting purposes.

The resulting digital states of 14- and 16-pin chips can be read quickly with the logic clip, an instrument previously mentioned in paragraph 3.10.7.

The instruction manual provided with the SST is thorough and simple. All the technician must do is determine what needs to be done to test the circuit, then set up the switches of the SST to give those digital levels.

318 □ Troubleshooting Software-Driven Circuits

Figure 8-14 The static stimulus tester allows the technician to manually operate the chips that the microprocessor normally commands, in 8-bit microprocessor circuits (Courtesy Creative Microprocessor Systems)

The static stimulus tester can be used to write to RAM memory and to read RAM and ROM. Any address within these chips is accessible to the technician. Input and output ports can also be read from or written to. The SST is able to find almost any problem in the memory and input/output circuits that the circuit may have. Its use will require that the technician think in binary terms, at the machine language level.

8.9.3 Look for Stuck, Missing Signals

The logic probe is valuable for single-point troubleshooting of microprocessor circuits. One of the best places to test to see if a chip is being used is to monitor the output enable (OE) or chip select (CS) lines. Without the proper signal on these lines, the chips cannot receive or transmit digital information to and from the microprocessor.

The data lines should be toggling back and forth with activity during the execution of a program. If one of these lines is stuck, the microprocessor circuit cannot operate properly.

The address lines should also have activity. The lower bits, in particular, should be toggling. If high memory is not being used by the program, the highest few bits may not have activity.

The control pins, read and write, memory and input/output, should have activity since they control the direction of data flow on the data lines.

8.9.3.1 Verify that Software Calls for a Signal Here

The lack of a signal on a specific line in a microprocessor circuit does not, by itself, indicate that a problem exists. Since the program determines which lines are active, it is up to the technician to determine if activity at this point is actually required. Detailed information on the software used may indicate this, but comparison with other equipment operating under the exact same software is probably easier to use for a check of whether or not the lack of activity is normal. If the logic probe then indicates a bad level (neither the high nor low LEDs light, and the pulse light does not flash) where activity should be present, the problem is near at hand. Any well-designed circuit will not leave inputs floating because of the problems that can be caused with noise. CMOS chips, in particular, cannot tolerate open inputs.

8.9.4 PROGRAM MACHINE CODE AVAILABLE? — If the program documentation is available, it is quite possible to determine where the hardware is failing to perform. Two options are available: using the existing software and inserting break points, or using a logic analyzer. The machine code referred to here is the complete program listing, with the hexadecimal codes included.

If the binary program is all that is available, a *disassembler* program can be used to build a more understandable picture of what the microprocessor is doing during the execution of that file.

8.9.5 Determine Where the Hardware Fails — With the program documentation, the actual address, data, and the controlling signals can all be traced and verified as follows.

8.9.5.1 THE LOGIC ANALYZER — The basic operation of the logic analyzer was covered in paragraph 3.10.17.1. Specific instructions on the use of the instrument must come from the logic analyzer operation manual. The use of this instrument requires an in-depth understanding of digital circuits and the signals needed to control them.

An important tip — The logic probe can be used effectively to troubleshoot most failure cases with the following tip, reserving the logic analyzer only for stubborn cases:

A big help in troubleshooting is a special program that exercises all parts of the hardware, over and over again. A program loop such as this can be run continuously and will produce predictable patterns that the logic analyzer and other instruments can recognize. This program can be stored in a ROM chip and inserted into the microprocessor circuit for troubleshooting. Using such a chip, the technician can use the logic analyzer to build a file of normal digital signals called a *correlation* file. Digital signals from a malfunctioning circuit can then be compared with these normal signals. The logic analyzer is particularly adept at comparing long listings of digital information. The point of departure from normal signals can be pointed out by the analyzer. Events leading up to the failure can be analyzed in detail and the problem traced to its cause.

Tests can be made of the system clock, reset, ROM, and many other IC pins with a simple, single-step program that puts the microprocessor into a tight loop of getting an instruction from ROM, resetting the program pointer and getting the same instruction, over and over with no way out of the loop. This program is called a "Here-Jump-Here" program. Depending upon the specific microprocessor used, the binary codes required for this program can be put into a ROM quite easily with an EPROM programmer.

8.9.5.2 SOFTWARE BREAKPOINTS AND INDICATOR — Software is developed for a specific purpose. The process of developing effective software may involve the technician. If so, a copy of the software must be made available to the technician for the purposes of debugging it, or finding errors or malfunctions.

One of the techniques available for finding software problems is to install temporary breakpoints in the software where they will allow the hardware to stop so that signal levels can be observed.

Some sort of indicator is useful to tell when the microprocessor has reached that point in the software. As an example, the Z80 microprocessor has one pin that goes low when the microprocessor is in a halt state. Monitoring this pin with a logic probe or an installed LED provides this indication.

□ Phase 8 Summary

This concludes a rapid overview of the options available when troubleshooting circuits that use software. The actual circuit, the hardware, is operated *entirely* under the control of the software. Neither is more important than the other. The digital technician must have a thorough understanding of the operation of the microprocessor and of the basics of programming in machine and assembly language. The technician able to use these tools is elevated to a higher professional status and becomes much more valuable to his employer. Specific troubleshooting techniques for microcomputers are given in Delmar's *Assembling and Troubleshooting Microcomputers* textbook.

Review Questions for Phase 8

1. Which is more important in microprocessor-driven circuits, the hardware or the software?
2. There are three basic causes for problems in software-driven equipment. What are they?
3. What is meant by incompatible software?
4. What is a "proven program"?
5. What is meant by "configuring" software?
6. What are the two principal parameters to consider when using a logic analyzer?
7. What is a breakpoint?
8. When troubleshooting a large software-driven system, what is the technician's first choice of troubleshooting methods?
9. How can most modern printers be quickly tested?

PHASE 9
TRACKING HUM AND UNWANTED SIGNALS

☐ **PHASE 9 OVERVIEW**

Hum in audio equipment, mysterious computer crashes, and radio interference all can be the result of unwanted signals. The origin and cure for many of the more common problems are given in this phase. Although the art of tracking radio interference could in itself be a separate book, the fundamentals are explained here.

9.0 Interview Operator, Get the Facts — In any reported case of interference, it is always a good idea to contact the person who has reported a problem. If possible, have that person demonstrate the problem to you. In a few instances, the problem may be just that the operator doesn't fully understand the total operation, and the technician can correct that with some instruction and a bit of tact.

The technician should first categorize the problem demonstrated. Generally speaking, audio equipment produces hum and squeals, computers mysteriously cease normal functioning, and radio equipment develops interfering signals, ranging from filtering problems and resulting hum or noise to the presence of radio signals where they don't belong. Categorizing the type of problem is the first step in finding the cause and cure.

9.1 A Computer Is Involved — Interference to a computer's normal operation can cause the computer to do strange things. Usually, the computer ceases to respond to keyboard input and accomplishes nothing at all. In severe cases, important data can be destroyed as the computer indulges in its own brand of "silicon insanity." There are two principal sources of interference to a computer: that conducted to it from the operator and that conducted into the computer from the power lines. The difference is that the computer can crash at any time, with or without an operator, if the interference is caused by power line problems. Static discharge from the operator, however, usually occurs when the computer is approached after having left it for a time. During the absence, the operator accumulates a substantial electrostatic charge, and touching the computer then provides a discharge path. The discharge of electricity then causes the computer to crash. Even small charges can affect the computer, charges so small that the operator may not be aware of the spark that causes the problem.

9.1.1 Conducted Power Line Problems — Power line problems take three main forms: totally missing power cycles, over- and under-voltage, and incoming transient pulses on the power line. These problems can sometimes be cured at the source, if known. Stubborn cases can be referred to the local power company, who is ultimately responsible for providing pure, no-break, clean power.

9.1.1.1 Pulses, Transients — These are two words for the same power line problem, momentary bursts of voltage that can be well above the normal line voltage. See Figure 9–1.

These pulses are commonly caused by electrical equipment that uses only a part of the whole line waveform for operation, for instance, light dimmers and variable speed electrical

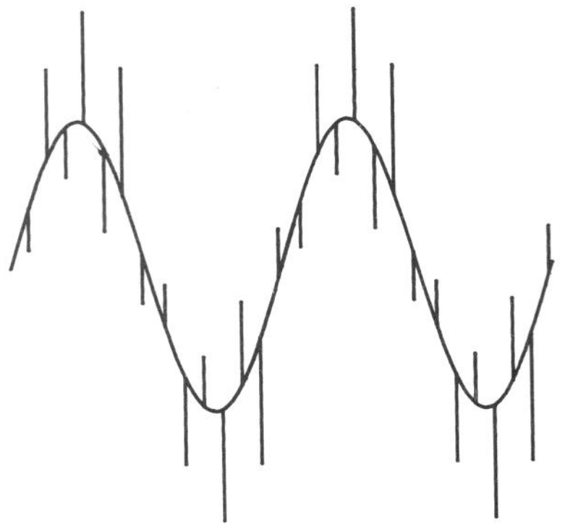

Figure 9-1 Graphic representation of power line voltage spikes

motor-driven appliances and tools. A variable speed electric drill is a good example of such a tool. The waveform produced on the power line by this kind of equipment might look much like that of Figure 9-1. Other sources of interference are electric arc welders and equipment that uses small "universal" motors, such as blenders and vacuum cleaners.

The cure for transients is to clamp them to a safe voltage with either of two methods, using zener diodes or metal oxide varistors (MOVs). These components are small, but they are able to absorb a great deal of energy if that energy doesn't last long. See Figure 9-2.

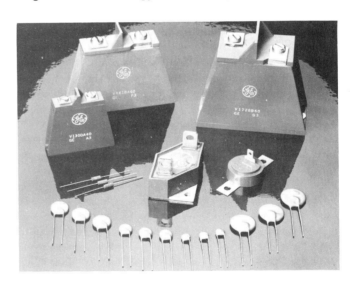

Figure 9-2 Metal Oxide Varistors (Photo reprinted with permission of GE Semiconductor, General Electric)

These components are often installed within the power supplies of computers. Refer back to Figures 7B-2 and 7C-4.

It is a good idea to have additional protection at the outlet supplying power for the computer. See Figure 9-3.

MOVs can be tested to be sure that they are providing the proper protection. A high-voltage DC source with a current limiting resistor can be used along with a voltmeter to determine the voltage at which the MOV clips the voltage. Since the MOV is effectively 2 zeners back to back in series, it can be tested in both directions as a zener, as shown in Figure 5-19.

9.1.1.2 Missing Cycles
Power may be interrupted for part of a cycle, several cycles, or the lights can just go out for a while. This interruption can cause a computer to forget everything in its RAM memory, data which is sometimes valuable, painstakingly entered by an operator.

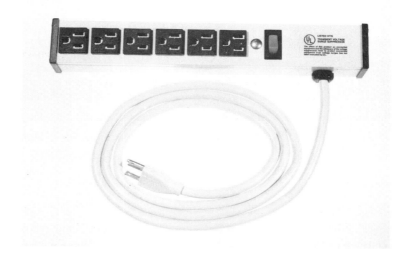

Figure 9–3 Commercial unit containing MOVs across the three wires of the outlet. (Photo courtesy of Brooks Manufacturing Company)

A partial hedge against such losses is to periodically copy the RAM data onto more permanent storage, storage that won't be affected by the loss of power. This leads to the recommendation that *information put into a computer should be saved frequently.* As this manuscript was being typed, it was automatically saved every 10 minutes, for example. This way, loss of power could cause a maximum loss of only 10 minutes of work.

9.1.1.3 Over-, Undervoltage

Momentary overvoltage and undervoltage on the power lines can also cause problems for computers. If severe enough, undervoltage can produce the same symptoms as missing cycles: a computer crash with the resultant loss of RAM data.

Overvoltage—cycles of more than normal amplitude—can cause computer failure by causing components such as capacitors and transistors to fail. The components that take the brunt of such stress are the power supply switching transistors, MOVs, and line filter capacitors. In extreme cases, a transformer could be made to fail by the puncturing of insulation.

It is interesting to note that a personal computer power supply can be designed to operate from 90 to 130 volts input. Voltages outside this range may well produce the symptoms just discussed.

9.1.1.4 Install No-Break Battery Backup Power Supply

The cure for all of these problems, including transient voltage problems, is the battery-operated, no-break power supply. See Figure 9–4.

The battery backup power supply comes in two general varieties, one which provides all power flow through the unit at all times, and the other which uses a high-speed relay to switch to the battery supply. See Figure 9–5.

The installation and operation of the supply is simple. Just put the battery supply between the computer and the power outlet. See Figure 9–6.

The battery is trickle-charged during normal operation and is thus in a standby status. The main power path is around the battery. If the power line voltage should fail, the battery takes over supplying all of the power needed by the computer. At the time of power line failure, most of these units also sound a small alarm signal to alert the operator. This alarm gives the operator time to put any valuable data in RAM into safe storage, often on a hard disk or floppy disk. The computer may only run a few minutes, typically 15, but this is quite enough time to save data. After the data is saved, the computer can be shut off until normal primary power is restored.

324 □ **Tracking Hum And Unwanted Signals**

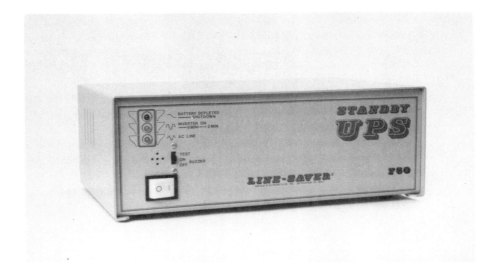

Figure 9-4 Commercial battery backup power supply for computer use (Courtesy of Kalglo Electronics Company Inc.)

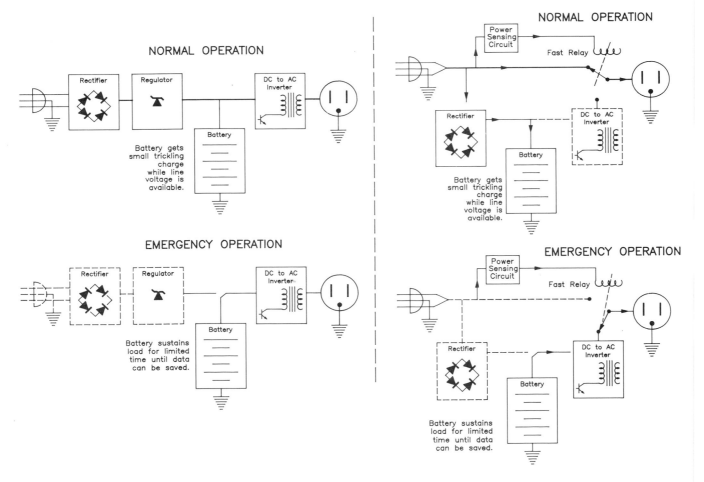

Figure 9-5 Block diagrams of the two basic types of battery-backup power supplies

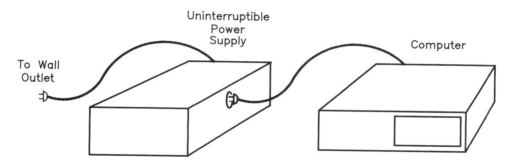

Figure 9-6 Block diagram of computer and battery backup supply

9.1.2 Computer Crashes if Touched

— Dry weather, synthetic carpets, and insulating soles on a person's shoes are a combination sure to cause computer problems. Handling styrofoam (including styrofoam cups) and many plastics, wearing nylon and rayon clothing, or simply moving around generates static. Static discharge can produce high currents for a very short time, currents that can actually weld the microelectronic circuitry within integrated circuits. Voltages as low as 30 volts can destroy or damage these tiny circuits. Even if the currents produced do no immediate damage, it has been determined that future failures can be caused by the cumulative effect of repeated static discharges. At the least, static discharge may be evident by the fact that the computer crashes when touched. This usually happens when the operator has left the computer for a few moments and has moved around the work area. Simply moving around can generate voltages up to 30,000 volts! Returning to the computer, a small static spark may or may not be felt as the keyboard is touched. The first indication of a malfunction comes when the computer refuses to respond to the keyboard—the computer is "locked up" or crashed. All data in RAM memory must now be "dumped," as the computer has to be restarted (rebooted).

9.1.2.1 INSTALL ANTISTATIC PRODUCTS

— Preventive measures against static discharge problems are aimed at preventing static voltage buildup, or by discharging existing static voltages harmlessly and directly to ground, through sufficiently high resistance to avoid a high peak current.

Prevention of static voltage buildup — Prevention of static voltage buildup is an effective way of avoiding the danger of discharges. Grounding of workers and the working surface is done to prevent static voltage buildup. Prevention products are shown in Figure 9-7.

Static-producing materials can also be eliminated from the work area to help in preventing static buildup. These materials include many kinds of plastics, styrofoam, and synthetic fabrics.

If the rug in the work area is not replaceable, yet seems to be the source of static electricity, it can be treated with special sprays that will prevent the buildup of excessive voltages.

Safe discharge of static voltages — Once static is present, it can be discharged safely if a path is provided to an earth ground through a high resistance. This also prevents the unpleasant sensation of a shock that is sometimes felt during the discharge of a high current static spark. A computer can be placed on a static control mat which is grounded through a high resistance to the third, grounded wire of the electrical outlet.

Delicate components should always be packaged in special material to shield them against static.

Installation of these anti-static products can enhance the computer operator's workplace in the prevention of computer crashes, and it can help the technician during servicing in preventing static-caused failures of microchips while equipment is open and very vulnerable to damage.

Figure 9-7 Commercial products available to reduce electrostatic discharge while operating a computer. Note the conductive floor mat connected to ground, the grounding chain from the chair to the floor mat, the conductive pad under the keyboard, and the anti-static spray product. (Photo courtesy of Plastic Systems, Inc.)

9.2 Radio Frequency Equipment Involved

— Unwanted signals sometimes involve radio frequency equipment, receivers mostly, but sometimes transmitters. Pulse and buzzing interference on transmitted signals can usually be solved with the same precautions and preventive measures as those described for automotive installations, paragraph 9.2.2.2. Interference in received signals takes three general forms: a general degradation of the receiver sensitivity, which may also accompany another type of interference; that of receiving signals not tuned for or desired; and that interference which is best described when a receiver makes noises sounding like a buzz, or a whine. Both of these are common problems when dealing with mobile radio installations.

9.2.1 Receiver Desense

— The desensitization of a receiver simply means that the receiver is not as sensitive to normal on-air signals as it should be. This kind of problem can occur when there is a powerful radio transmitter near the receiving equipment. Persons at the receiving site may or may not be aware of the strong signal, as its frequency may be far from the resonant frequency of the receiver. A radio frequency (RF) signal that is strong enough to force its way past the first tuned stages of a receiver and into the first RF amplifier stage can produce a DC voltage that tends to cut off the first amplifier stage. See Figure 9–8. The result is that the undesired signal is rejected by the remaining stages of the receiver, but sensitivity to the proper signal suffers.

Field effect transistors are often used in the first amplifier stages of receivers because they are less prone to this kind of problem. They are less prone because they do not have a junction in the input circuit to produce the DC-bias upset.

9.2.1.1 Install an RF Trap

— Other than replacing the entire receiver with one that can handle the overloading signal better, there is the possibility of installing an RF trap to reduce the strength of the interfering signal. The RF trap should be made using a coil of high Q and a high-quality RF capacitor, either air or Mylar are good choices. Either the capacitor or the inductor should be made adjustable to tune the trap exactly to the interfering signal. Also, either a series or a parallel circuit, or both, may be used to help trap the signal. See Figure 9–9.

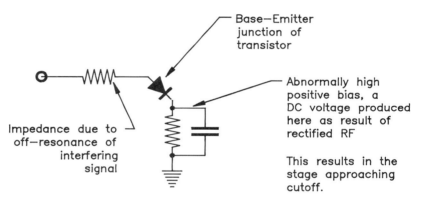

STRONG SIGNAL INPUT SEES DIODE TO GROUND

Figure 9-8 Schematic of a receiver front end, with result of strong off-channel signal producing bias in the base circuit

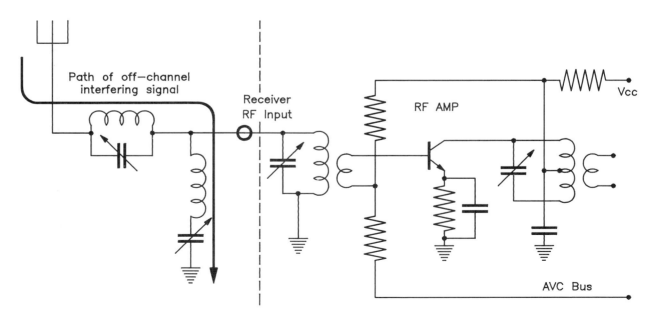

Figure 9-9 Parallel and series RF traps will help eliminate strong off-channel signals that may be causing desensitizing of the receiver

9.2.2 Received Pulse or Buzz Interference
— When a receiver receives buzzing or pulsed interference, it has a ragged sound. Such interference is common in mobile radio receiver installations. This kind of interference may be constant, or it may vary in intensity from time to time.

9.2.2.1 Identify Source
— A general rule applies to most interference problems: *It is most effective to eliminate interference at the source.* The first step is to identify where the noise is originating. It may be very local, such as interference produced by a vehicle in which the radio is installed, covered in paragraph 9.2.2.2, or it might be similar noise received by a stationary radio. A stationary radio installation sometimes receives noise from unusual sources, covered in paragraph 9.2.2.1.1.

Vehicle noise is easily identified—it stops when the vehicle engine is stopped and the vehicle is not moving. The ignition switch is the principal factor involved when tracking noise. If the noise stops instantly when the ignition switch is turned off, the ignition system is causing the problem. If the noise seems to "wind up and down" with the RPM of the engine and sounds something like a police siren, the alternator is probably the cause.

Stationary sources can be a bit more difficult to track down. A portable radio can be used to track the interfering signal to its source. Help in tracing interfering signals is sometimes available through local radio clubs. Amateur radio and citizen band radio clubs are often enthusiastic about helping with these problems. These clubs have contests based on the members' abilities to accurately track phony interfering signals. See Figure 9–10.

Figure 9–10 A group of amateur radio operators enjoying a "rabbit hunt," a search for a hidden transmitter, to sharpen their RF tracking skills and to enjoy a day in the open (Photo courtesy of the American Radio Relay League, Inc., Newington, CT)

9.2.2.1.1 Other Noise Sources
— Noise sources other than from vehicles can be difficult to track down. A portable radio will help find the source. The nearer the radio comes to the

problem, the louder the signal should get. This is simple enough with signals that are amplitude-modulated, as noise and buzzing signals usually are. When using a small AM portable radio for this purpose, it should be noticed that signals arriving at the receiver seem louder with a certain horizontal orientation. This variation in strength is due to the directional qualities of the small ferrite antenna used in these radios. See Figure 9-11.

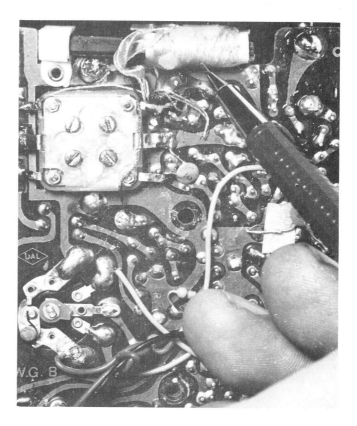

Figure 9-11 A small AM radio receiver can be used in some cases to help find RF noise sources. Note the use of the ferrite antenna, which has directional qualities

9.2.2.1.2 Find the Source —
The directional qualities of the antenna used in a portable AM receiver can be used to advantage in finding an interfering noise source. First, find the most sensitive direction of reception for the receiver. Since the antenna is probably mounted horizontally within the receiver, signals arriving from two directions should be much weaker than any other direction. This direction of greatest sensitivity should be when the antenna is broadside to the incoming signal. Use a distant signal whose direction is known to verify the sensitivity pattern of the receiver antenna. See Figure 9-12.

The weak point of reception is used because it is a far sharper indicator than the maximum points of reception. This results in a more definite direction to the interfering source.

The next step is to take a bearing on the unknown source from a single location. Using a map, draw a line indicating the direction of the signal. Draw this line for quite a distance on the map.

Next, move a considerable distance, at least several blocks, to another point from which a reading can be obtained. Again, take a reading and mark the direction on the map with another line.

The final step is to make a third move, in a direction 90 degrees from the first leg of the journey. Take another reading, drawing the indicated direction for a third time.

After these steps have been done, the three lines should approximately meet at a common point. See Figure 9-13.

330 ☐ Tracking Hum And Unwanted Signals

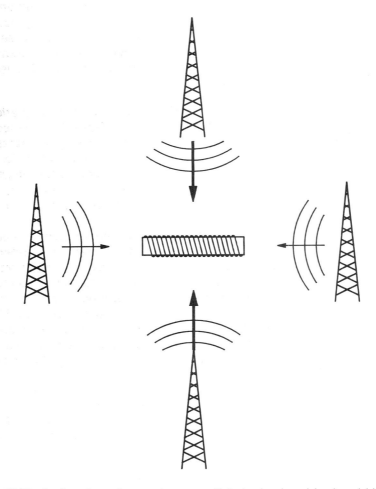

Figure 9-12 The ferrite antenna has maximum sensitivity to signals arriving broadside to the loop

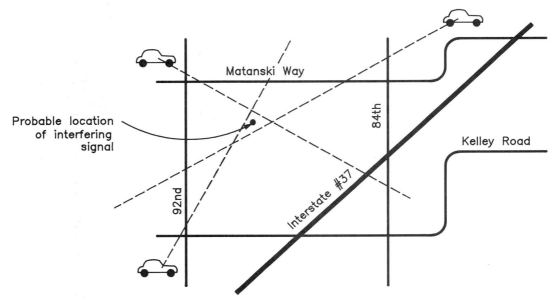

Figure 9-13 How three lines of position might indicate the source of an interfering signal

If all of the lines seem to lie in the same bearing and do not cross at a large angle, longer legs of the journey must be taken to make a good, clean cross of the lines.

Common sources for interference of the buzz, pulse, or noise interference types are power line insulators leaking high voltage during damp weather, bad power line connections on power poles, electric fences, and almost anything that causes an arc, including some electric motors and arc welding equipment.

9.2.2.1.3 Seek a Mutual Solution

Once the interfering signal is found, the persons involved should be contacted. An electric fence, for instance, could be causing a great deal of interference if it is not properly installed. Using a great deal of tact, explain the situation to the person responsible. Usually, people are glad to help, if it doesn't cost them anything. Offer what help you can, but be careful not to be too helpful and leave yourself open to criticism or worse. Often, a simple cleaning of insulators on an electric fence can cure the problem, or cutting down high weeds contacting the fence, which can cause leakage. Other situations will require appropriate cures.

If the person responsible for the interfering signal is not cooperative, the Federal Communications Commission (FCC) may be consulted for further assistance in any specific case.

The author was recently involved in a search for an intermittent interfering noise signal in a residential area. With the help of a portable marine radio direction finder receiver, the residence which was producing the heavy noise problem was easily located. The persons involved, however, were not at all cooperative. When the FCC was asked to intervene, the house was found to have a burned-up attic ventilation fan that was arcing during hot weather as the motor attempted to run. Little did the people in the house know that they were in very real danger of losing their house, partly because they refused to cooperate in cleaning up an RF noise problem!

9.2.2.2 Automotive Receiver Noise

There are two major sources of vehicle buzz and pulse noise interference; the ignition system and the charging system. Both sources of interference are present in and caused by the vehicle itself, and are therefore relatively easy to identify.

9.2.2.2.1 Alternator Noise

Alternator noise sounds like a whine or a siren. It may come through loudly into the receiver audio system. This kind of noise is being conducted into the radio on the incoming 12 VDC power line from the alternator. See Figure 9–14.

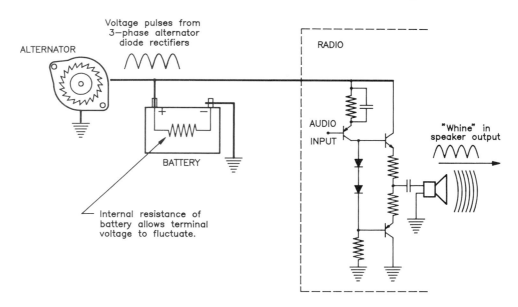

Figure 9–14 How the alternator causes audio noise on the vehicles electrical system. A major contributor is a battery with high internal resistance (worn out battery)

Alternator noise is most pronounced when the alternator is producing heavy charging current. More current is demanded of the alternator just after starting, when the battery must be charged back to normal voltage quickly. Alternator noise is also more noticeable when extra load is placed on the electrical system, as when the air conditioner and lights are turned on.

Alternator output is relatively high audio frequency, higher than that of 60 or 120 hum normal in 120 VAC-operated power supplies. The output of the alternator comes from a three-phase, full-wave bridge. See Figure 9–15.

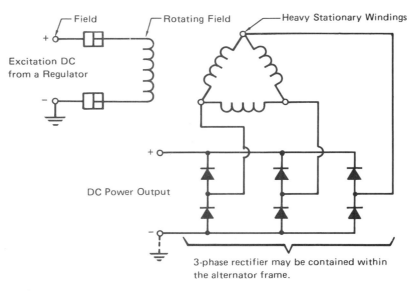

Figure 9–15 Schematic of an alternator with the six rectifying diodes mounted internal to the alternator

If any of the diodes fail by opening, the output waveform will have missing cycles, a fault that will increase the noise produced by the unit.

A storage battery with high internal resistance will also contribute to the production of noise, because it will not act as a huge filter capacitor as well, allowing the voltage on the electrical system to go higher and lower than normal, thus producing a higher voltage interference signal. Replacement of the battery may help the situation. See Figure 9–14.

9.2.2.2.2 Install Inductive, Capacitive Filtering — If the condition of the alternator and the battery are above reproach, additional filtering of the electrical system before it enters the receiver circuits may be necessary. See Figure 9–16.

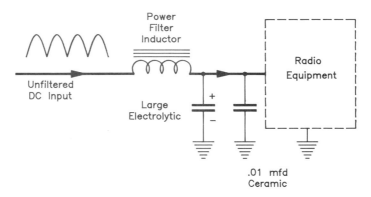

Figure 9–16 How incoming DC power can be filtered for audio and RF interference, using inductance and capacity to help reduce alternator whine

9.2.2.3 Ignition Noise
— The ignition system normally produces very high voltage that is used to produce a spark that ignites the explosive vapors within the engine. This high voltage spark does not need to have a high current to be effective in igniting the vapors, but the spark has to have sufficient voltage to jump two series gaps, within the distributor and the spark plug. If the peak current during the actual spark is high, there will be electromagnetic radiation from the spark plug wires. See Figure 9-17.

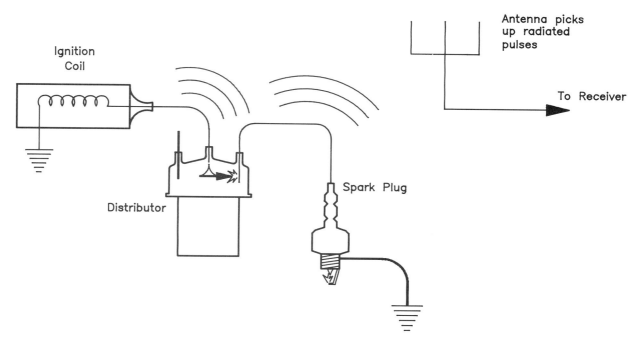

Figure 9-17 Radiation caused by the high pulsed currents carried by the high-voltage wiring of the ignition system

9.2.2.4 Install Resistor Plugs, HV Wiring, Install Shielding —
If the current in the high-voltage circuit can be reduced without seriously affecting the spark at the plug, the engine will still run as well as it did before, but the radiation from the wiring will be greatly reduced. One way to reduce the peak current is to insert resistance distributed within the wiring. This requires a special kind of wire made specifically to have resistance. This wire is sometimes an insulating fabric core saturated with carbon granules.

Another method of inserting resistance is to use spark plugs that have a resistor in series with the incoming high voltage. See Figure 9-18.

As a last resort, the high voltage and the primary or low-voltage side of the ignition coil can be shielded to prevent the radiation of interference. Although common tinfoil can be wrapped around the wiring to see if this method would help, commercial kits are available to shield wiring in critical applications.

Stubborn cases of ignition interference might be caused by noise pulses being conducted into the radio by the primary wiring. See Figure 9-19.

Conducted RF noise problems can be cured by proper RF filtering of the incoming power lead. See paragraph 9.2.3 for the principle of this kind of filtering. Filtering for RF rather than the audio components covered next will mean that the filter components should reject RF, rather than audio, on the primary winding. The values used for inductance and capacitance will be much less than those for audio filtering.

9.2.3 Transmitted Noise
— Transmission of noise, such as buzzes and whines, is often noticed after installing a new radio transmitter in a vehicle. The same sounds that are received on the radio equipment could also be transmitted, particularly the conducted noise

334 ☐ **Tracking Hum And Unwanted Signals**

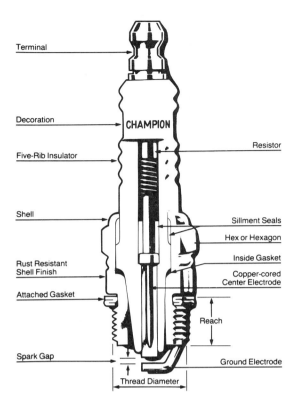

Figure 9-18 Internal structure of a resistor spark plug, designed to reduce RF interference by reducing the peak current during the spark (Photo courtesy of Champion Spark Plug Company)

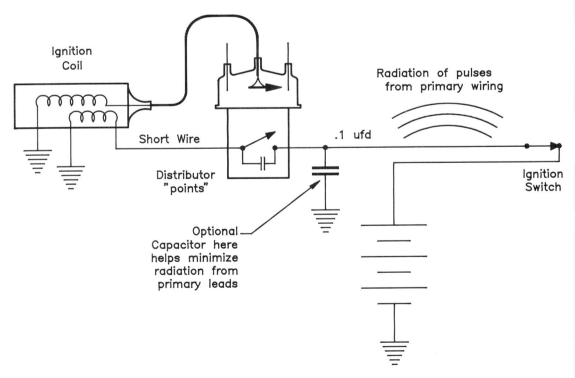

Figure 9-19 How ignition noise might be conducted into equipment via the wiring from the ignition switch

of the alternator whine. See paragraph 9.2.2.2 for suggestions on how to prevent this kind of noise problem. The cure is the same as that for received noises.

9.2.4 Interfering RF Signals Being Received — The reception of improper signals by radio receivers can come from two principal sources: those that are radiated and thus received into the receiver via the antenna, and those that are conducted into it via the incoming power lines. In either case, "extra stations" are being received.

9.2.4.1 Tune to the Interference, Remove Antenna — Tuning to the interfering station and removing the antenna is a simple test that will tell if the signal is coming into the receiver via the antenna. If the signal is still there after the antenna is removed, the signal is being conducted into the receiver via the power lines. Usually, though, the interfering signal will no longer be heard.

9.2.4.1.1 Conducted Interference — Interfering signals conducted into radio equipment on the power lines are rare, but they can happen. The interfering signals will have to be filtered out of the incoming power lines. RF bypass capacitors to ground on the incoming power line may help. See Figure 9-20. Note that the chassis of the equipment must have a good RF ground.

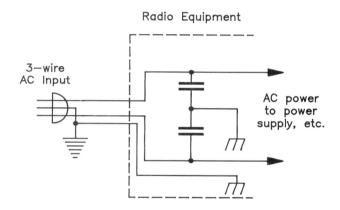

Figure 9-20 The installation of RF bypass capacitors on the incoming power lines can help reduce conducted RF interference

9.2.4.1.2 Power Company May Help — It is the responsibility of the power company to provide reliable, clean power. Stubborn cases of conducted interference can be referred to your local power company for assistance. Be sure to explain the details and what you have done to filter the problem, and your results in so doing.

9.2.4.2 Radiated Interference — If the removal of the receiver antenna also removes the interfering signal, then the interference must be coming into the radio through the antenna.

9.2.4.3 Use Better Receiver on Same Frequency — If there are very strong signals in the vicinity, there may be problems with the receiver as noted in paragraph 9.2.1, resulting in the desensitizing of the receiver. Unintended reception of signals due to receiver design problems are sometimes called spurious responses, or "spurs" for short. A simple way of determining if the receiver is at fault is to substitute a receiver of a superior design, to see if the problem is still there. This will also indicate if the signal is actually coming down the antenna, or if there is a strong signal in the area causing the receiver to internally produce the interfering signals. A superior receiver will have an FET first amplifier which will produce less "spurs" than a receiver having a bipolar amplifier.

Interference problems are often experienced by TV technicians. The nearby radiation of signals far from the normal tuning range of the TV receiver can easily overload the TV amplifiers and cause reception problems. See Figure 9-21.

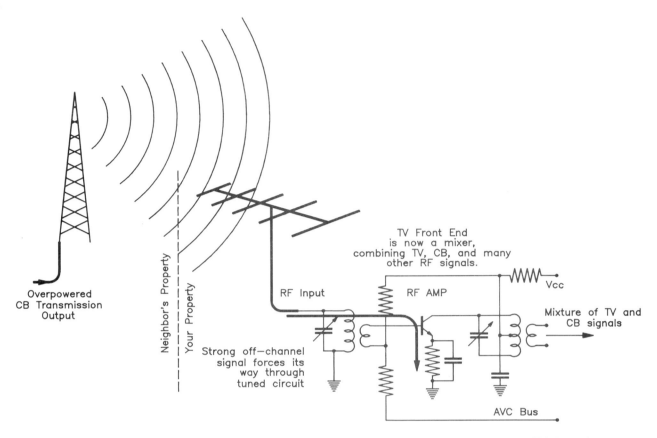

Figure 9-21 How a TV might receive CB signals, which are normally far removed from normal TV channels

9.2.4.3.1 Intermodulation Problem Indicated — A diode junction can produce sum and difference radio frequencies if an RF field is present that is strong enough to take the junction into forward conduction and if there is some means of radiating it, an antenna. Once these conditions are met, any other RF signal, strong or weak, will be mixed with the strong signal. Two additional frequencies will then be generated, the sum and difference of the two original signals. Adding a third and fourth signal will produce additional sum and difference signals in addition to all of the other signals involved. Interfering RF signals generated by this means are called intermodulation products or *intermod* for short.

9.2.4.3.2 Find the Cause, Cure the Problem — The one item that can often be controlled in an intermod interference situation is the diode junction. In free space, signals do not mix. A junction is required for the generation of the sum and difference frequencies. The junction can be located in the front end of a receiver, such as the base-to-emitter junction of a transistor, it may be external to the receiver, the result of poor bonding of tower guy lines, or it may even be generated within a transmitter final amplifier.

Signal mixing in the receiver front end — A receiver of good design can tolerate strong off-channel signals without degradation of performance. Such a receiver will have one or more sharply tuned (High "Q") circuits before the first amplifier. The amplifier component used is the critical component in generating interfering signals. It should not have the junction effect that produces interference. An FET is the best kind of first RF amplifier to use in a receiver because there is no diode junction action between an FET gate and the source element of the component.

A bipolar transistor, however, has a very definite junction, that of the base to emitter. See Figure 9-22.

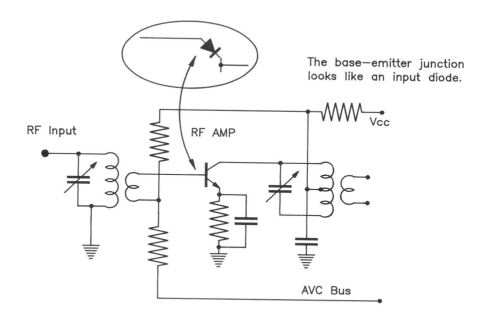

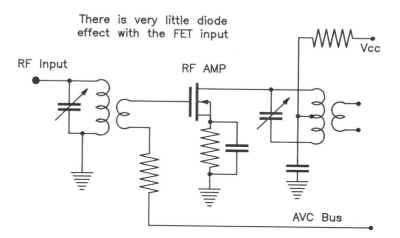

Figure 9-22 Comparison of the input circuits of the bipolar and FET RT amplifiers

If even a very high-quality receiver does not cure the interference problem, two recourses are available: install expensive, high "Q" traps, or find the cause of the generation of specific interfering frequencies.

If the receiver in question has a bipolar first RF amplifier, little can be done without redesigning the receiver. The strong RF signal that is causing the receiver to generate extra signals might be filtered out of the first stage by installing an RF trap between the antenna and the receiver.

High "Q" traps for the receiver — High "Q" traps are difficult to make for low RF frequencies. At VHF and UHF frequencies, however, such filters become reasonable. VHF and UHF traps are called resonant cavities. The dimensions of the cavity make it seem like inductance and capacitance to the RF signals. See Figure 9-23.

Figure 9-23 Resonant cavities such as these are very high "Q" tuned circuits that are used to "notch out" interfering signals where they cannot be avoided otherwise (Photo courtesy of Motorola Communications & Electronics)

These cavities find frequent use in mountain-top radio relay installations. Such installations may have many RF transmitters and receivers, often within the same building. See Figure 9-24.

Figure 9-24 This mountain-top radio installation makes use of many RF cavities to avoid interference between the various transmitters and receivers (Photo courtesy of Motorola Communications & Electronics)

A small modification may help — If the interfering signal is so strong that it can force its way into the front end of a receiver and cause problems, it is possible that adding a resistor may help the problem. This trick used to work with vacuum tubes when the incoming signal was rectified by the grid of the tube. The idea is to prevent rectification of the strong RF, which may cause desensing of the receiver (paragraph 9.2.1). We may be able to literally "throw away" the strong RF field by allowing it to drop harmlessly across a resistor rather than cause a current flow in the input circuit of the RF amplifier. See Figure 9-25.

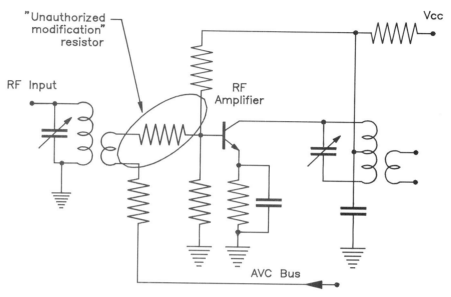

Figure 9-25 The installation of an "unauthorized modification" in the form of an extra resistor may cure intermod problems that are generated in the first RF amplifier of a receiver

The resistor used should be selected for each individual situation. Select a resistance value that is high enough to prevent the generation of the interfering signals, yet not too high in value to cause the receiver sensitivity to unduly suffer.

Signal mixing caused by poor bonding — Generation of additional frequencies can be accomplished with a simple nonlinear junction and an antenna connected to it, in the presence of a strong RF signal. See Figure 9-26.

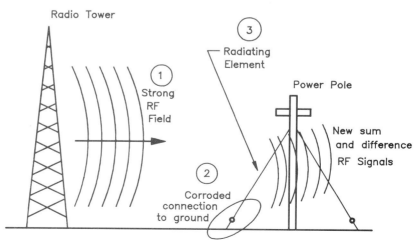

Figure 9-26 The three requirements for generating extraneous signals: a strong RF field, a nonlinear junction, and a radiating element connected to it

Since there will probably be little that can be done about the presence of the strong RF field, the next steps will be to find any source of radiation (antenna in the vicinity) and then closely examine it for nonlinear junctions.

A nonlinear junction is any junction between two conductors that conducts differently in the two possible directions. Junctions like this can occur when guy wires are put up in a sloppy manner, twisting dissimilar wires together rather than using an insulator between them. Such a junction might be a copper wire wound around an aluminum wire, for instance. Given a little time, *particularly in a marine environment with its salt-air*, corrosion problems arise. This makes a crude semiconductor between the two dissimilar metals, nonlinear by virtue of the fact that the applied voltage and current are not proportional. In this sense, the poor connection becomes a *detector*, rather than a good, low-resistance joint.

The cure for poor joints is make them 100 percent good joints or to completely isolate the metals involved with a good insulator. Environments with a high level of air pollution will also have a high percentage of problems because of the increased corrosion produced by such pollution.

Signal mixing caused by transmitter final amplifier
— A normal transmitter unavoidably produces low level "extra" signals at even multiples of the fundamental frequency of operation. Thus, a transmitter of 1250 kHz will have transmitted frequencies at 2500 kHz and 3750 kHz. The higher the multiple, the lower the power at those frequencies. These frequency amplitudes are carefully watched by the transmitting station technicians and are kept to an absolute minimum.

The generation of new additional frequencies within the transmitter final amplifier is possible under special circumstances. Mountain-top VHF and UHF radio installations are particularly susceptible to these problems because numerous transmitters are located within mere feet of each other. The transmitted signal from one transmitter is an extremely strong signal that easily enters the final amplifier of another transmitter via its antenna. See Figure 9-27.

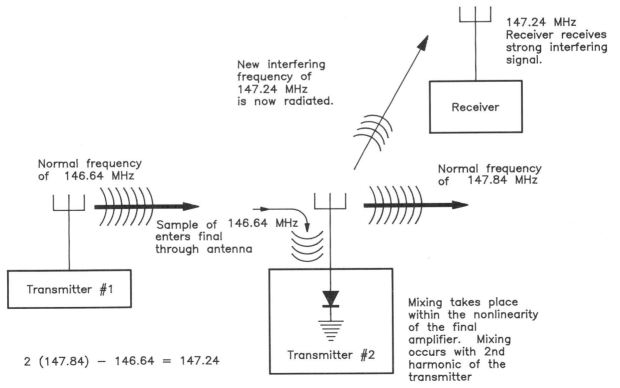

Figure 9-27 How two transmitters could cause the generation of new frequencies because of mixing within the final amplifier

The final amplifier is itself a large diode junction, since it is usually operating Class C. Now we have all three elements necessary to generate sum and difference signals: the strong RF field, the junction, and an antenna to radiate the signals.

One cure for such problems is to install very high "Q" traps to shunt RF signals traveling down the transmitting antenna to ground, not allowing them to mix in the final amplifier. Another is to install one-directional devices that allow RF to travel only one way in the transmission line, thus preventing the entrance of strongly received signals back into the final amplifier.

9.2.4.4 Poor Receiver, Get Better One — If substituting a more costly, better designed receiver cleans up the reception of unwanted signals, the cure is obvious—buy a better receiver, one that is able to survive in that relatively "dirty" RF environment. Additional external RF filtering, using tuned traps, might be effective, but the time spent in clearing up such a problem may be more expensive than a better receiver.

9A Tracking Hum and Unwanted Signals

9.3 Audio Frequency Equipment Involved — Three possibilities exist for interference problems in audio amplifiers, depending on what the sound of the interference might be: hum, squeal, or RF interference. Accurate identifiction of the kind of interference is necessary to trace the problem further.

9.3.1 Audio Amplifier Hum (60/120 Hz) — Hum interference does not change in frequency. It is a constant tone that comes, by one route or another, from the incoming power line. While 60 Hz interference can be introduced into the equipment by several means, 120 Hz interference is the result of a power supply filtering problem.

9.3.1.1 Filter Capacitors Ineffective — The filter capacitors of the power supply are supposed to charge during the peaks of the incoming voltage pulses from the rectifiers, holding a relatively constant voltage when the peak subsides. If the capacitor is not filtering because of an open circuit or because it is drying out internally and thus losing capacity, it can no longer hold a constant voltage when the charging pulse drops in amplitude between pulses. The output waveform from the power supply is no longer pure DC, but has an AC component: hum. See Figure 9–28.

9.3.1.1.1 Parallel with Good Capacitors — Connecting a good capacitor directly in parallel with a suspected open one will cause an immediate cessation of the hum if the original capacitor was indeed open. This is a rare troubleshooting instance where holding a good component in parallel with a bad one is effective. If this temporary connection removes the hum, the fix is obvious—replace the old component. Phase 5 is the next step.

9.3.1.1.2 Select Different Category — If paralleling a suspected filter cap with a good one does not substantially reduce the hum problem, then the original capacitor is doing its job. Another solution must be sought for the hum problem.

9.3.1.2 Power Supply Overloaded — If there is an excessive load on the power supply, the filter capacitors cannot sustain the normal voltage to the circuit between pulses from the rectifier. Temporarily installing a paralleling capacity across the installed filter capacitors may help, but it will not cure the problem.

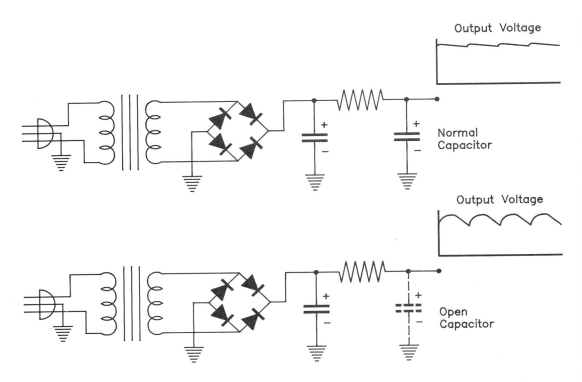

Figure 9-28 Normal and abnormal DC outputs from a power supply due to an open filter capacitor

An overload on the power supply will usually blow the fuse. Partial overload might merely cause the introduction of hum, however. *Be sure that the installed fuse is of the correct size!* Under overload conditions the transformer will probably run hot if operation continues for more than perhaps a half hour.

The cure for this kind of problem is to identify and correct the cause of the overload. Help is available beginning at point P of Chart 7A.

9.3.1.3 Open Shield on Cable

The high-gain input circuits of audio equipment are provided with a grounded shield to prevent the pickup of stray hum and noise. Any break in the ground connection of this shield immediately makes the whole cable an antenna that will gather great amounts of hum and introduce it directly into the amplifier.

The cure is to repair the cable or replace it. Repair will usually mean disassembling the connector at either end and resoldering a poor or broken ground connection.

9.3.1.3.1 Trace to Source by Shorting Signal

A special technique is used to trace unwanted signals in audio amplifiers: short the offending signal to ground to determine where it is coming from within the circuit.

Remember that a bipolar transistor can be rendered inactive without damage if the base of the transistor is directly shorted to its emitter. Shorting the base to the collector, on the other hand, will destroy the transistor in most circuits because of the excessive base current that would flow. Be sure to positively identify the transistor leads before using this technique! Since the power supply output can be either positive or negative and both PNP and NPN transistors are used, it is not safe to assume that the base of a transistor can always be shorted to chassis ground!

While using this shorting technique, keep in mind that if the hum stops while shorting a stage, that stage was necessary in passing the hum, and is therefore identified as being in the path of unwanted signal flow. The idea is to find, by going back through the circuitry, the stage where shorting the signal no longer makes a difference. See Figure 9-29.

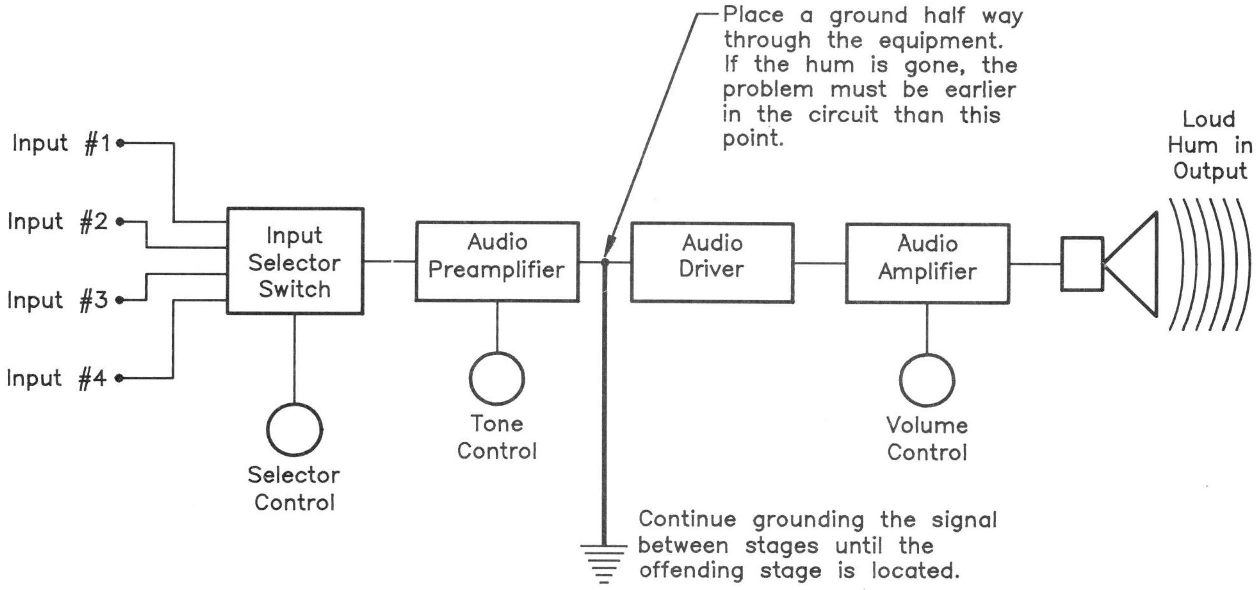

[Diagram: Block diagram showing Input #1–#4 feeding an Input Selector Switch (with Selector Control), then Audio Preamplifier (Tone Control), Audio Driver, Audio Amplifier (Volume Control), and speaker with "Loud Hum in Output."]

- Place a ground half way through the equipment. If the hum is gone, the problem must be earlier in the circuit than this point.
- Continue grounding the signal between stages until the offending stage is located.

Figure 9-29 Technique of shorting a signal to determine the stage from which it comes

9.3.1.4 Too Many Ground Points in a High Gain Circuit —
This sort of problem sometimes occurs in new high gain circuits or installations. Grounding the shield on a low-level signal cable at more than one point sets up a loop circuit that will be very sensitive to stray magnetic fields from power transformers. See Figure 9-30.

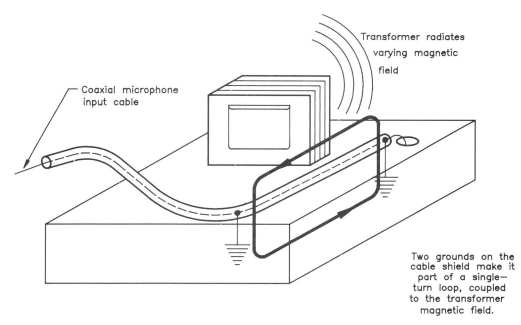

Two grounds on the cable shield make it part of a single-turn loop, coupled to the transformer magnetic field.

The single-turn loop has high current flow since it has very low resistance. This couples hum into the microphone amplifier circuits.

Figure 9-30 Example of how two grounds on a shielded cable can produce heavy currents in the shield which could then couple hum into an amplifier input

The cure for any installation that has too many grounds is to simply eliminate all but one of the grounds.

9.3.2 Audio Amplifier Squeal
— Unintentional squealing of an audio amplifier can be caused by two means: through audio feedback via a microphone, or internally to the amplifier equipment because of poor filtering in the power supply. The easiest way of determining which is causing the problem is to disconnect the microphone. Obviously, if there is no microphone in the system, the second cause must be the problem.

9.3.2.1 Acoustic Feedback
— Acoustic feedback is a term meaning that sound waves are involved in the oscillation of the amplifier system. The observer hears a squealing sound. Movement of the microphone or placing the hand over the microphone affects the oscillation frequency and intensity.

There are several things that can be done to alleviate this problem. One is to simply lower the gain of the amplifier. When the gain is low enough, the amplifier has insufficient gain to drive the oscillating loop of amplifier, sound path, microphone, and back to the amplifier. Similar effects can be accomplished by lengthening the acoustic path by moving the speakers farther from the microphone or vice-versa. Another solution is to use a microphone that is relatively insensitive in one direction, the cardioid pattern microphone. Placing the speakers within the area of least sensitivity of the microphone will enable the amplifier to operate at higher gain settings before the system oscillates.

9.3.2.2 Filter Capacitors Open; Test, Replace
— If an amplifier continues to squeal when all inputs are removed, the oscillation is occurring within the amplifier. Such a case often has an accompanying hum, too. An oscillation path of a positive feedback phase relationship exists between the power output stages and the input stages in some amplifiers. Under normal operation, oscillations cannot exist because of the filtering provided by the power supply output filter capacitors. Any fluctuation of the power supplied to the power amplifier stages cannot be allowed to enter the earlier stages of amplification. Failure of any of the filters contained in the earlier stages of the amplifier, usually caused by open or dried out electrolytic capacitors, can lead to oscillation. The first test is to parallel capacitors with new ones of a similar value and voltage rating, until the oscillation stops. The paralleled capacitor is the open one. The main filter capacitors on the power supply should also be included in a search for the open capacitor.

Once the oscillation ceases when a capacitor is paralleled, replace the original capacitor with a new one.

9.3.3 Radio Signals Entering Audio Equipment
— Audio equipment is not supposed to respond to RF signals. Very strong RF signals, however, can be received by such equipment if the interfering RF signal is of sufficient amplitude. A very strong RF signal meeting a diode junction such as the input to a bipolar transistor can produce a voltage which, when used as an input to a high gain amplifier, will be detected and produce the amplitude modulated waveform of the RF signal. This occurs in much the same manner as that shown in Figure 9-8.

If the RF source can be removed from the vicinity of the audio equipment, the problem is resolved... at least for the time being. Prevention of these problems in the future can be done either of two ways, or by a combination of both. One cure for the problem is to install small capacitors at the input of the amplifier to bypass the RF to ground, but of such a small capacity value so as not to affect the normally lower frequencies of the audio signal applied.

A second choice to eliminate RF interference to audio equipment is to install the "small modification" mentioned in paragraph 9.2.4.3.2. A series resistor in the input signal path should not affect the overall audio gain as much as it would help to reduce or eliminate the sensitivity to interfering RF signals.

Phase 9 Summary

Unique to troubleshooting texts, this phase has shown the technician some of the subtle causes of radio frequency interference. The unique short-it-out method of tracking hum and the causes and cures for audio amplifier squeal were explained. Causes of computer crashes due to external interference and the prevention of further problems was discussed. Armed with this information, the technician can confidently tackle problems that might appear to be equipment problems and which are, in reality, not problems within the equipment at all. They may be problems caused by something *entirely external* to the malfunctioning equipment.

Review Questions For Phase 9

1. Name two common sources of interfering signals to a computer.
2. What is the best protection for a small computer system for over- and undervoltage, transients and power loss?
3. What is radio receiver desense?
4. What is an RF trap?
5. What is the most effective way to eliminate noise interference problems?
6. What are the two most common sources of noise interference in a mobile radio?
7. Name three methods of reducing ignition-type pulsed interference.
8. What simple method can be used to see if interference is coming into a radio receiver via the antenna?
9. By what means other than the antenna might an interfering signal be introduced into a receiver?
10. What elements are necessary to produce intermodulation problems?
11. How might an inexpensive receiver produce intermodulation problems?
12. You suspect that the filter capacitors of an audio amplifier are not filtering as they should. What simple test can be made to see if a different capacitor might help?
13. What special troubleshooting technique can be used to find the source of an unwanted squeal in an amplifier?
14. How can an audio amplifier receive RF radiation?

PHASE 10
REPAIRING PROFOUND FAILURES

☐ PHASE 10 OVERVIEW

Profound failures are those where the damage is more extensive than the failure of one or two components. Reading the major headings of the flowchart will show examples of some of these big problems, and what the technician can do about them. The technician must remember that because of the amount of time required to make repairs, it is often advisable to discard the equipment rather than to attempt repair.

10.1 IS IT WORTH YOUR TIME TO REPAIR?

— The failures discussed in this phase can, by their very nature, involve a great deal of damage and thus require much of the technician's time to correct. In these modern times, it is often more economical to discard rather than to repair. Multiply this by the number of failures that might be present because of a major disaster to one piece of equipment, and the result is that only rarely will a technician go to the great lengths required to repair the equipment involved.

If it is deemed necessary for any reason that the equipment be repaired anyway, the technician should begin with a careful visual inspection and the replacement of any obviously damaged parts. The troubleshooting then begins in earnest.

10.1.1 Fire and Smoke Damage

— While smoke by itself is not very damaging to electronic equipment, the high temperatures reached during a fire can of course ruin it. Some indication of the amount of heat sustained can be indicated by the condition of any plastic parts on the equipment. If plastic parts are still intact, there is a chance the technician may be able to salvage the equipment. If the plastic has melted or is gone, there is very little hope of restoring the equipment to service with any reasonable amount of work.

Most smoke can be removed using a household cleaning agent such as "409," applying a small amount to a cloth and carefully wiping down the equipment. Severe cases may require more prolonged exposure to the cleaner.

10.1.2 Dropped Equipment

— The mechanical portions of equipment suffer most when equipment is dropped. As a general rule, the electronics parts are not much affected. A portable radio, for instance, may work quite well after being dropped, but the cabinet may be a mess. Equipment that is largely mechanical is more likely to be damaged beyond repair. Audio tape recorders and video cassette players have intricate mechanisms and are therefore easily ruined by dropping. Purely electronic items such as calculators and computers can survive surprisingly heavy blows without affecting the operation of the equipment.

The heaviest components of equipment are the ones most likely to be broken loose when dropped. Batteries and transformers are candidates for breaking loose and perhaps causing additional damage while adrift. Repair of this sort of damage is obvious—just remount the components. Plastic mountings can sometimes be adequately repaired by using plastic or "super" glue.

10.1.3 Overvoltage or Reversed Polarity Applied to Circuit —

The application of too much voltage to a circuit is often caused by a technician while working on circuitry. Inattention to the settings of a power supply before turning it on and the slip of a dull probe during troubleshooting are two common causes of this sort of major damage. Little can be said about the application of external power to equipment, other than *be careful to avoid the application of too much voltage or reversed input.* It is up to the technician to be very careful to avoid this kind of expensive "accident."

There is little excuse for using dull test probes, too. A dull test probe, in accordance with Murphy's law, will slip when it can do the most damage. If there is 12 V and 5 V used in the same equipment, for instance, a slip and accidental short between them can apply the 12 V source to components designed only for 5 V. The moral to this story is *use only test probes with sharp points* that will not slip when applied to the circuit.

10.1.3.1 Determine Likely Damaged Components —
If overvoltage or reversed polarity is applied to a circuit, certain components are more likely to be damaged than others. Semiconductors of any kind, including integrated circuits, and electrolytic capacitors are generally the first components to go. Look for these components and then look for other components that may have protected them. For example, a transistor with high values of base and collector resistance to the normal Vcc supply will not be as readily damaged as those that are connected without such buffering. See Figure 10-1.

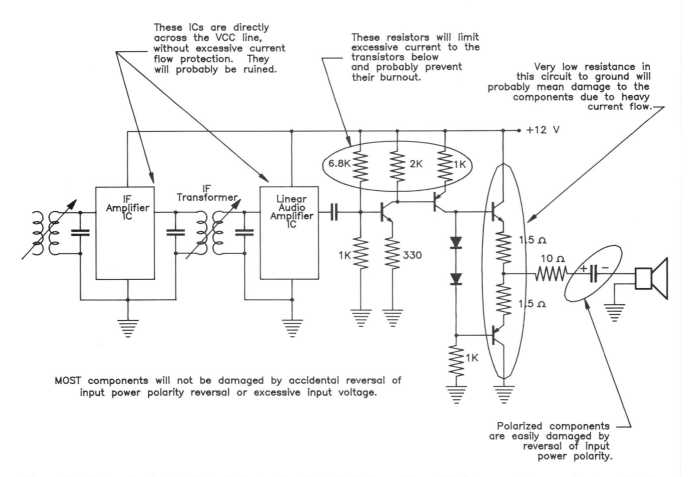

Figure 10-1 Schematic of a typical circuit showing the effect of buffering resistors which help prevent damage due to accidental overvoltage or the application of reversed voltage polarity

Based upon the likelihood of damage, use an ohmmeter or solid state tester to test suspected components for damage while the components are still in the circuit. See Phase 6 for further assistance.

10.1.3.2 Get as Much Documentation as Possible

— The amount of work involved to repair a multitude of problems demands the availability of proper documentation for the equipment. This is not the time for guessing. It may be necessary to trace signals and voltages, over and over, through the equipment.

10.1.3.3 Use Methods Beginning with Phase 6

— Phase 6, Dead Circuit Troubleshooting, is the starting point to begin troubleshooting the circuit. Check only the major components, replacing as necessary.

10.1.3.4 Use Methods Beginning with Phase 2

— Once the major components of the board are eliminated as possible problems, it is time to fire up the circuit and begin troubleshooting it, stage by stage, until the bad components are identified. Reviewing Phase 2 several times may be required before all of the problems are resolved.

10.1.3.5 Test, Replace One at a Time

— As each defective component is discovered and replaced, the overall operation of the equipment must be evaluated to see if there is further damage. If so, the troubleshooting again reverts to the beginning of Phase 2.

Once all of the problems are resolved, the end of Phase 5, point "N," gives details to complete the job.

10.1.4 Lightning Damage

— Lightning damage can cause any or all of the components in any piece of equipment to be damaged completely. A nearby hit of lightning on a power pole, for instance, can ruin all of the equipment in the immediate vicinity because the tremendous voltage surge is conducted into all of the nearby buildings without substantial attenuation. Any equipment on the circuits must be written off as damaged due to an "act of God." Insurance is the only practical help in such a case, assisting in the complete replacement of the equipment.

Lightning strikes farther away may do relatively little damage. The MOVs in the equipment may help, but there is a good chance that they have destroyed themselves by shorting. In this case, they will cause the equipment fuses to blow. Replacement of the MOVs and the fuse is indicated, of course.

Other damage to equipment due to a nearby strike can be traced using the standard troubleshooting techniques beginning with Phase 2.

10.1.5 Submerged Equipment

— Contrary to first insight, sinking electronic equipment in water doesn't necessarily damage it. It takes time for water to seep into some of the components, and two other major factors have a great deal to do with the resulting amount of damage: whether or not the power is turned on while underwater, and the kind of water involved, fresh or salt water.

10.1.5.1 Stop Further Damage, Rinse in Warm Fresh Water

— If the equipment is submerged in salt water, the very first thing to do is to thoroughly rinse the equipment in fresh water. *It is important to get the salt water off the equipment as quickly as possible, because salt-water corrosion progresses even if the equipment is apparently dry.*

If the fresh water rinse can be made warm, so much the better, as warm water is a better solvent to carry away the salts on the equipment. Batteries, if any, should be immediately removed and probably discarded. At least a few hours should be allowed for the salts to dissolve. Change the water now and then to help carry them away. Occasional agitation of the water is also a help in dislodging salt particles.

After thorough rinsing, the equipment should be allowed to drain and to dry.

Evaluation of the damage done by the immersion should consider at least these factors:

1. Did the equipment have power applied during immersion?
2. How long was the equipment submerged?
3. What kind of water—fresh or salt—was involved?
4. Results of a visual inspection
5. Original cost of the equipment
6. Value of equipment after depreciation
7. Can future failures because of submersion be tolerated?

10.1.5.2 Consult Owner, Insurance Company — The owner and the insurance company involved will make the decision at this point as to whether or not the equipment will be repaired or if it will be scrapped and replaced with new equipment. The technician should not be too anxious to attempt repair until the appropriate decisions are made.

10.1.5.3 Replace Components Susceptible to Water Damage — Most of the components used today are relatively immune to the direct effects of immersion. The components that are likely to be damaged beyond repair are meters, switches of all kinds, potentiometers, connectors, and mechanical assemblies. Plated assemblies such as waveguides are easily corroded and will probably require replacement. Transformers must be thoroughly dried out before applying power.

Phase 10 Summary

This phase has provided guidance in deciding whether or not to attempt the repair of a probable series of individual problems stemming from a common mishap. Once the decision to go ahead with the repair has been made, Phase 2 provides the principal entry point for further repairs. Several loops through Phase 2 may be necessary to revive severely damaged equipment.

Review Questions for Phase 10

1. What is most important about repairing equipment that has been dropped into salt water?
2. What component is most likely to be damaged, if present, during a nearby lightning strike?
3. What kind of damage is most likely when solid-state equipment is dropped?
4. What might be used as an indicator of the amount of heat sustained by equipment in a fire?

APPENDIX A
THE FRUSTRATION LIST

This checklist may help the technician nearing "the end of the rope" on a job. Sometimes, it is best to stand back and look at the whole picture and consider things you may be overlooking. Use this list when all else fails, and you may come up with the solution.

1. Have the batteries of either the equipment under test or of the test instrument gone dead? Are the battery connectors clean and making positive contact to the batteries? Beware of an intermittent battery connector!
2. Check the test equipment test leads and/or RF cables for intermittent opens that may have occurred since starting the job.
3. Did you change ranges on some of the equipment under test or the test instruments and forget to change them back?
4. Try substituting known good test instruments for the test instruments in use at the moment. The test instrument may have failed. Try the same with the equipment under test. Remove the suspected unit and substitute a known good piece of equipment to verify the symptoms you are getting.
5. If you have access to known good equipment of the same type as the defective one, try comparing readings from good to bad unit to help in isolating the problem in the bad. Apply power to both units and take readings back and forth between them.
6. The problem you are looking for in a DC circuit may not be evident as a DC problem at all — it may be a totally unexpected RF problem that will not be noticeable with DC readings. This principle might apply to low frequency audio problems when unintentional RF is present.
7. Get away from the problem for a while. Sometimes you get too involved and begin making mistakes.
8. When you go back to the job, make notes on paper as you go. This will assist you in keeping things straight in your head. This is a particularly good idea when you are taking many voltage readings and have been trying to remember them and they've become confusing.
9. Consider asking the help of another technician. No one has all the answers, no matter how experienced. You could be overlooking something very simple that another person could point out.
10. Start your troubleshooting all over again. Something may have changed as you were working and that could scramble any logical approach you are trying to maintain.
11. Consider the possibility that the book you are using is wrong. Trying to trace a problem that isn't there is very nonproductive!

APPENDIX B
ELECTRONIC SCHEMATIC SYMBOLS

NAME OF DEVICE	CIRCUIT SYMBOL	COMMONLY USED JUNCTION SCHEMATIC	ELECTRICAL CHARACTERISTICS	MAJOR APPLICATIONS
GE-MOV® Varistor		WIRE LEAD / ELECTRODE / INTERGRANULAR PHASE / ZINC OXIDE GRAINS / ELECTRODE / EPOXY ENCAPSULANT / WIRE LEAD	When exposed to high energy transients, the varistor impedance changes from a high standby value to a very low conducting value, thus clamping the transient voltage to a safe level.	Voltage transient protection High voltage sensing Regulation
Diode or Rectifier	ANODE / CATHODE	ANODE / n / p / CATHODE	Conducts easily in one direction, blocks in the other.	Rectification Blocking Detecting Steering
Tunnel Diode	POSITIVE ELECTRODE / NEGATIVE ELECTRODE	POSITIVE ELECTRODE / p / n / NEGATIVE ELECTRODE	Displays negative resistance when current exceeds peak point current I_p.	UHF converter Logic circuits Microwave circuits Level sensing
Back Diode	ANODE / CATHODE	ANODE / n / p / CATHODE	Similar characteristics to conventional diode except very low forward voltage drop.	Microwave mixers and low power oscillators
n-p-n Transistor	COLLECTOR / BASE / EMITTER	COLLECTOR / n / p / n / BASE / EMITTER	Constant collector current for given base drive.	Amplification Switching Oscillation
p-n-p Transistor	COLLECTOR / BASE / EMITTER	COLLECTOR / p / n / p / BASE / EMITTER	Complement to n-p-n transistor.	Amplification Switching Oscillation
Unijunction Transistor (UJT)	BASE 2 / EMITTER / BASE 1	BASE 2 / EMITTER / BASE 1	Unijunction emitter blocks until its voltage reaches V_p; then conducts.	Interval timing Oscillation Level Detector SCR Trigger
Complementary Unijunction Transistor (CUJT)	BASE 1 / EMITTER / BASE 2	BASE 1 / EMITTER / BASE 2	Functional complement to UJT	High stability timers Oscillators and level detectors
Programmable Unijunction Transistor (PUT)	ANODE / GATE / CATHODE	ANODE / GATE / CATHODE	Programmed by two resistors for V_p, I_p, I_v. Function equivalent to normal UJT.	Low cost timers and oscillators Long period timers SCR trigger Level detector
Photo Transistor	COLLECTOR / BASE / EMITTER	COLLECTOR / n / p / n / BASE / EMITTER	Incident light acts as base current of the photo transistor.	Tape readers Card readers Position sensor Tachometers

Reprinted with permission Microwave Products Department, General Electric Company, Owenburo, Kentucky

NAME OF DEVICE	CIRCUIT SYMBOL	COMMONLY USED JUNCTION SCHEMATIC	ELECTRICAL CHARACTERISTICS	MAJOR APPLICATIONS
Opto Coupler 1) Transistor 2) Darlington Outputs			I_C vs V_{CE} curves $I_{F1}, I_{F2}, I_{F3}, I_{F4}$	Output characteristics are identical to a normal transistor/Darlington except that the LED current (I_F) replaces the base drive (I_B).	Isolated interfacing of logic systems with other logic systems, power semiconductors and electro-mechanical devices. Solid state relays.
Opto Coupler SCR Output			$I_{ANODE}(+)$ vs $V_{ANODE}(+)$	With anode voltage (+) the SCR can be triggered with a forward LED current. (Characteristics identical to a normal SCR except that LED current (I_F) replaces gate trigger current (I_{GT}).	Isolated interfacing of logic systems with AC power switching functions. Replacement of relays; micro-switches.
AC Input Opto Coupler			I_C vs V_{CE} curves $I_{F1}, I_{F2}, I_{F3}, I_{F4}$	Identical to a "standard" transistor coupler except that LED current can be of either polarity.	Telecommunications — ring signal detection, monitoring line usage. Polarity insensitive solid state relay. Zero voltage detector.
Silicon Controlled Rectifier (SCR)	ANODE, GATE, CATHODE	ANODE, CATHODE, I_g, GATE	ANODE I vs $V_{ANODE}(-)$ / $V_{ANODE}(+)$	With anode voltage (+), SCR can be triggered by I_g, remaining in conduction until anode I is reduced to zero.	Power switching, Phase control, Inverters, Choppers
Complementary Silicon Controlled Rectifier (CSCR)	ANODE, GATE, CATHODE	ANODE, GATE, CATHODE	ANODE I vs $V_{AC}(-)$ / $V_{AC}(+)$	Polarity complement to SCR	Ring counters, Low speed logic, Lamp driver
Light Activated SCR*	ANODE, GATE, CATHODE	ANODE, CATHODE, I_g, GATE	ANODE I vs $V_{ANODE}(-)$ / $V_{ANODE}(+)$	Operates similar to SCR, except can also be triggered into conduction by light falling on junctions.	Relay Replacement, Position controls, Photoelectric applications, Slave flashes
Silicon Controlled Switch* (SCS)	ANODE, CATHODE GATE, ANODE GATE, CATHODE	ANODE, CATHODE GATE, ANODE GATE, CATHODE	ANODE I vs $V_{ANODE}(-)$ / $V_{ANODE}(+)$	Operates similar to SCR except can also be triggered on by a negative signal on anode-gate. Also several other specialized modes of operation.	Logic applications, Counters, Nixie drivers, Lamp drivers
Silicon Unilateral Switch (SUS)	ANODE, GATE, CATHODE	ANODE, GATE, CATHODE, R_B	$I_{ANODE}(+)$ vs $V_{ANODE}(+)$	Similar to SCS but zener added to anode gate to trigger device into conduction at ~ 8 volts. Can also be triggered by negative pulse at gate lead.	Switching Circuits, Counters, SCR Trigger, Oscillator
Silicon Bilateral Switch (SBS)	ANODE 2, GATE, ANODE 1	ANODE 2, GATE, ANODE 1, R_B	$I_{ANODE\ 2}$ vs $V_{ANODE\ 2}(-)$ / $V_{ANODE\ 2}(+)$	Symmetrical bilateral version of the SUS. Breaks down in both directions as SUS does in forward.	Switching Circuits, Counters, TRIAC Phase Control
Triac	ANODE 2, GATE, ANODE 1	ANODE 2, GATE, ANODE 1	I vs $V_{ANODE\ 2}(-)$ / $V_{ANODE\ 2}(+)$	Operates similar to SCR except can be triggered into conduction in either direction by (+) or (−) gate signal.	AC switching, Phase control, Relay replacement
Diac Trigger			I vs V	When voltage reaches trigger level (about 35 volts), abruptly switches down about 10 volts.	Triac and SCR trigger, Oscillator

Reprinted with permission Microwave Products Department, General Electric Company, Owenburo, Kentucky

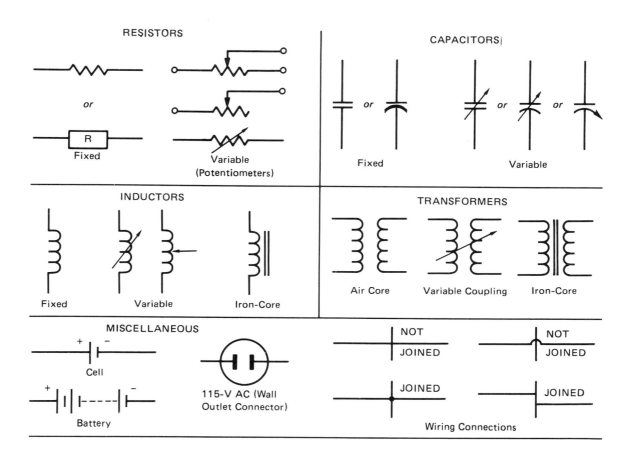

356 ▢ Electronic Schematic Symbols

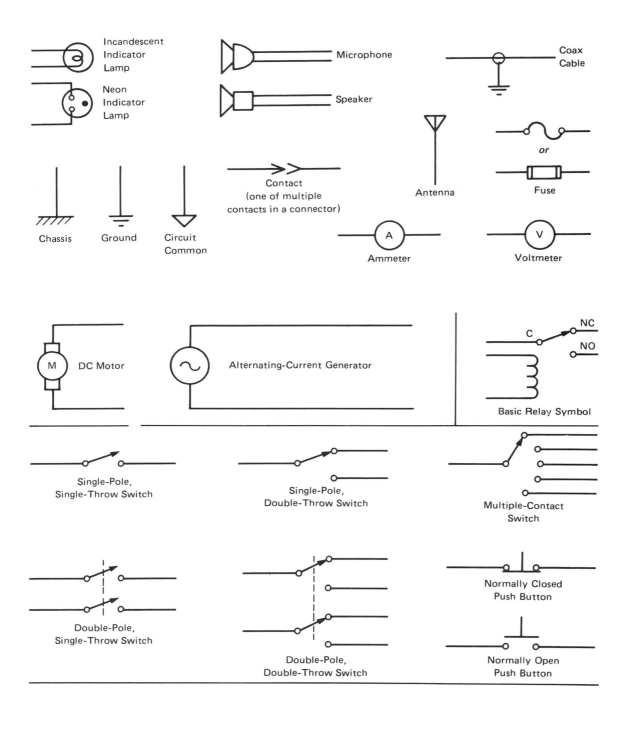

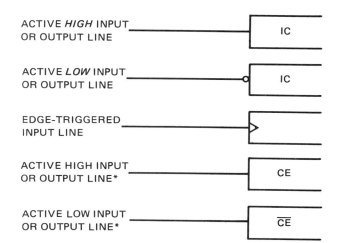

DIGITAL SYMBOLS

ACTIVE *HIGH* INPUT OR OUTPUT LINE	IC
ACTIVE *LOW* INPUT OR OUTPUT LINE	─o| IC
EDGE-TRIGGERED INPUT LINE	▷
ACTIVE HIGH INPUT OR OUTPUT LINE*	CE
ACTIVE LOW INPUT OR OUTPUT LINE*	$\overline{CE}$

*CE (chip enable) is an example of a lettered IC pin designation. The presence or absence of a bar over the letters determines the active state of the pin.

Gate	OR GATE	A	B	Out
		0	0	0
		0	1	1
		1	0	1
		1	1	1

Gate	AND GATE	A	B	Out
		0	0	0
		0	1	0
		1	0	0
		1	1	1

Gate	INVERTER	A	Out
		0	1
		1	0

Gate	NOR	A	B	Out
		0	0	1
		0	1	0
		1	0	0
		1	1	0

Gate	NAND	A	B	Out
		0	0	1
		0	1	1
		1	0	1
		1	1	0

Gate	EXCLUSIVE-OR (X-OR)	A	B	Out
		0	0	0
		0	1	1
		1	0	1
		1	1	0

Gate	EXCLUSIVE-NOR (X-NOR)	A	B	Out
		0	0	1
		0	1	0
		1	0	0
		1	1	1

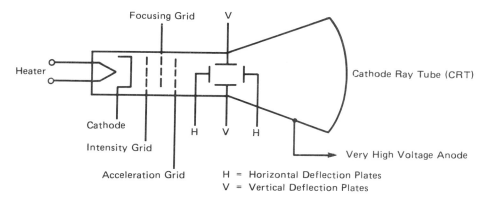

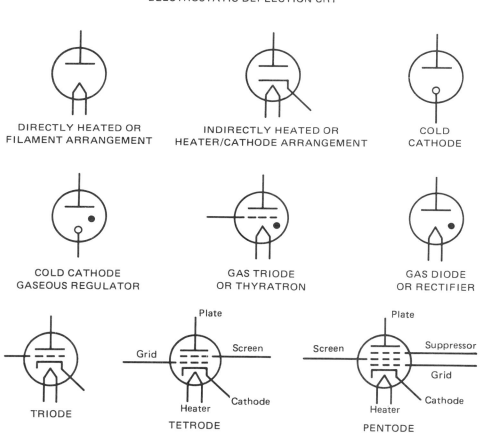

APPENDIX C Part 1
PART IDENTIFICATION PREFIXES

These part-number prefixes may help you in determining the manufacturer or supplier of original parts. The information contained here is believed to be correct, however, no liability to its correctness is assumed by RCA Corporation.

Prefix	Manufacturer
001-	Hammond
010-	Ampex
019-	Hallicrafter
020-	Scott
050-	Hitachi
051-	Automatic
057-	Hitachi
075-	Hammond, Heathkit
1A	Automatic Radio
1N	Diode or rectifier, JEDEC number.
1S	Japanese Rectifier or Diode
1T	Japanese Rectifiers
1W, 1X, 1Z	Farfisa
2A	See G.E. 4JX2A Series
2D	Kimball
2N	Transistor or SCR, JEDEC number.
2S, 2SA	Japanese Transistors
2SB, 2SC	Japanese Transistors
2SD, 2T	Japanese Transistors
3L4-	Philco
3N	Field-Effect Transistor (FET)
4-	Simpson Sears
4C, 4D	General Electric
4JX	General Electric
5E	See General Electric 4JX5E Series
6-, 6A, 6D-, 6L	Automatic Radio
6MC	Automatic Radio
6RS	General Electric
7-	Standel
7A	General Electric, Sherwood
8-	Blaupunkt
8D, 8H, 8L, 8P	Automatic Radio
13-	Sylvania
14-	Electrophone, Philco
16-	Symphonic
16A-, 16E-	See General Electric 4JX16A/16E Series
20-	Philco
20A	Standel, Gibson
24T-	Standard Kollsman
28-	Electrophone
31-	Philco
32-	Philco, Sylvania
33H	Philips
34-	Philco, Marantz
35-	Marantz, Autolite
40-, 41-	Acoustech
46-	Philco, Simpson Sears
48A, B, C, D, K, P, S	Motorola
51S-	Motorola
53A, 53B, 53C, 53E	Magnavox
56, 57-	Heathkit, Admiral
58	Admiral
62-, 63-	RCA
63B, 63C, 63J	Magnavox
66F, 69SP	Fleetwood
69A, 69SP	Standel
73C	General Electric
76	Philco
77-, 78-	Clairtone
79F	Fleetwood
86	Fleetwood, Warwick
85-5000 Ser.	Thomas
88A/B/D, 89B	Automatic
90-, 91-	Imperial
91A	Standel
93	Admiral
93SE-	Clairtone
96-	Blaupunkt
101-	RCA
103-	Zenith
104-	RCA
115-, 118-	RCA
117-	Heathkit, RCA
120-	Hallicrafter
121-	Zenith
132-	McIntosh
156	Westinghouse
173	General Electric
207A	Argos
209-	Dual
212-	Zenith
221-	Zenith
229-	Philco
236-	Allen Organ
247-	Heathkit
294-	Fanon
295	Westinghouse
296	Fanon, Westinghouse
297	Westinghouse
324-, 325-, 344-	Philco
353-	Sony
378-	Dynakit
417-	Heathkit
421-	Zenith
431-	Harmon Kordon
442-, 443-	Heathkit
462-, 464-, 465, 469-	Nordmende
571-, 572-, 574-	Dynakit
624-	Sylvania
690-	Westinghouse
800-	Clairtone, Zenith
880-	Clairtone

Electronic Troubleshooting □ 361

Part Identification Prefixes

919-	Lowery, Farfisa, Maestro	50,000 Series	Conn, Ford	801,000 Series	Gulbransen
964-	Zenith	56,000 Series	Leece Neville	960,000 Series	Fleetwood
991-, 992-	Lowery	58,000 Series	Leece Neville	980,000 Series	Fleetwood
1410-, 1412-, 1414-	Rogers	60,000 Series	Ford, RCA	1,810,000 Series	Simpson Sears, Sears
2057B	Admiral	80,000 Series	RCA	2,000,000 Series	Bendix
		93,000, 94,000, 95,000 Series	Eico		
2093B	Admiral	100,000 Series	RCA	2,300,000 Series	Hitachi
2400 Series	RCA	161,000 Series	Toshiba	4,000,000 Series	Bendix
3400 Series	RCA	171,000 Series	Toshiba	7,570,000 Series	Ampex
3500 Series	RCA	199,000 Series	Toshiba	23,000,000 Series	Toshiba
3600 Series	RCA	530,000 Series	Fleetwood		
3900 Series	RCA	570,000 Series	Simpson Sears, Sears, Dynaco		
4822-	Philips	600,000 Series	Sharp		
7000 Series	RCA	612,000 Series	Magnavox		
30,000 Series	RCA	800,000 Series	Emerson		
40,000 Series	RCA				

A	See 2SA	CDT	Clevite	KR	Craig
A	General Electric Rectifier	CS	Continental, Simpson Sears	LA, LD, LH LM, LP	Sanyo IC
A	Amperex, Baldwin, Fisher, Nivico	D	General Electric, Sylvania	LM	National IC
A514-	Baldwin	DHD	Japanese Rectifier	M4000 Series	Motorola
AA	European Diodes	DM	National IC	M5000 Series	Mitsubishi Sony IC
AC, AD	European Transistors	DS, DT	General Motors	M7000 Series	Motorola
AF, AI, AL	European Transistors	DTG, DTS	General Motors	M8000 Series	Toshiba
AMD	Philips	E	Fleetwood, Nivico	M9000 Series	Toshiba
AN	Panasonic, Matsushita	ECG	Sylvania	MC, MU, MM, MDA	Motorola
AR, AQ	Philco	EA, ER, ET, EV	General Electric	MFC	Motorola
AT	Philco	EP, ES, EV	General Electric	MM	Motorola, Toshiba
AS, AU, AX	European Transistors	EX	Fisher	MP	Motorola, Philco
		FA	Hitachi IC	MRF	Motorola
AW01	Hitachi Zener Diode	FJ	Philips IC	MT	General Electric
		FR	Japanese Rectifier	MV	Motorola
B	See 25B	FS	Fairchild- See S No.		
B	Bendix Farfisa			N	Signetics IC
		FCS	See CS	OA	European Diode
BA, BB	European Diodes	GB	Philco	OC	European Transistors
BC, BD, BF	European Transistors	GC	Texas Instrument		
		GE	General Electric	P	Lowery
BCM	Philco	GD	Philco	PA	Fisher, Philco, General Electric IC
BF	European Transistors	GM	Japanese UHF Transistors		IC
BN	Maestro, Lowery	GP	Texas Instrument	PC	Craig
		GT	General Instrument	PD	General Electric IC
BR, BS	European Transistors	HA, HB, HC	Simpson Sears, Lloyds, Sanyo Hitachi	PE, PL	Philco IC
BT	European SCR			PM	Motorola
BU	European Transistors	HD, HE	Admiral, Lloyds	PR	See Philco AR
BY	European Rectifiers	HEP	Motorola	PET	See Philco
BZ	European Zener Diodes	HF, HR	Philco	PH	European Rectifiers
		HS	Lloyds	PT	European
		HT	Admiral	QA, QP, QRS	Scott
C	See 2SC	HV	Admiral, Philco	R	General Electric, Texas Instrument
C	General Electric SCR	HX	Lloyds		
CA, CD	RCA IC. G.E.	J	Philco	RC	Raytheon
CD, CJ, CS	Simpson Sears, Scott	JC	Lloyds	RD	Japanese Diodes
		JT	Philco	RE	Raytheon
CDC	Monarch	KB	RCA		

RM	Raytheon	SGB	Standel	TAA, TAD	Philips, SGS IC
RR, RS	General Electric, Craig	SIS, SID, SIT	Fisher	TD	General Electric, Toshiba
S	Fairchild, Syntron	SH	Japanese Rectifier	TE, TH	Sanyo
SA, SB, SD	Japanese types— See 2SA, 2SB, etc.	SK	RCA, Motorola, Texas Instrument	TI, TIS, TIP, TIM, TIX	Texas Instrument
SA	Standel	SKA	Texas Instrument	TQ	Sanyo
SAJ	ITT, IC	SL	Motorola	TR, TRO	Fisher
SC	Motorola, Texas Instrument	SM	Texas Instrument	TS	
SC	General Electric SCR	SN	Texas Instrument IC, Japanese Rectifier	TV	Sanyo, Philco
SD	Japanese Rectifiers, Philco	SP, SPS, SPF	Motorola	TVC	Sprague
SDT, SES	Solitron, Standel	SR, SS, SL	Motorola	UA, UL	Fairchild IC
SE	Fairchild	STK	Sanyo IC Modules	ULN, ULX	Sprague IC
SF	Texas Instrument	SV	Transitron	UT	Panasonic IC
SFC	Motorola	SX	Texas Instrument	TZ	Sprague
SFT	Farfisa, Simpson Sears	T	Philco, Motorola	V	Benco, Fisher
SG	Japanese Rectifier, Telefunken	TA	RCA, Toshiba	VS	Japanese Rectifier
				XB, XC	Clairtone, Motorola
				X16	General Electric
				ZR	Fisher

APPENDIX C Part 2
MANUFACTURER'S LOGO DIRECTORY

LOGO	MANUFACTURER/ADDRESS	LOGO	MANUFACTURER/ADDRESS
ADVANCED ANALOG A Division of intech	Advanced Analog 2270 Martin Avenue Santa Clara, CA 95050 (408) 988-4930		Data General Corp. 4400 Computer Drive Westborough, MA 01580 (617) 366-8911
AM	Advanced Micro Devices P.O. Box 3453 Sunnyvale, CA 94088 (408) 732-2400	DATEL	Datel 11 Cabot Boulevard Mansfield, MA 02048 (617) 339-3000
TFK	AEG-Telefunken Corp. P.O. Box 3800 Somerville, NJ 08876 (201) 722-9800	EG&G RETICON	EG&G Reticon Corp. 345 Potrero Avenue Sunnyvale, CA 94086 (408) 738-4266
ANALOG DEVICES	Analog Devices One Technology Way Norwood, MA 02062-9106 (617) 329-4700		EXAR Integrated Systems 2222 Qume Drive San Jose, CA 95161 (408) 732-7970
ANALOGIC	Analogic Corporation 8 Centennial Drive Peabody, MA 01961 (617) 246-0300	FERRANTI	Ferranti 87 Modular Avenue Commack, NY 11725 (516) 543-0200
BECKMAN	Beckman Instruments 2500 Harbor Boulevard Fullerton, CA 92634 (714) 773-8603	FUJITSU	Fujitsu Microelectronics, Inc. 3545 North First Street San Jose, CA 95134 (408) 922-9000
BURR-BROWN BB	Burr-Brown P.O. Box 11400 Tucson, AZ 85734 (602) 746-1111	GENERAL ELECTRIC SOLID STATE	General Electric Solid State 724 Route 202, P.O. Box 591 Somerville, NJ 08876-0591 (201) 685-6000
CALIFORNIA MICRO DEVICES MICROCIRCUITS DIVISION	California Micro Devices 2000 W. 14th Street Tempe, AZ 85281 (602) 968-4431	GENERAL INSTRUMENT MICROELECTRONICS	General Instrument 600 W. John Street Hicksville, NY 11802 (516) 933-9000
CHERRY SEMICONDUCTOR	Cherry Semiconductor Corp. 2000 South County Trail East Greenwich, RI 02818-0031 (401) 885-3600	GoldStar GOLD STAR CO.,LTD.	Goldstar Semiconductor Ltd. 1130 E. Arquez Avenue Sunnyvale, CA 94086 (408) 737-8576
C=	Commodore Semiconductor Group 950 Rittenhouse Road Norristown, PA 19403 (215) 666-7950	GOULD Electronics	Gould Semiconductors 2300 Buckskin Road Pocatello, ID 83201 (208) 233-4690
	Cybernetic Micro Systems P.O. Box 3000 San Gregorio, CA 94074 (415) 726-3000		Harris Semiconductor P.O. Box 883 Melbourne, FL 32901 (305) 724-7000
CYPRESS SEMICONDUCTOR	Cypress Semiconductor 3901 N. First Street San Jose, CA 95134 (408) 943-2600		Hitachi America, Ltd. 2210 O'Toole Avenue San Jose, CA 95131 (408) 435-8300

Electronic Troubleshooting □ 365

LOGO	MANUFACTURER/ADDRESS
HONEYWELL	Honeywell 1150 E. Cheyenne Mountain Blvd. Colorado Springs, CO 80906-4599 (303) 576-3300
HUGHES AIRCRAFT COMPANY	Hughes Aircraft 500 Superior Avenue, P.O., Box H Newport Beach, CA 92658-8903 (714) 759-2349
ʮ	Hybrid Systems 22 Linnell Circle Billerica, MA 01821 (617) 667-8700
HYUNDAI ELEXS AMERICA	Hyundai Elexs America 166 Baypointe Parkway San Jose, CA 95134 (408) 286-9800
inmos	INMOS Corp. 1110 Bayfield Road Colorado Springs, CO 80906 (303) 630-4000
intel i	Intel 3065 Bowers Avenue Santa Clara, CA 95051 (408) 987-8080
INTERSIL	Intersil 2450 Walsh Avenue Santa Clara, CA 95051 (408) 996-5000
ITT semiconductors	ITT Semiconductors 470 Broadway Lawrence, MA 01841 (617) 688-1881
Jameco ELECTRONICS	Jameco Electronics 1355 Shoreway Road Belmont, California 94002 (415) 592-8097 FAX 415-592-2503
λ	Lambda Semiconductor 121 International Drive Corpus Christi, TX 78410 (512) 289-0403
LT LINEAR	Linear Technology 1630 McCarthy Boulevard Milpitas, CA 95035-7487 (408) 432-1900
3M	3M/Electronic Products Division P.O. Box 2963 Austin, TX 78769-2963 (512) 834-1800
MICRON TECHNOLOGY, INC.	Micron Technology Inc. 2805 E. Columbia Road Boise, ID 83706 (208) 383-4000
MN	Micro Networks 324 Clark Street Worcester, MA 01606 (617) 852-5400
M	Micro Power Systems 3151 Jay Street, Box 54965 Santa Clara, CA 95054-0965 (408) 727-5350
⊕	Mitel Corporation 350 Lagget Drive, P.O. Box 13089 Ontario, Canada K2K 1X3 (613) 592-5630
▲	Mitsubishi Electronics America 1050 E. Arques Avenue Sunnyvale, CA 94086 (408) 730-5900
MMI	Monolithic Memories P.O. Box 3453 Sunnyvale, CA 94088 (408) 970-9700

LOGO	MANUFACTURER/ADDRESS
Ⓜ	Motorola 5005 E. McDowell Road Phoenix, AZ 85008 (602) 244-7100
NS	National Semiconductor 2900 Semiconductor Drive Santa Clara, CA 95051 (408) 721-5000
NCR	NCR 8181 Byers Road Miamisburg, OH 45342 (513) 866-7217
NEC	NEC Electronics, Inc. 401 Ellis Street Mountain View, CA 94039-7241 (415) 960-6000
OKI JAPAN	OKI Semiconductor, Inc. 650 N. Mary Avenue Sunnyvale, CA 94086 (408) 720-1900
MATSUSHITA	Panasonic (Matsushita) 1 Panasonic Way Secaucus, NJ 07094 (201) 348-7000
PLESSEY	Plessey Semiconductor 1500 Green Hills Road Scotts Valley, CA 95066 (408) 438-2900
PMI	Precision Monolithics, Inc. 1500 Space Park Drive Santa Clara, CA 95052-8020 (408) 727-9222
Raytheon	Raytheon Semiconductor 350 Ellis Street Mountain View, CA 94039-7016 (415) 968-9211
Rockwell	Rockwell International 4311 Jamboree Road, P.O. Box C Newport Beach, CA 92658-8902 (714) 833-4700
SAMSUNG Semiconductor	Samsung Semiconductor, Inc. 3725 North First Street San Jose, CA 95134 (408) 434-5400
SANYO	Sanyo Semiconductor Corp. 7 Pearl Court Allendale, NJ 07401 (201) 825-8080
seeQ	SEEQ Technology, Inc. 1849 Fortune Drive San Jose, CA 95131 (408) 432-7400
SGS	SGS Semiconductor 1000 E. Bell Road Phoenix, AZ 85022 (602) 867-6100
SIEMENS	Siemens Components 1900 Homestead Road Cupertino, CA 95014 (408) 725-3531
signetics	Signetics 811 E. Arques Avenue Sunnyvale, CA 94088-3409 (408) 991-2000
SG Silicon General	Silicon General 11861 Western Avenue Garden Grove, CA 92641 (714) 898-8121
⊕	Silicon Systems 14351 Myford Road Tustin, CA 92680 (714) 731-7110

LOGO	MANUFACTURER/ADDRESS
	Siliconix 2201 Laurelwood Road Santa Clara, CA 95054 (408) 988-8000
S-MOS SYSTEMS	S-MOS 2460 North First Street San Jose, CA 95131 (408) 922-0200
S	Solitron Devices, Inc. 1177 Blue Heron Boulevard Riviera Beach, FL 33404 (407) 848-4311
② SPRAGUE	Sprague Electric 115 N.E. Cutoff Worcester, MA 01613-2036 (617) 853-5000
SPRAGUE SOLID STATE	Sprague Solid State 3900 Welsh Road Willow Grove, PA 19090 (215) 657-8400
	Standard Microsystems 35 Marcus Boulevard Hauppauge, NY 11788 (516) 273-3100
STS THOMSON MICRO ELECTRONICS	STS Thomson Micro Electronics 1310 Electronics Drive Carrollton, TX 75006 (214) 466-6000
	Supertex 1350 Bordeaux Drive Sunnyvale, CA 94089 (408) 744-0100
TELEDYNE PHILBRICK	Teledyne Philbrick 40 Allied Drive Dedham, MA 02026-9103 (617) 329-1600
	Teledyne Semiconductor 1300 Terra Bella Ave., Box 7267 Mountain View, CA 94039-7267 (415) 968-9241

LOGO	MANUFACTURER/ADDRESS
	Texas Instruments P.O. Box 655474 Dallas, TX 75265 (214) 995-2011
TRW	TRW Semiconductors P.O. Box 2472 La Jolla, CA 92038 (619) 457-1000
	Thomson-CSF Components Corp. 6203 Variel Avenue, Unit A Woodland Hills, CA 91365 (818) 887-1010
Toshiba	Toshiba America, Inc. 9775 Toldeo Way Irvine, CA 92718 (714) 455-2000
UNITED MICROELECTRONICS CORPORATION	United Microelectronics Corp. 3350 Scott Boulevard #57 Santa Clara, CA 95054 (408) 727-9239
VITELIC CORP.	Vitelic Corp. 3910 North First Street San Jose, CA 95135-1501 (408) 433-6000
WESTERN DIGITAL	Western Digital 2445 McCabe Way Irvine, CA 92714 (714) 863-0102
Xicor	Xicor 851 Buckeye Court Milpitas, CA 95035 (408) 432-8888
Zilog	Zilog, Inc. 210 Hacienda Avenue Campbell, CA 95008 (408) 370-8000

APPENDIX D

ANSWERS TO ODD-NUMBERED REVIEW QUESTIONS

ANSWERS TO REVIEW QUESTIONS FOR PHASE 1

1. Any indicator lights on, internal cooling fan running, transformer is humming slightly, equipment warms up, etc.
3. Substitute another, known good cable.
5. There must be at least two wires. This would mean that four wire ends must be showing. Lack of a wire end probably means a wire has become disconnected.
7. A fast rise to a plateau, then another smaller rise to a second final plateau.
9. Installed indicators.
11. Substitution.
13. Reinstall it. The problem should reappear.
15. See paragraph 1.1.3.2.
17. The intermittent problem must be present long enough to effect a repair.
19. A problem that will not recur in the presence of the technician. The customer insists there is a problem, but it is not evident to the technician.

ANSWERS TO REVIEW QUESTIONS FOR PHASE 2

1. Classroom discussion. Include personnel grounding through wrist strap, grounding of equipment to true earth, providing leakage paths for accumulated charge.
3. Using a three-wire cord to provide a true earth ground for the chassis, use of a polarized two-wire cord, use of an isolation transformer.
5. The much higher voltages required, typically 150 to 300 volts for vacuum tube circuits as compared with the 5 to 12 volt circuits of today.
7. The dead man stick.
9. Make a careful visual check of the interior.
11. Apply power and test all power supply internal outputs.
13. To orient the technician as to signal flow, what each of the major circuits do, and how the signals interrelate. In short, to orient the technician to the equipment.
15. Test points have been chosen by the manufacturer as being good points at which to conveniently determine where a problem might lie.
17. Reduction of noise pickup, less circuit loading, improved bandwidth characteristics, no cause for ringing of digital signals.

ANSWERS TO REVIEW QUESTIONS FOR PHASE 3

1. Depending on the individual situation, there may be power supplies without regulation of any kind, voltage-regulated, or variable voltage-regulated types available. These may also be current-regulated or variable current-regulated supplies. Current regulation may be of the reducing-voltage or the fold back type.
3. See end of paragraph 3.2; index entries of each for additional help.
5. A defect in one circuit shows up as abnormal readings in all stages within the loop.
7. The DC blocking capacitor keeps any DC at the test point of the equipment from flowing back into the signal generator output. A typical value may be anywhere from 1 to 10 ufd.
9. To physically identify the location of a component selected from the schematic drawing.
11. Call the manufacturer for help. Order instruction book. Draw the schematic yourself. Use the techniques of Phase 6.
13. Classroom on-going assignment.
15. See Figure 3–28.
17. The OR gate. When the circuit would be easier to understand.
19. The timing of the circuit signals.
21. Reverse engineering.
23. Classroom discussion: To include advantages, disadvantages, complexity, cost, definitions of each.
25. Switch bounce will provide many 1's and 0's for each mechanical actuation of the switch.
27. Seven:

HIGH LED	LOW LED	ACTIVITY	
OFF	OFF	OFF	Indeterminate level; disconnected circuit; TTL input open circuit.
OFF	OFF	ON	Normal digital signal, frequency of operation too high for high and low LED circuit.
OFF	ON	OFF	Constant low digital level.
OFF	ON	ON	Normal digital signal, very low duty cycle.
ON	OFF	OFF	Constant high digital level.
ON	OFF	ON	Normal digital signal, very high duty cycle.
ON	ON	OFF	Not a valid reading.
ON	ON	ON	Normal digital signals.

29. There is no constant overwriting of the same line on the display.
31. The corners of the waveform will be rounded off. See Figure 3–51.
33. About 0.8 volt.
35. As a logic high, even though the level is about 1.6 volt.
37. The output of the CMOS chip will be erratic.
39. Yes. The timing pin of the multivibrator chip is not a digital signal.
41. No. The timing of the signal may be defective.
43. An open inside the chip, on an active-high input line.
45. See paragraph 3.11.
47. See Figures 3–67, 3–68, and 3–82.
49. Class C
51. Class A
53. Class C
55. The push-pull Class B stage or the complimentary-symmetry circuit.

57. The answer would calculate to be a peak amplitude of 2 volts, except that the oscilloscope is operating at its bandwidth frequency. The actual amplitude of the waveform would be 1.41 × 2 or 2.82 V peak.
59. AVC (Automatic Volume Control) bus.
61. Between the AVC test point and the antenna input.
63. By measuring the amount of negative bias produced by the oscillations.
65. 0° or 180°.
67. The time required for a waveform to go from 10% to 90% of the way from minimum to maximum.
69. The oscilloscope.
71. Implosion and high voltage.
73. Realignment of the circuit if applicable.
75. That there is an intermittent problem.

ANSWERS TO REVIEW QUESTIONS FOR PHASE 4

1. Check for proper Vcc and ground connections.
3. Substitution of a good chip, one at a time until the bad chip is identified.
5. Good activity at the output chip pin, indeterminate level without activity at the input pin of the following chip.
7. The semi-short and the hard-wire short.
9. Use the current tracer to find the high current path.
11. The chip probably has an open-collector output.
13. The greater risk of slipping and accidentally shorting circuitry when using dull test leads.
15. Fragile; inaccurate; hazardous ohmmeter currents; loads circuit on DC scales.
17. Estimate first, measure second. This way the technician is not swayed by suggestion.
19. The one farthest from normal, the 50% reading.
21. No. It could be open.
23. Yes (if there is no paralleling path around the resistor).
25. To force a more equal division of current between the two transistors than would be provided by the internal resistance of the transistors alone.
27. Like a short or very low resistance.
29. Like a very low resistance, almost a short.
31. 0.3 or 0.7 V (germanium, silicon).
33. An open circuit; transparent.
35. Diode; current-controlled resistor.
37. With and without the bias stabilizing resistor.
39. The gate must have a polarity opposite that of the drain, with respect to the source lead.
41. Where the depletion type FET requires no special biasing arrangements, the enhancement type FET requires a source of forward-biasing to conduct.
43. Very near either one of the rails.
45. The voltage *difference* between the two inputs.
47. The circuit may have dangerously high voltages on one or both sides of the isolator.
49. The use of screws made of insulating material such as nylon for mounting and the obvious avoidance of chassis mounting screws by the ground bus trace of the PC board.
51. Classroom discussion.
53. Probably not. Complete failures usually produce a higher percentage of DC voltage shift from normal than a mere 5%.

55. 1) The voltage source is too low. 2) The meter is defective. 3) The meter is loading the circuit. 4) The circuit (paralleling the meter) is overloaded. (Too little resistance.) 5) An increase in the resistance between load and source.
57. Usually short, sometimes leaky, seldom open or out of tolerance.
59. A bad capacitor.
61. Always turn off all power and discharge all high voltage or high value capacitors before using an ohmmeter or solid state tester.
63. Insufficient current is used on the ohmmeter scales to forward-bias a junction reliably.
65. All Vcc to ground paths in the circuit, read through the stabilization resistor in series with them. The power supply bleeder, if any, would also be included.
67. Very definitely, yes. This is one of its strongest points, particularly if used for comparison tested of good and bad circuit.
69. 60 Hz. Accuracy is often maintained to about 1000 Hz for audio circuit measurements.
71. The oscilloscope ground lead is connected to earth ground via the third wire of the instrument power cord. Circuit damage can result if the oscilloscope ground is connected to "hot" points within a circuit under test.
73. There is much arbitrary twisting of knobs in the hope that the display will behave.
75. AC, DC.
77. $T = RC = (1 \times 10^{-8})(1 \times 10^5) = 1 \times 10^{-3} = 1$ millisecond.
79. DC, average.
81. The integrator.
83. The circuit oscillations are dampened in pulsed applications.
85. 1) Use stagger-tuning of the stages. 2) Add damping resistors to the LC circuits. 3) Use overcoupled transformers for interstage coupling.
87. The amplifier will go into saturation during part of the input cycle, causing distortion.
89. To load the inductive kick voltage spike and ringing voltage produced when the DC current is suddenly interrupted.

ANSWERS TO REVIEW QUESTIONS FOR PHASE 5

1. Discharge all of the large-value or high-voltage capacitors in the circuit.
3. NEVER by filing! Repeated applications of fresh rosin-core solder and immediately wiping the tip on a wet sponge is the only acceptable way to rejuvenate an oxidized tip.
5. Installation of a socket.
7. An initial low-resistance value, tapering to a much higher or open value. Small-value capacitors may show an open indication almost immediately. Large values may take an excessively long time to approach an open condition.
9. Compare with a known good, identical part; consult the parts list for the DC resistance.
11. No. The resistance should be zero.
13. Yes.
15. Substitution of a known good device into the circuit.
17. Substitution of a known good EC into the circuit.
19. A line, vertical, horizontal, or slanted, depending on the value of the resistor and the range used.
21. A checkmark, inverted or not.
23. The Huntron tracker, a solid state tester with special triggering circuitry for these components.
25. Yes. Input and output stage failures can be detected. In addition, some types of pre-failure discrepancies can be seen, thereby possibly averting operational failure later, while in service.

27. No. Look for primary or secondary failures and eliminate them before reapplication of power.
29. Capacity value; voltage rating; capacitor type; current rating; physical size and mounting.
31. The transformer will draw too much current, probably opening a fuse or burning up.
33. Common (C); normally open (NO); normally closed (NC).
35. This is a book that converts specific semiconductor identification numbers to a manufacturer's line of generic replacement semiconductors.
37. Higher current can be drawn from the bank of batteries than that of an individual battery. The voltage remains the same as a single battery.
39. Ni-Cad, Lead-acid, Gel cell.
41. Burning rosin flux.
43. Test for proper operation; do required paperwork.

ANSWERS TO REVIEW QUESTIONS FOR PHASE 6

1. The ohmmeter and the solid state tester.
3. The edge connector, then the major components.
5. The high-current, high voltage, heat-dissipated components.

ANSWERS TO REVIEW QUESTIONS FOR PHASE 7 THROUGH 7D

1. To avoid slipping and the resulting risk of shorting circuitry, thus causing additional damage to a circuit.
3. Charge with a constant current of 1/10 the Ah rating, or in this case 400 mA, for a period of 14 hours.
5. The location of the common or ground connection.
7. Whether or not the fuse blows tells whether the failure is of the short or open type.
9. By disconnecting-testing-connecting then disconnecting again, beginning at the input end of the supply, progressing component by component down through the circuit until the fuse blew.
11. The type of Vcc bus structure used on the board. This will determine where cuts must be made for efficient troubleshooting.
13. No. The 4.7 K resistor limits the current to "normal" levels, even though the transistor may be shorted.
15. The possible cause of the first diodes' opening.
17. They should never be operated without a load.
19. The switching transistor is shorted, sending DC into the transformer primary.
21. The oscilloscope test lead is grounded to earth ground and may cause damage if connected to a "hot" circuit.
23. The output voltage is too high and not adjustable.
25. The pass transistor within the IC chip is shorted.

ANSWERS TO REVIEW QUESTIONS FOR PHASE 8

1. Neither. Both must be in perfect working order for the results to be correct.
3. The program was written for different hardware and will therefore not operate on any other hardware configuration.
5. Modification of the software to work with a specific set of hardware parameters.

7. A halt in the operation of a microprocessor to allow logic levels and registers to be examined for appropriate content.
9. Run the internal self-test diagnostics.

ANSWERS TO REVIEW QUESTIONS FOR PHASE 9

1. Static discharge to the computer and conducted noise and transients via the power line.
3. The reduction in receiver sensitivity caused by strong off-channel signals nearby.
5. Eliminate them at the source.
7. Use resistor spark plugs, resistive HV wiring, and ignition system shielding.
9. Via the power circuit, conducted interference.
11. In the presence of a strong off-channel signal, the receivers first RF stage may rectify incoming signals, producing sum and difference interfering signals.
13. Short the signal progressively from one stage to another to identify the source.

ANSWERS TO REVIEW QUESTIONS FOR PHASE 10

1. Remove batteries, if any, and stop further damage by thorough rinsing.
3. Mechanical damage, mechanisms jam, etc. Large, heavy components torn loose.

INDEX

Active   73
Acoustic Feedback   344
Alignment   124, 206, 248
Alternator Noise   331
Ammeter   17, 276
Amplifier
    Alignment   124
    Balanced   109
    Classes of   105
    Complementary-symmetry   112
    Darlington   112
    DC   111
    Low frequency   113, 341
    Radio Frequency   113, 117
    Squeal   344
    Transistor Configurations   106
    Unbalanced   108
Analog Circuits
    Alignment   124
    Amplifier Configuration   106
    Classes of Operation   105
    DC amplifier   111
    Locating Defective stages   103
    Stage Functions   104
Antennas, Testing   10
Aperiodic Waveforms   88
Application of Power   53
Applying Input Signals   54
Assembly Language   308
Automatic Volume Control (AVC)   117

Backplane   36, 316
Balance Control   205
Batteries
    Alkaline   262
    Backup Supply   323
    Carbon-zinc   262
    Charging with Bench Supply   50
    Gel-Cell   264
    Nickel Cadmium (Ni-Cad)   264
    Lead-Acid   263, 332
    In Series   242
    In Parallel   242
    Primary Cells   262
    Secondary Cells   263
    Substitution of   242
    Testing   228, 261
Battery Backup Supply   323
Bench Power Supply   46
Block Diagram
    Equipment   37
    Stage   129
    System   12
Bonding   339
Bounceless Switch   83
Break Point   308, 310, 319
Bridge Rectifier   267
Built-in Indicators   10

Cables, Testing
    Coaxial   6
    Multiple Wire   3
Calibration   205, 248
Capacitors
    DC Blocking   55
    Filter   189, 272, 278, 289, 341, 344
    High Capacity   211
    High Voltage   28
    In DC Circuits   153
    In AC Circuits   187
    Open Emitter Bypass   205
    Removing   221
    Parallel   237, 341, 344
    Series   237
    Shorting   30
    Substitution   237
    Testing with Ohmmeter   230
    Testing with SST   230
    When DC signal tracing   64
    With Diode & Large Signal   198
    With Inductors & Sinewaves   196
    With Inductors, Resistors & Sinewaves   197
    With Resistors & Sinewaves   188
    With Resistors & Pulses, squarewaves   190
Capacitor Bridge   222
Cavities, Resonant   338
Chassis   95
    Defined   271
    Use of Voltmeter   149, 169

Circuit Symbols
　See Component Designations
Clamper　204
Class A Amplifier
　Operation　105
　Distortion Problems　202
　IGFET Circuits　163
　In RF Stages　115
　JFET Circuits　163
　Open Emitter Bypass　205
　Operating Curve　106
　Transistor Biasing for　111, 160, 162
Class B Amplifier
　Crossover Distortion　203
　Distortion Problems　202
　In Complementary-Symmetry Stage　112
　Operating Curve　107
　Operation　106, 160
Class C Amplifier
　DC Voltages In　145
　Biasing　160
　Distortion Problems　204
　In Pulsed Stages　124
　In RF Stages　117
　In Digital Circuits　139
　Operating Curve　119
　Operation　106
Classes of Operation
　By use　159
　See Also Individual Classes; A, B, and C
Clear　73
Clipping　202
　See Also Distortion
Clocked Logic　82
Closed Loop Circuits
　Troubleshooting　52
Complementary Metal Oxide Semiconductor (CMOS)
　Defined　73
　Erratic　65
　Logic Levels　86, 93
　Open Input　96, 136
　Output Configuration　76
　Soldering　221
　Stuck Bus　137
Coaxial Cables　3
　Safety Hazards　27
Color Coding
　Analog Schematics　103
　Digital Schematics　73
Common　95
　Defined　271
Common Base Amplifier　106
Common Collector Amplifier　106
Common Emitter Amplifier　106
Comparison
　Circuit　65
　Component　233, 288
Complementary-symmetry Amplifier
　Operation　112
　DC Voltages in　161

Component Designations
　List of　68
Configuring
　Software　306, 308
　Switches　312
Crossover Distortion　203
Crowbar Circuit　270
Current Limiting　298
Current Tracer
　Notes on Using　139
　On Stuck Bus　139, 140
　On Vcc Short　281
Cutoff
　Class A Amplifier　202
Cut-Trace-and-Try Method　280

Darlington Amplifier　112
DC Amplifier　111
DC Voltages, Testing　169
Dead Circuit Troubleshooting　253
Debounce　142
Dedicated Test Instruments　56
Defective
　Card, Finding　25
　Equipment, Finding　1
　Stage, Finding　45
Desense, Receiver　326
Diagnostic Software　305, 313
Differentiator　192
Digital Circuits
　Equivalent Gates　74
　IC Output Configurations　76
　Locating Defective Stage　71
　Non-standard Gates Symbols　76
　Notes on Discrete
　Symbology　74
　Terminology　73
　Timing　78
Digital References　71
Diode
　As a Clamper　204
　Failures　157
　In DC Circuits　157
　Infrared　168
　Power Supply Rectifier　284
　Removing　221
　Testing In-Circuit　178
　Testing with Ohmmeter　224
　Testing with SST　230, 231
　With Capacitor, Large Signal　198
　Zener—See Zener Diode
Distortion
　Feedback to reduce　53
　Notes on　201
　Tracing　42
Divide-and-conquer
　See Half-split
Documentation
　Defined　56
　Ordering　66
　Using Similar　66

Dropped Equipment Damage 347
Duty Cycle 121
Dynamic
    Defined 73
    Component Effects 146

Equivalent Gates 74
Electromagnetic Interference (EMI) Filter 283, 290
Electrostatic Discharge (ESD) 30, 54, 325
Erratic Intermittent
    Defined 18
    During Testing 65
Estimating
    AC Voltages 187
    DC Voltages 146
Extender Boards 50

Field Effect Transistors (FETs)
    Erratic 65
    In DC Circuits 163
    Protecting 31
    Simplified 164
    Soldering 221
    Testing with Ohmmeter 227, 228
Fire Damage 347
Filter
    See Also Capacitor
    Alternator Electrical Noise 332
Floating Output—*See* Three-state
Floppy Diskettes 306
Fractured Knee 233
Freezer Method 281
Freon 18
Frequency Counter 124
Frequency Response 198
Front Panel Controls
    During Equipment Troubleshooting 37
    Presetting for Alignment 169
Frustration List—Appendix A
Fuses 15
    As Indicator 275
    Feed it 276
    Holders 17
    In DC Circuits 156
    In Phasing Supply 295
    In Switching Supply 288
    In Transformer/Rectifier Supply 275
    Pico 286
    Substitution 238
Fuse Holder 17

Generic Parts 235
Ground
    Bench Power Supply 50
    Defined 25, 271
    On ICs 130
    Open in 342
    Oscilloscope 185
    Oscilloscope Probe 39
    Power Supply 269
    Return Line 95
    Too Many 343
    Virtual 167
    Voltmeter 149, 169

Half-Split Method 62, 276, 277, 280
Hard Disk 306
Hardware 303
Hard-wire Short 138
Here-Jump-Here Program 319
High Voltage Probe 123
Hot Box 19
Hum, Tracking 321
Huntron Tracker
    See Solid State Tester
    Photo 253
Huntron Switcher 257
Hydrometer 264

IC—*See* Integrated Circuit
IC Extender Clip 88
IGFET—*See* FET
Ignition Noise 333
Ignored Inputs
    In Digital Circuits 97, 138
Indeterminate Levels 86
Inductors 37
    DC Signal Tracing 64
    Filters 193, 282
    In AC Circuits 193
    In DC Circuits 154
    Measuring Resistance In-Circuit 182
    Testing with Ohmmeter 225
    Testing with SST 230
    With Capacitors & Sinewaves 196
    With Capacitors, Resistors & Sinewaves
    With Resistors & Sinewaves 194
    With Resistors & Pulses, Squarewaves 194
Installation, Mechanical 247
Instruction Book—*See* Documentation
Isolation Transformer 25
Integrated Circuits
    Burned 33
    Desoldering 206
    Failures, Kinds 132
    Failures, with Logic Probe 132
    Inputs Ignored 97
    Open Collector 77, 142
    Output Configurations 76
    Pin Numbers 96
    Removing 221
    Schmitt Trigger Gates 94
    Shorting Output to Ground 83
    Substitution 242
    Testing with Ohmmeter 228
    Testing with SST 232
Integrator 190
Interference, RF 328, 335
    Tracking 329
    Conducted 335

Interlocks   10, 285
Intermittents   17
Intermodulation   336
Interrogator   315
Investigator   314

Junction Field Effect Transistor (JFET)
    Biasing   163
    In DC Circuits   162
    Simplified   162
    Testing with Ohmmeter   227

Load IC   131
Layout Diagram—See Parts Layout Diagram
Lightning Damage   349
Limiters   117
Lip, Upper   35
Logic Analyzer
    Notes on Using   98, 308
    Using Correlation Data   82
Logic Clip   82, 317
Logic Monitor—See Logic Clip
Logic Probe   85, 318
    Finding Open Trace   135
    Indications   135
    On Analog Signals   144
    On Discrete-to-IC Stages   142
    On IC-to-IC Stages   132
    On IC-to-Discrete Stages   139, 141
Logic Pulser
    As Signal Source   83
    Notes on Using   94, 138
    Overriding Digital Signals   97
    To Find Vcc Short   281
Loose Connections   32
Low Frequency   113

Mechanical Intermittent   18
Mechanical Problems   316
Metal Oxide Semiconductor Field Effect Transistor (MOSFET)
    See FET
Metal Oxide Varistor (MOV)   21, 283, 290, 322
Meters
    Substitution   244
Microphones
    Substitution   245
Missing Cycles   322
Mixer Stage   104
Mockup   56
Mother Board   37, 316
Multimeter
    Dangers of a VOM
    In DC Circuits   145
    Signal Tracing With   184
Multiple Wire Cables, Testing   3

Negative Feedback
    Operational Amplifier   165
    To Reduce Distortion   53

Neutral   25
    Defined 271
Noise
    Alternator   332
    Ignition   333

Odds of Component Failure   177
Ohmmeter   15, 37
    Diode Scale   179
    In-Circuit Limitations   178
    Out-of-Circuit Testing   222
    Reading on Schematics   103
    Testing In-Circuit   259
    Versus Solid State Tester   182
Open circuits
    Review of   271
    Troubleshooting   85
Open Collector Output
    Defined   77
    Troubleshooting   142
Operational Amplifiers
    In DC Circuits   163
Operating Controls   1
Operator Error   1
Optical Isolators   291
    In DC Circuits   168
Oscillator Stage   104
    Radio-Frequency   118
Oscilloscope
    AC/DC Coupling   41
    Bandwidth   90, 117
    Ground   50, 149, 185
    Holdoff Control   89
    On Switching Power Supply   296
    Probe   39, 88, 115
    Response   115
    Risetime   90
    Signal Tracing   39, 185
    Transistor Driver Circuits   141
    Triggering, Digital Circuits   87
    Triggering, Pulsed Circuits   123
    Use in Digital Circuits   87
    Using Trigger as Input   100
    Unusual Waveforms   103
    Voltage Limitations   121
Overcoupling   198
Overtemperature Stressing   19
Overvoltage Damage   347
Overvoltage Stressing   19

Paint
    Calibration Locking   33
Panic—See Appendix A
Paperwork   249
Parts Layout Diagram
    Using   64
    Drawing   67
Parts List   60
Pass Transistor   298
Periodic Waveforms   88

Perozzo 5-step Method 172
    Voltage Too Low 172, 272
    Voltage Too High 175
Phasing Power Supply 270
    Troubleshooting 293
Phase Shift 189
Phasing Dots 120
Photo Isolators
    See Optical Isolators
Pico Fuse 286
Polarity
    of Components 221
Positive Feedback
    Operational Amplifiers 163
Power Line Problems 321
Power Supply
    Bench 46
    Checking Outputs 34
    Constant Current 49
    Constant Voltage 49
    Current Limiting 46
    Foldback 50
    Hard 46
    Input Circuit 266
    Output Circuit 267
    Remote Voltage Sensing 46
    Stiff 46
    Soft 46
    Transformer/Rectifier 269
    Troubleshooting 261
    Types 266
Preset 73
Primary Component Failure 234
Profound Failures 347
Programmable Logic Array (PAL) 235
Proven Program 305
Pseudo-Zener Load 278
Pulse Amplifiers 119
    Notes on Stages 124
Pulse Circuits 190
    Capacitors, Resistors in 188
    Capacitors, Inductors in 198
    Inductors, Resistors in 194
Pulse Stretcher 86
Pulse Terminology 120
Pulse Timing 120
Push-Pull
    DC Voltages In 161

Radio Frequency 113
Radio Frequency Oscillator, Testing 118
Rails 165
Random Access Memory (RAM) 304
Read Only Memory (ROM) 304
Re-alignment
    See Alignment
Reboot 305
Recording Voltmeter 21
Regulators, Voltage
    Bench Power Supply 46

Discrete Semiconductor 298
    Variable Voltage 46
    Current Limiting 46
    Remote Voltage Sensing 46
    Switching 297, 300
    Troubleshooting 297
Relays
    Substitution 243
Removing Components 220
Removing Covers 32
Replacing Components 211
Reset 73
Resistors
    Across Series Capacitors 151
    Burned 33
    Calculating Current Through 274
    In DC Circuits 149
    In Transistor Base Circuit 139
    Isolating Shorts 178
    Measuring Resistance In-Circuit 179
    Pullup 77, 142
    Smell of burned 34
    Substitution 236
    Swamping 198
    Testing with Ohmmeter 222
    Testing with SST 229
    With Capacitor & Sinewaves 188
    With Capacitors, Pulses & Squarewaves 190
    With Inductors 194
    With Inductors, Capacitors & Sinewaves 197
    With Paralleled Transistors 151
Resonance 197
Return Line 95
Reverse Engineering 80
Reversed Polarity Damage 348
Ringing
    Defined 198
    Diode to Stop 205

Safety 25
    Cathode Ray Tube 124
    Dead Man Stick 28
    Measuring Voltage 28
    Power Supply 261
    Using Proper Voltmeter Probes 144, 170
    Vacuum Tubes 28
    When Testing Inductors with Ohmmeter 225
    Working Alone 30
Saturation
    Class A Amplifier 202
Secondary Component Failure 234
Self-Check—See Diagnostics
Sequential Troubleshooting 62
Semiconductors
    See Individual Types, FET, SCR, etc.
Semi-Short
    Defined 138
Sensing
    Remote Voltage 46
Set 74

Schematic Diagram
    Drawing   67
    Preparing Analog   103, 148
    Preparing Digital   72
    Using   62, 169
Schmitt Gates   94, 142
Shock Exciting   198
Short Circuits
    Between Digital Inputs   134, 138
    Isolating   178
    Reviewed   271
    Troubleshooting   85
Signal Generators
    Applying Signals   54
    Signal Tracing with   183
    Using   37
Signal Paths   62
Signal Splitter Stage   104
Signal Tracing
    In Defective Stage   200
    In Power Supplies   271
    Notes on   58
    Through DC Amplifier   108
    Through Low-frequency Amplifier   113
    Through RF Amplifier   113
Silicon Controlled Rectifier (SCR)
    In AC Circuits   199
    In DC Circuits   168
    In Phasing Power Supply   276, 293
    Testing with Ohmmeter   227
    Testing with SST   232
Similar Parts   235
Simple Things, Checking   3
Sinewave Circuits
    Capacitors, Resistors in   188
    Inductors, Resistors in   194
    Inductors, Capacitors in   196
    Inductors, Capacitors, Resistors in   197, 198
Single-Ended Amplifier
    See Unbalanced Amplifier
Single Stepping   82
Smoke Damage   347
Smoke Out Method   282
Soft Fuse   276
Software   303
    Configuring   306, 308
Software Driven Circuits
    Troubleshooting   303
Solder   215
Solder Bridge   278
Soldering
    Making Connections   247
Soldering Iron   212
Soldering Tools   216
Solid State Tester (SST)   45
    Comparison Method
    Edge Connector   255
    In-Circuit Testing with   182, 254
    Out-of-Circuit Testing With   229
    Pattern Recognition   254
    Recorded Patterns   73, 103

Source Code   308
Source IC   131
Speakers
    Substitution   244
Specialized Test Equipment   10
Squarewaves
    Capacitors, Resistors With   192
    Inductors, Capacitors With   198
    Inductors, Resistors With   196
Squeal, Amplifier   344
SST
    See Solid State Tester
Stagger-Tuning   198
Static
    Component Effects   146
    Digital Troubleshooting   84
    Static Analog Troubleshooting   145, 169
    Without Motion   74
Static Stimulus Tester   317
Storage Oscilloscope   19
Stuck Bus   137, 318
Stuck Signals   95
Submerged Equipment Damage   349
Substitution
    Basic Technique of   12
    of Batteries   242
    of Capacitors   237
    of Cards   36, 65, 315
    of Chips
    of Equipment   15, 311
    of Fuses   238
    of Integrated Circuits   242
    of Meters
    of Microphones   245
    of Relays
    of Resistors   236
    of Semiconductors   241
    of Speakers   244
    of Switches   240
    of Transformers   239
    Resistor, for Load   298
Sulphation   264
Switches
    Bounce   83
    Conditioning—*See* Debounce
    In DC Circuits   157
    Substitution   240
    Testing with Ohmmeter   226
    Testing with SST   232
Switching Power Supply   270
    Troubleshooting   287
    Versatile Input Circuit   289
Switching Transistors   139, 288, 289, 291
Symbols, Electronic—*See* Appendix B
System Board   316
System Troubleshooting   10

TDR—*See* Time Domain Reflectometry
Temperature Probe   34
Termination Resistance   10

Test Instruments
    See Individual Instruments
Test Point   39, 60
Testing
    Antennas   10
    Coaxial Cables   6
    DC Voltages   169
    Multiple-wire Cables   3
    RF Oscillator   118
    Vcc on Digital Circuits   94
    Wired Circuits   3
Thermal Intermittent   18
Theory of Operation   58, 79
Three-State Output   78, 82, 305, 317
    Indeterminate Level   86
    Waveforms on Oscilloscope   89
Time Constant
    Capacitor, Resistor   190
    Inductor, Resistor   194
Time Domain Reflectometry   6
Timing Diagram   79
Timing Problems
    Digital Circuits   131, 138
Toothpick Test   219
Totem-Pole Output
    Defined   76
    Logic Levels   133
TP
    See Test Point
Transformer
    Buck/Boost AC Voltages   19
    Flyback   35
    In DC Circuits   156
    Open Winding   272, 285
    Over-coupling RF   198
    Pulse   291
    Removing   221, 282, 292
    Shorted   283
    Substitution   239
    Testing with Ohmmeter   226
    Testing with SST   231
    Variable AC   19, 275
Transformer/Rectifier Power Supply   269
    Troubleshooting   275
Transients
    In Switching Supply   288
    On Power Line   321
Transistor, Bipolar
    Base-to-Emitter Test   170
    Biasing   111, 160, 202
    Configurations of   106
    In DC Circuits   158
    Photo-Transistors   168
    Producing Intermodulation   336
    Removing   221
    Simplified   159
    Switching   139
    Testers   222
    Testing with Ohmmeter   224
    Testing with SST   231

Transistor-Transistor Logic (TTL)
    Families   92
    Logic Levels   86, 92
    Open Inputs   96, 136
    Output Configuration   76
    Stuck Bus   137
Trap, RF   326, 227
Triacs
    In AC Circuits   199
    In DC Circuits   168
    In Phasing Supply   270, 293
    Testing with Ohmmeter   227
    Testing with SST   232
Troubleshooting Charts
    Examples   12
    Using   35
Truth Tables   76
TTL
    See Transistor-Transistor Logic

Ultra-High Voltage Probe   123
Unbalanced Amplifier   108
Unclocked Logic   82
Unijunction Transistor (UJT)
    In AC Circuits   200
    In DC Circuits   169
    In Phasing Supply   270, 293
    Testing with Ohmmeter   227, 228
Unwanted Signals, Tracking   321
Usual Repairs   2

Vacuum Tubes   28
Vcc   130
Vcc Bus Structures   280
Virtual Ground   167
Visible Faults   129
Visual Inspection   32
Voltage Regulator—See Regulator, Voltage
Voltmeter
    AC, in Power Supplies   283, 284, 285
    In Analog Stage   169
    In RF Circuits   117
    In Wired Circuits   3
    Recording   21
    Safety   28
    Test Probes   144, 170, 348
    Testing Vcc in Digital Circuits   94
    Too Sensitive   5
    Using DC SCales   145
    With Fuses   15
Voltage Reference   271
Vss   132

Warranty   33
Wattmeter   7, 10
Waveshape Problems
    See Distortion
Wired Circuits, Testing   3

Zener Diodes
    Regulator   297, 299
    Spike Suppression   322
    Testing with ohmmeter   227, 228
    Testing with SST   232

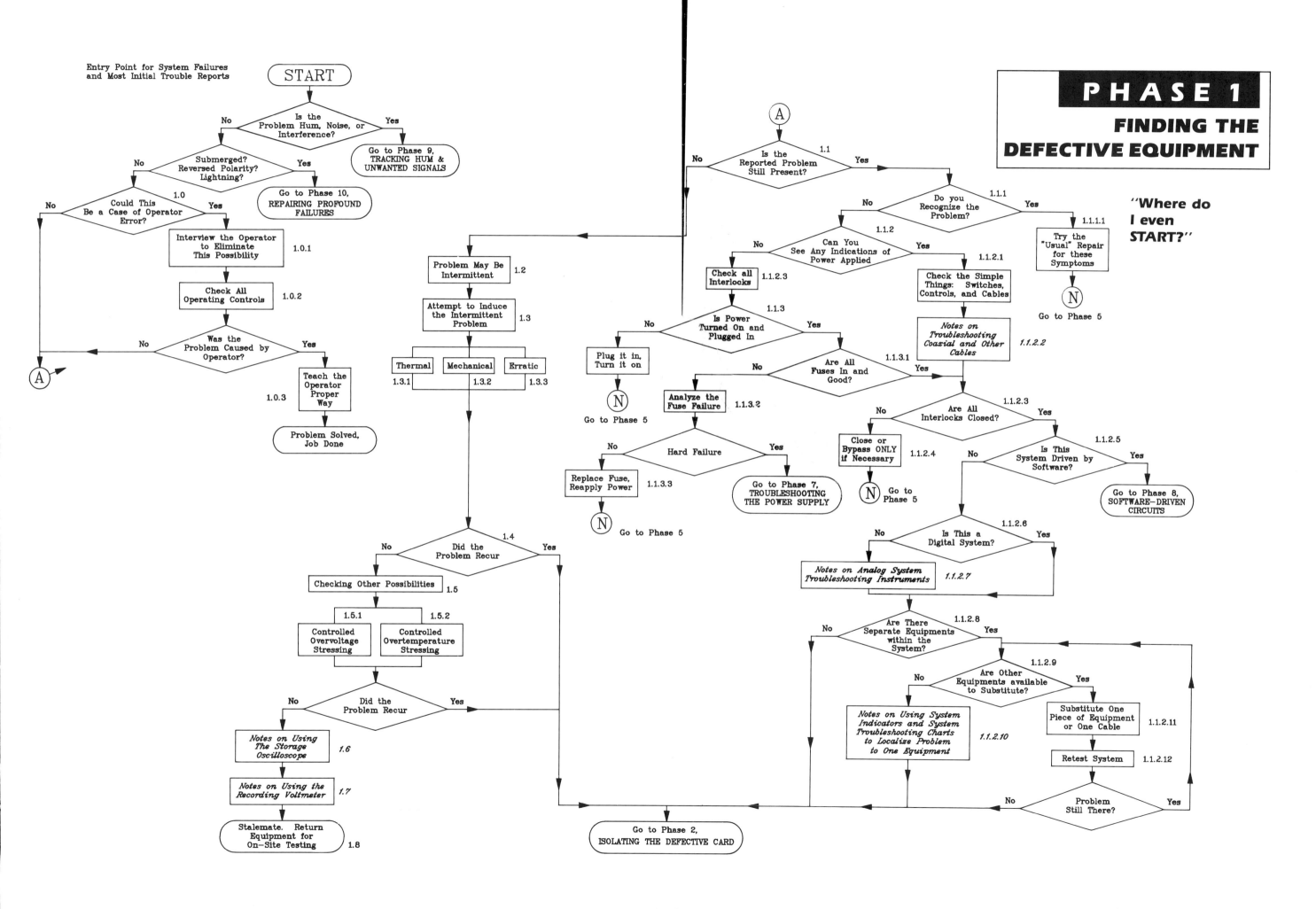

PHASE 2
ISOLATING THE DEFECTIVE CARD

"I KNOW the problem is in here SOMEWHERE...!"

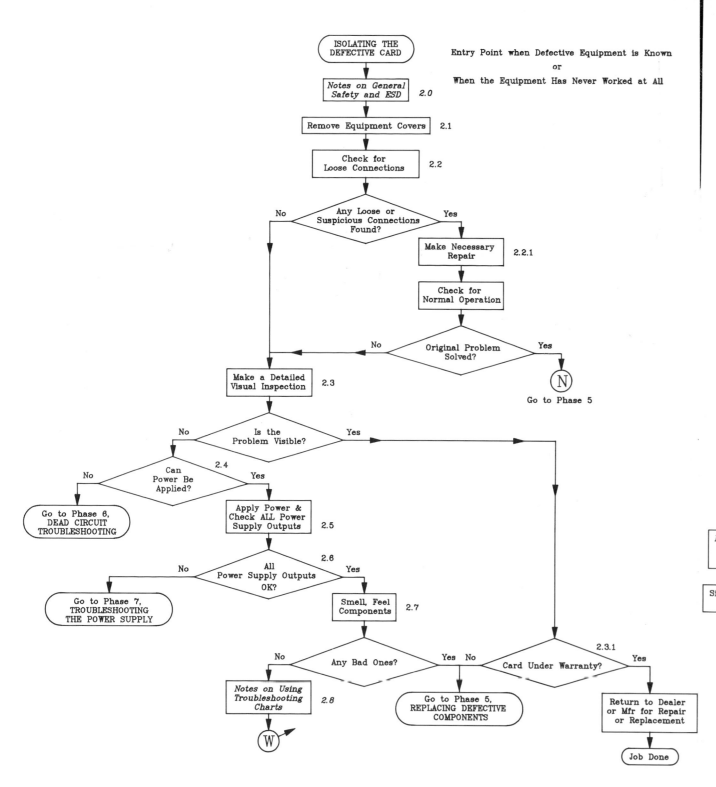
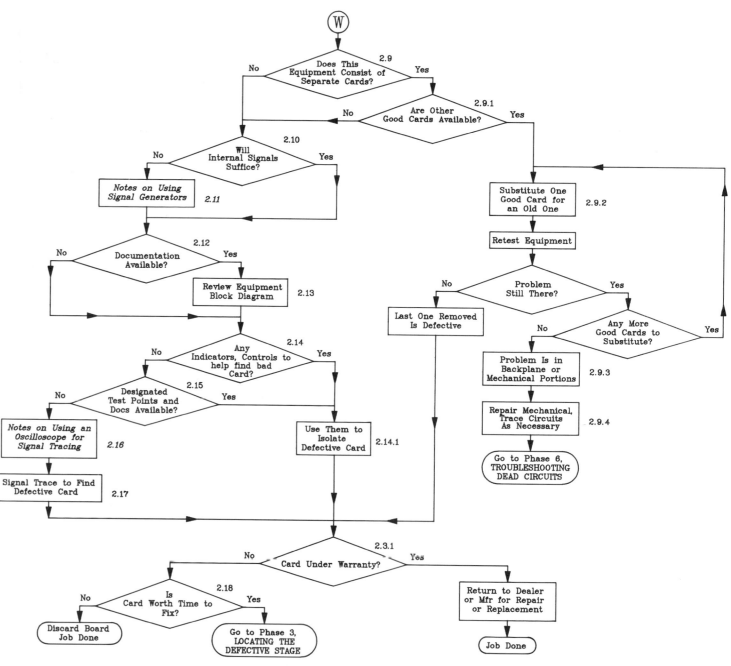

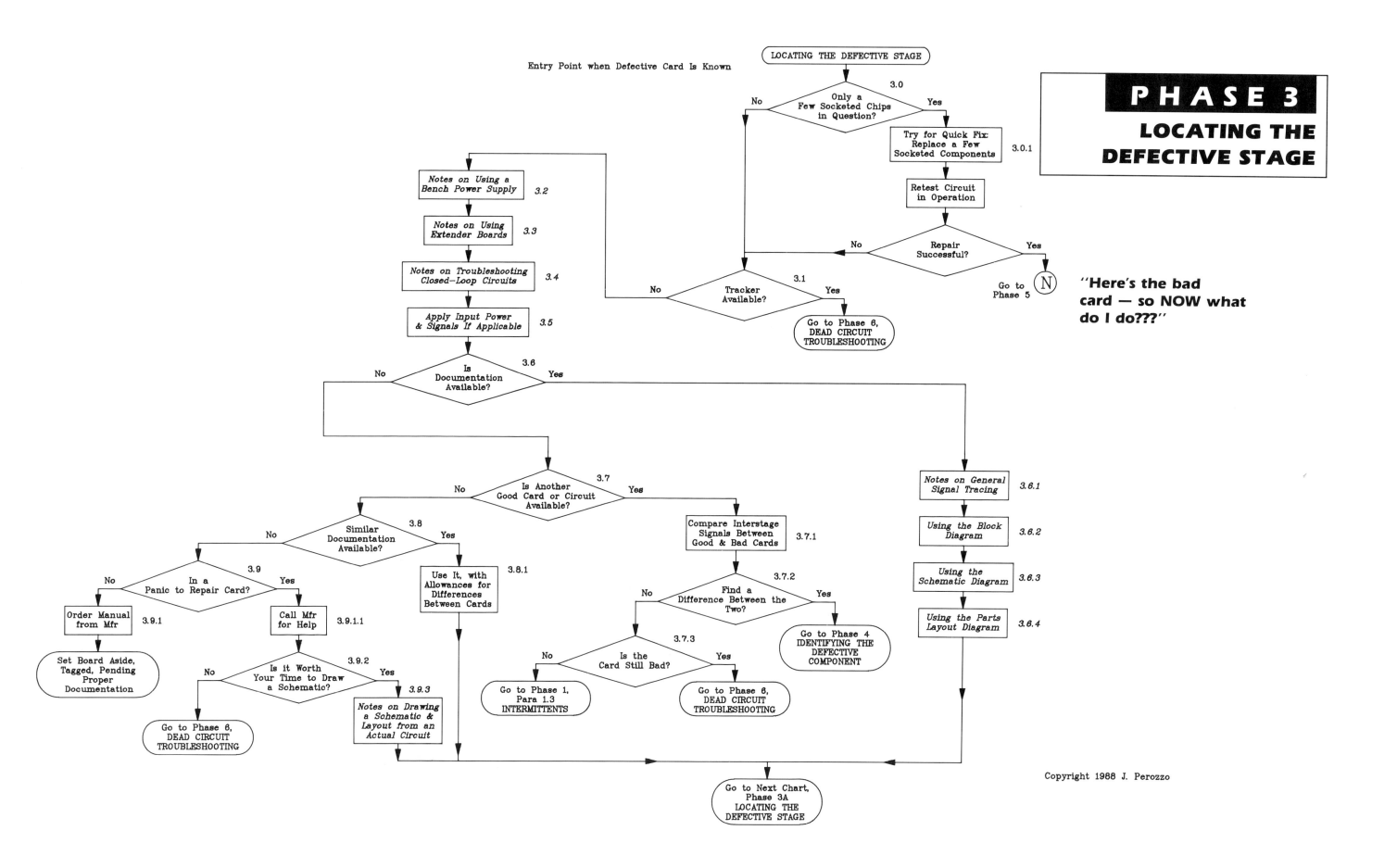

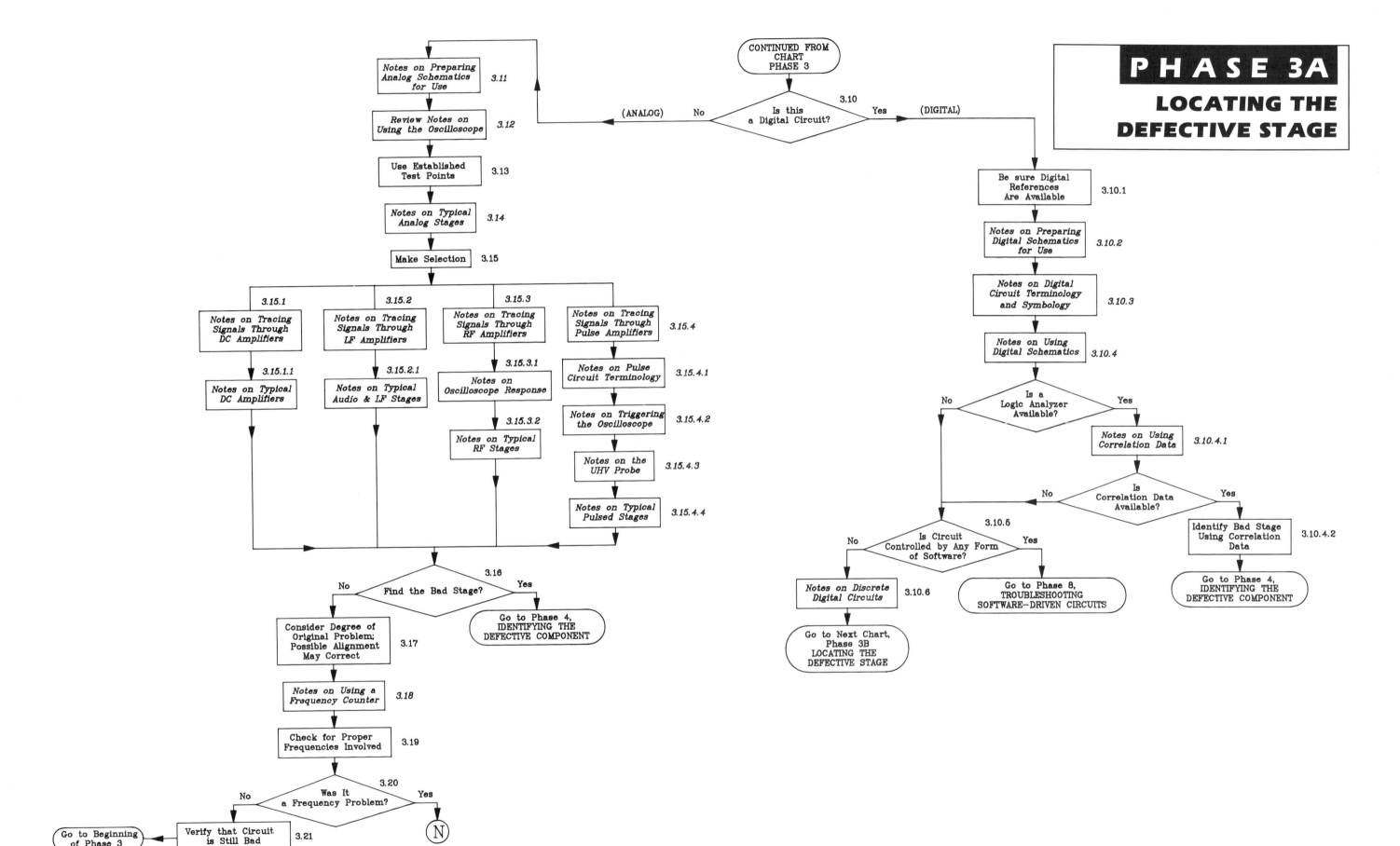

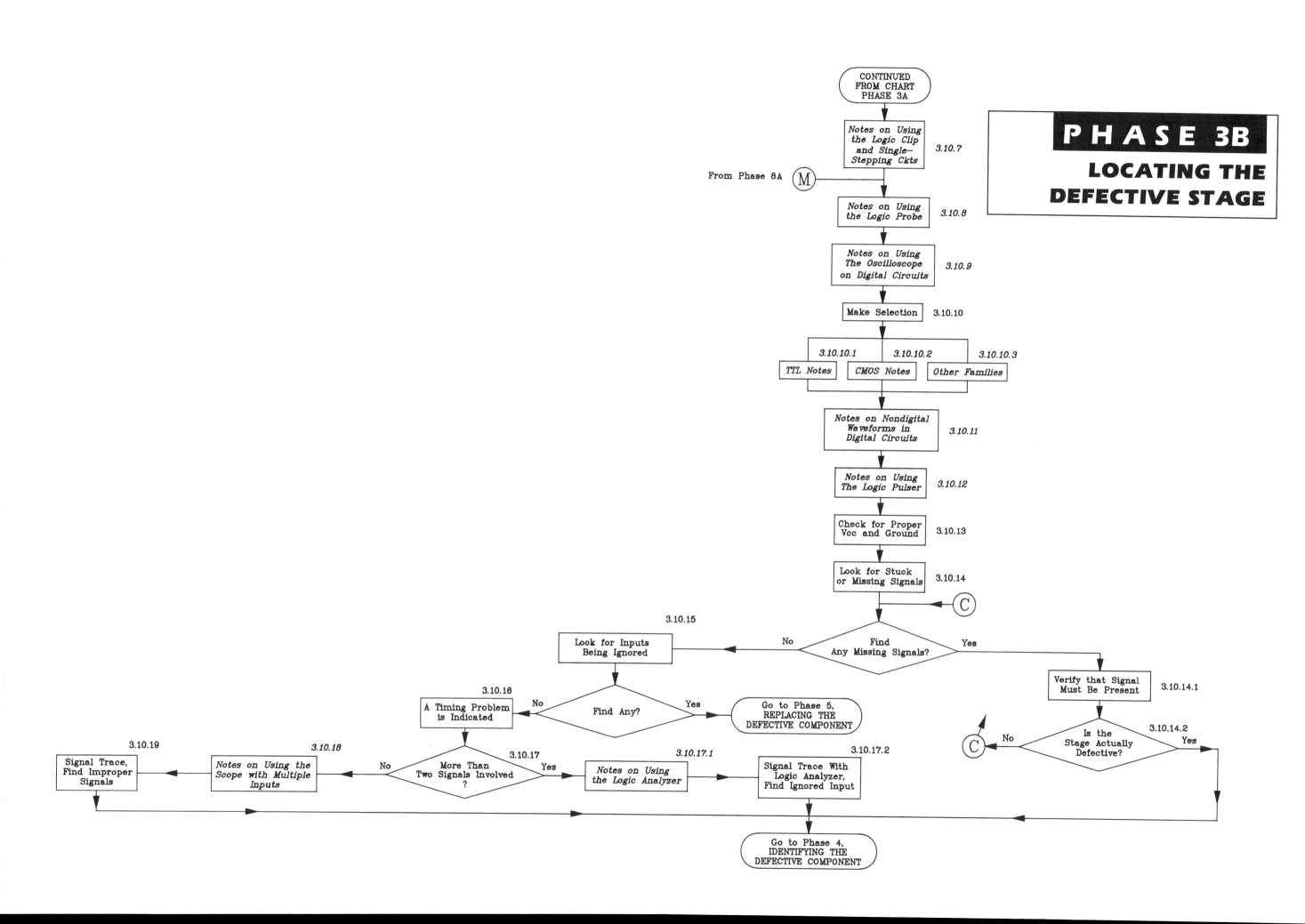

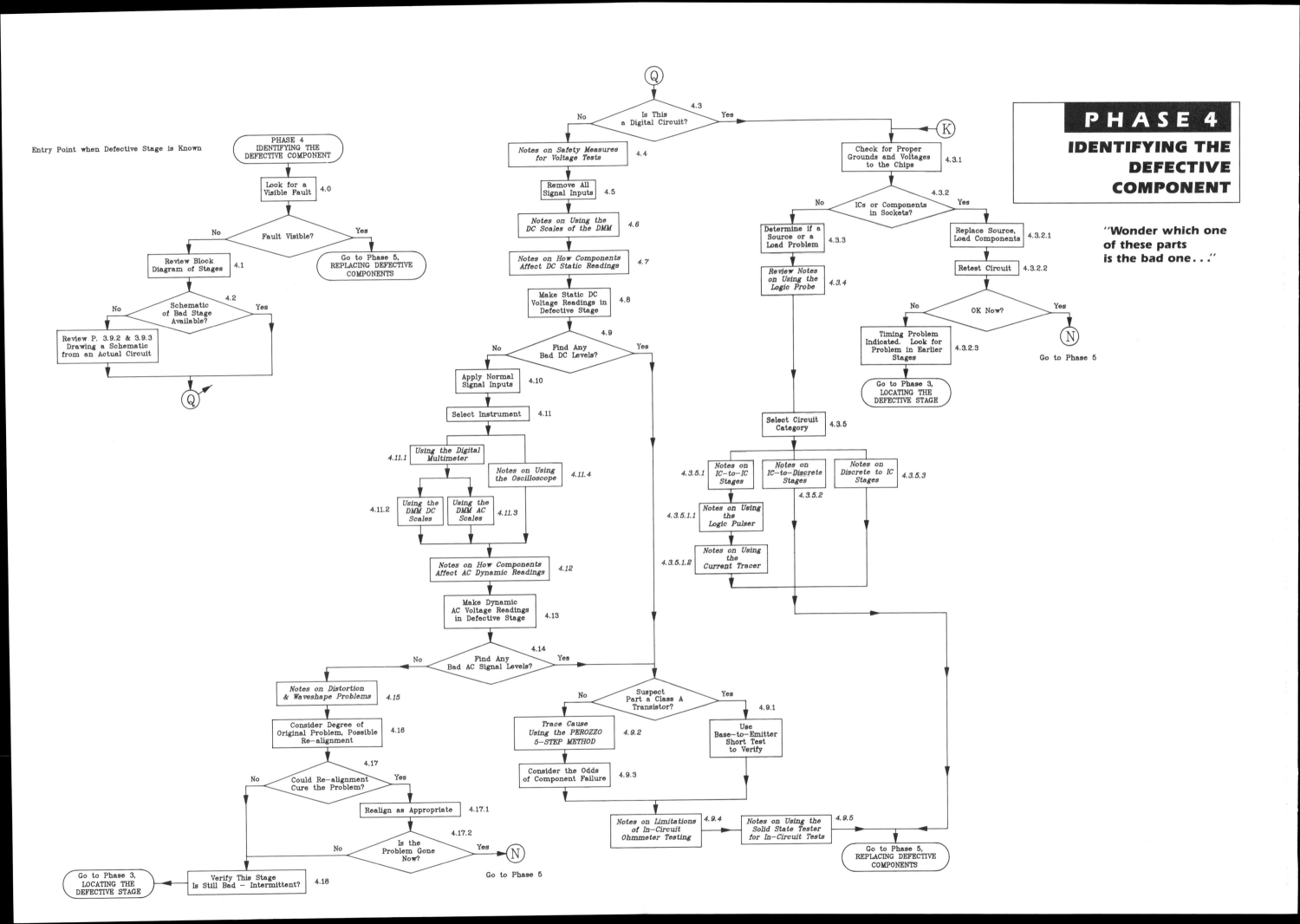

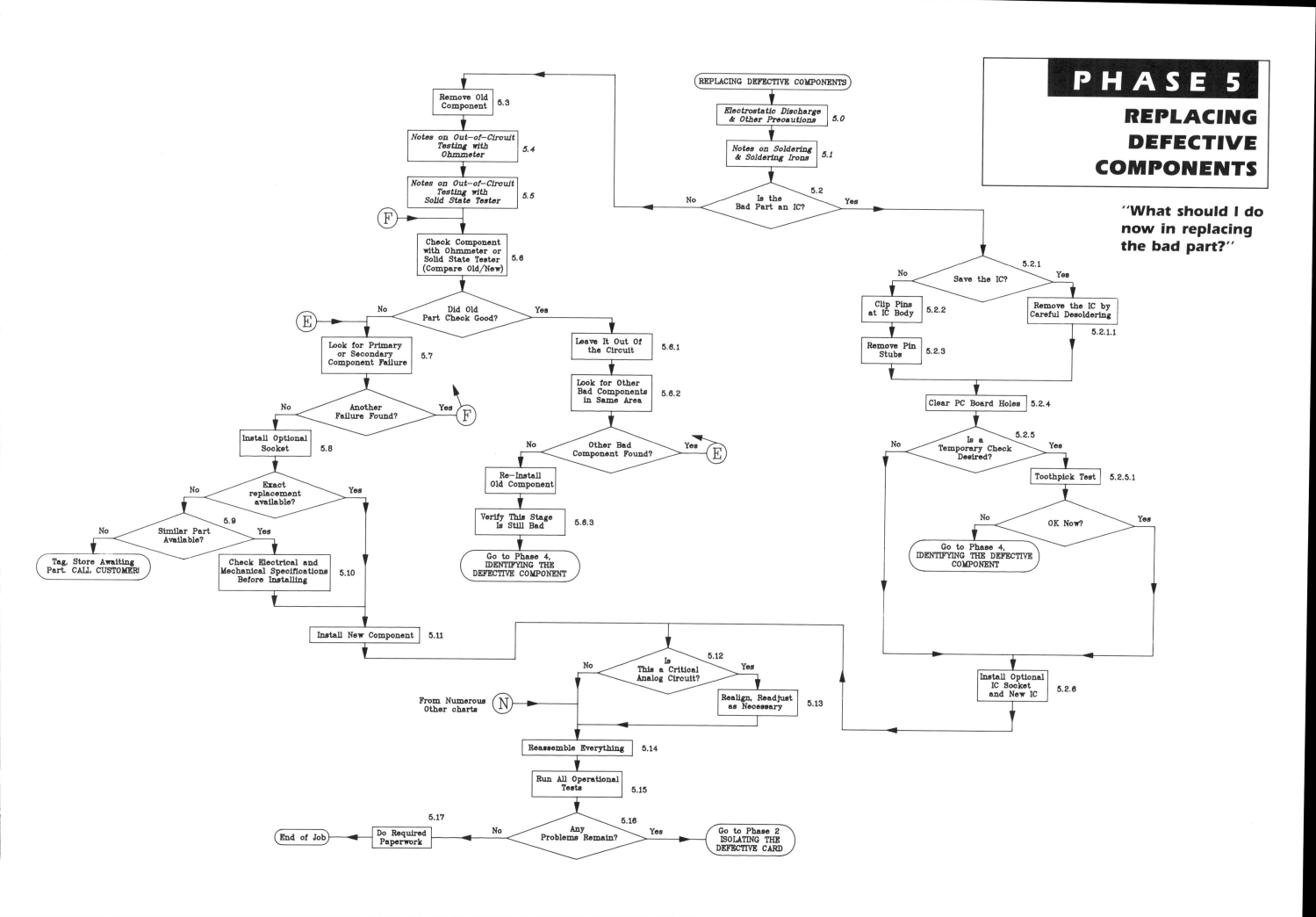

PHASE 6
DEAD CIRCUIT TROUBLESHOOTING

"I've never seen one of these before, can't turn it on, don't have any information about it, but I've still got to fix it!"

Flowchart: Dead Circuit Troubleshooting

Entry Point for "Dead" Circuits

- **DEAD CIRCUIT TROUBLESHOOTING** → Turn Off Equipment, Discharge Filter Capacitors
- **6.0** Is a Solid State Tester Available?
  - **Yes** → Using the Solid State Tester for In-Circuit Tests (6.0.1)
    - Use Pattern Recognition or Documented Waveform Methods (6.0.2)
    - **6.0.3** Is There an Edge Connector?
      - **No** → (H)
      - **Yes** → (G) → Test Edge Connector (6.0.4)
        - Are the Edge Connector Waveforms OK?
          - **No** → Follow PC Board Trace Back to Defective Component (6.0.5) → Remove Component Review Para. 5.3 → Test Component Review Para. 5.5 → Component Check Good?
            - **No** → (back to trace)
            - **Yes** → Re-Install Old Part → (H)
          - **Yes** → (H) → Test Power and Major Components In-Circuit (6.0.4.1)
            - These Components Test OK?
              - **Yes** → Are You Using the Comparison Method Now? (6.0.4.1.1)
                - **Yes** → (I)
                - **No** → Is a Good Identical Board Available? (6.0.4.1.2)
                  - **Yes** → Use Comparison Method (6.0.4.1.2.1)
                    - Are Multiple ICs Involved?
                      - **Yes** → Use Huntron Switcher If Available (6.0.4.1.2.2)
                      - **No** → (G)
                  - **No** → (I)
              - **No** → Evaluate Time Required vs Probability of Repair (6.0.4.1.3)
                - Is the Card Worth your Time?
                  - **No** → Discard Board
                  - **Yes** → Check Every Component on the Board (6.0.4.1.4)
                    - Minor Components Test OK?
                      - **Yes** → Discard Board or Order Docs, Save for Later Repair
                      - **No** → Go to Phase 5, REPLACING DEFECTIVE COMPONENTS
  - **No** → **6.1** Is an Ohmmeter available?
    - **No** → Set Aside until Proper Instruments available or Discard Board
    - **Yes** → Use Ohmmeter for testing. Review Paragraph 4.9.4 (6.2) → Test Power and Major Components In-Circuit (6.3)
      - All OK?
        - **Yes** → (I)
        - **No** → Remove Component Review Para. 5.3 → Test Component Review Para. 5.4 → Do the Components Check Good?
          - **Yes** → (I)
          - **No** → Go to Phase 5, REPLACING DEFECTIVE COMPONENTS

Copyright 1988 J. Perozzo

PHASE 7
TROUBLESHOOTING THE POWER SUPPLY

Entry Point for Known Defective Power Supply

```
TROUBLESHOOTING
THE POWER SUPPLY
        │
        ▼
┌─────────────────────┐
│   Notes on Safety   │  7.0
│ for You and the Supply │
└─────────────────────┘
        │
        ▼
      ╱ Are ╲
  No ╱ Batteries the Power ╲ Yes
    ╲    Source?    ╱        7.0.1
     ╲            ╱
        │
┌──────────────┐         ┌──────────────────┐
│ Review of Three │ 7.0.2   │ Care, Feeding, and │
│ Power Supply Types │      │  Troubleshooting  │
└──────────────┘         │    of Batteries    │
        │                └──────────────────┘
        ▼                         │
┌──────────────┐                  ▼
│ General Notes on │         ┌──────────────┐
│  Power Supply   │ 7.0.3   │ Test, Maintain │
│ Troubleshooting │         │  or Replace as │
└──────────────┘         │    Required    │
        │                └──────────────┘
        ▼                         │
  ┌───────────┐                  ( N )
  │ Select Type │              Go to Phase 5
  └───────────┘
```

Transformer/Rectifier	Switching Supply	Phase Controlled
Go to Chart 7A	Go to Chart 7B	Go to Chart 7C

"This power supply blows fuses one after another... I wonder if the problem will be easy to fix."

Copyright 1988 J. Perozzo

PHASE 7A
TROUBLESHOOTING THE TRANSFORMER/RECTIFIER POWER SUPPLY

TROUBLESHOOTING THE TRANSFORMER/RECTIFIER POWER SUPPLY

Entry point when power supply is known to be defective

CAUTION! Remove All power and discharge filter capacitors before disconnecting any components or using a TRACKER or Ohmmeter

CAUTION! Do NOT use an oscilloscope on the input side of the transformer!

7.1 Is the Power Supply Blowing Fuses?

— No → An Open Circuit Is Indicated
— Yes → A Short Circuit is Indicated ← (T) From Chart 7B

Open Circuit Branch

7.1.1 Measure AC at ALL Secondaries, Point C

Normal AC?
- No → 7.1.2 Check AC Input at Point A
  - Normal AC?
    - Yes → 7.1.2.1 Transformer has an Open Winding. Replace → Go to Phase 5, REPLACING DEFECTIVE COMPONENTS
    - No → 7.1.3 There is an Open in the Primary Wiring. Check Interlocks, fuses & Switches
      - 7.1.4 Check with AC Voltmeter for Opens to Line Input
        - 7.1.1.4 Circuit is Open Between Transmitter Secondary and Filter Input
- Yes → 7.1.1.1 Check All Power Supply Output Voltages between Filter and Regulator, Point E
  - Normal?
    - Yes → 7.1.1.1.1 Check all Regulator Outputs, Point F
      - Normal?
        - No → Go to Chart 7D, TROUBLESHOOTING VOLTAGE REGULATORS
        - Yes → 7.1.1.1.2 Trouble Is Not in Power Supply → (A) Go to Phase 1
    - No → Remove Power Input
      - 7.1.1.2 Check All Rectifiers for Opens
        - An Open Found?
          - No → 7.1.1.3 Check Filter Resistor/Inductor for Open Circuit
            - An Open Found?
              - Yes → Replace Open Component → Go to Phase 5, REPLACING DEFECTIVE COMPONENTS
          - Yes → 7.1.1.2.1 Replace All Rectifiers in Circuit → Go to Phase 5, REPLACING DEFECTIVE COMPONENTS

7.1.0.10 Check PC Board for Shorts with Ohmmeter → Repair as Necessary → (N) Go to Phase 5

Short Circuit Branch

7.1.0.1 Select Option
- 7.1.0.2 Use Variable Transformer
- 7.1.0.3 Use Soft Fuse
- 7.1.0.4 Feed Fuses

(R) From Chart 7C

7.0.1.5 Disconnect All After Filter at Point E

7.1.0.6 Apply Power

Overload Gone?
- No → Reconnect Point E
  - 7.1.0.7 Disconnect Secondaries at Point C
    - Short Gone? ← From Chart 7C (S)
      - No → 7.1.0.8 Disconnect Primary at Point A
        - Short Gone?
          - No → Reconnect Point A & Point C
            - 7.1.0.9 Short Is Between Power Input and Transmitter Primary
          - Yes → 7.1.0.8.1 Transformer Is Shorted → One or more Diodes Is Shorted → Go to Phase 5, REPLACING DEFECTIVE COMPONENTS
      - Yes → Reconnect Point C Disconnect Point D
        - Short Gone?
          - No → 7.1.0.7.1 Filter Capacitor or Inductor Shorted to Ground
          - Yes → 7.1.0.7.2 Determine Which Component is Defective With Ohmmeter → Go to Phase 5, REPLACING DEFECTIVE COMPONENTS
- Yes → Is there a Regulator Installed?
  - No → Troubleshoot Regulator → Go to Chart 7D
  - Yes → 7.1.0.6.1 Reconnect Point E Disconnect Point F
    - Overload Gone? ← (P) From Charts 7C, 7D, and 9
      - Yes → 7.1.0.6.2 Load or Output Filter Capacitor Is Shorted
      - No → 7.1.0.6.3 Identify Shorted Component
        - 7.1.0.6.4 Does the Load Have a Semi-Short?
          - No → Select Method
            - 7.1.0.6.5 Cut-Trace-and-Try
            - 7.1.0.6.6 Current Tracer & Logic Pulser
            - 7.1.0.6.7 The Freezer Method
            - 7.1.0.6.8 Removing Likely Components
          - Yes → 7.1.0.6.9 "Smoke Out" Method Optional
        - → Go to Phase 5, REPLACING DEFECTIVE COMPONENTS

(P)

Circuit diagram: INPUT AC → EMI FILTER → Transformer (A, A, C, C) → Diodes (D) → Filter caps (E) → Resistor → Reg → Load (F)

Copyright 1988 J. Perozzo

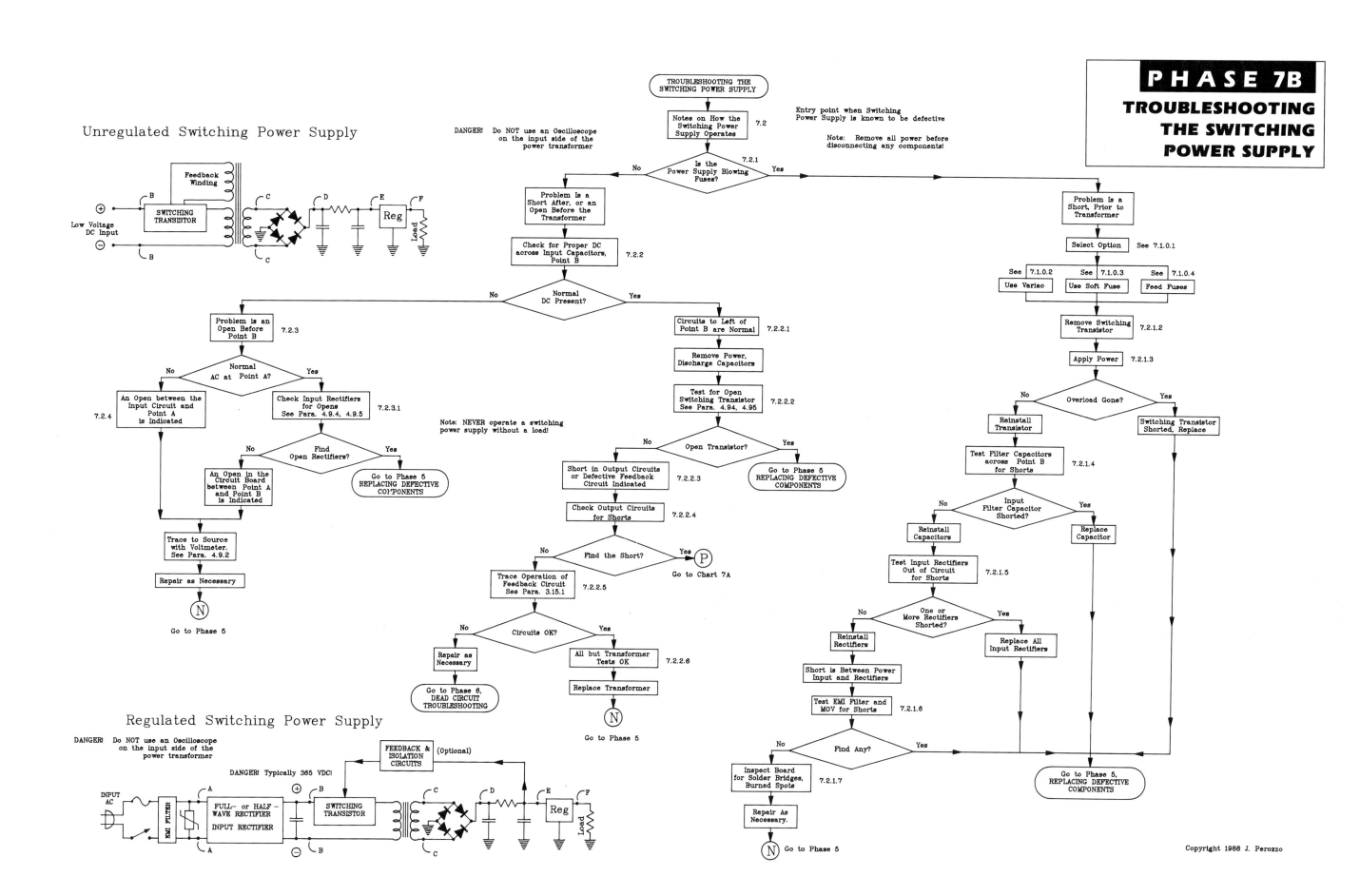

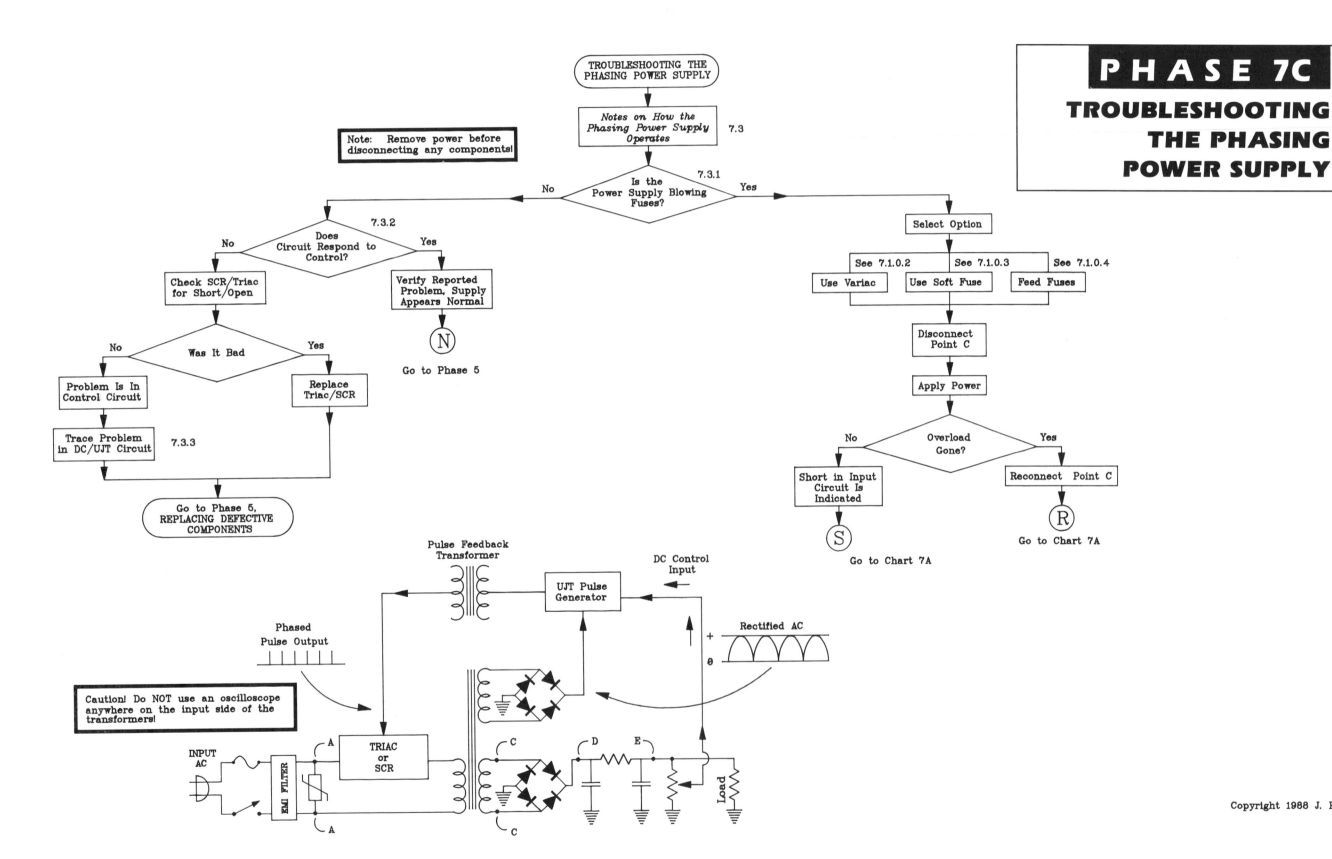

PHASE 7D
TROUBLESHOOTING VOLTAGE REGULATORS

*"Let's see now...
Plenty of Gizinta,
but no Gizouta..."*

Entry point for defective regulator

Caution! Remove All power and discharge filter capacitors before disconnecting any components or using a solid state tester or ohmmeter

DISCRETE REGULATOR
Input DC → PASS TRANSISTOR → Shunt → Output DC
Current Monitor
Amplifier
Voltage Comparator
Voltage Reference

SWITCHING REGULATOR
Input → SWITCHING REGULATOR IC → Output

ZENER DIODE REGULATOR
Input → Series Resistance → Output

INTEGRATED CIRCUIT REGULATOR
Input → REGULATOR IC → Output

Flowchart:

- TROUBLESHOOTING VOLTAGE REGULATORS
- Notes on Regulator Types — 7.4
- Apply Input Power — 7.4.1
- Input Voltage OK? — 7.4.2
  - **No** → Disconnect Regulator from Power Supply — 7.4.3
    - Power Supply Output Voltage OK now?
      - **No** → Problem is in the Power Supply — 7.4.4 → Go to Phase 7
      - **Yes** → Regulator or Load has a Short — 7.4.3.1 → Reconnect Power Supply Input — 7.4.3.2
  - **Yes** → Disconnect Load and Filter Cap from Regulator — 7.4.2.1
    - Output Voltage of Regulator OK Now? — 7.4.2.2
      - **Yes** → Substitute Resistor for Load — 7.4.2.3
        - Output Voltage of Regulator OK Now? — 7.4.2.4
          - **No** → A Current Limiting Problem is Indicated — 7.4.2.5
          - **Yes** → Go to Chart 7A (P)
      - **No** → Is this a Shunt Zener or an IC Regulator — 7.4.2.6
        - **No** → (V)
        - **Yes** → Replace Regulator

From (V) Select Type:
- Switching Regulator — 7.4.2.7
  - Output Voltage Too Low or zero?
    - **No** → Output Voltage Is Too High — 7.4.2.8
    - **Yes** → Disconnect Load Including Filter Capacitor — 7.4.2.7.1 → Use Substitute Filter Capacitor
      - Output Voltage OK Now?
        - **No** → Check Inductor for Open — 7.4.2.7.2
          - Was It Open
            - **No** → Regulator IC Defective — 7.4.2.9 → Replace Regulator
            - **Yes** → Replace Inductor
        - **Yes** → Go to Chart 7A (P)
- Discrete Semiconductor Regulator — 7.4.2.6.1
  - Check Pass Transistor — 7.4.2.6.2
    - Was it OK?
      - **No** → Replace as Required
      - **Yes** → Check Voltage Reference Component — 7.4.2.6.3
        - Was it OK?
          - **No** → Replace as Required
          - **Yes** → Problem Is in Error Amplifier Stages — 7.4.2.6.4 → Trace DC Signal to Find Cause → Go to Phase 3, LOCATING THE DEFECTIVE STAGE

Go to Phase 5 REPLACING DEFECTIVE COMPONENTS

Copyright 1988 J. Perozzo

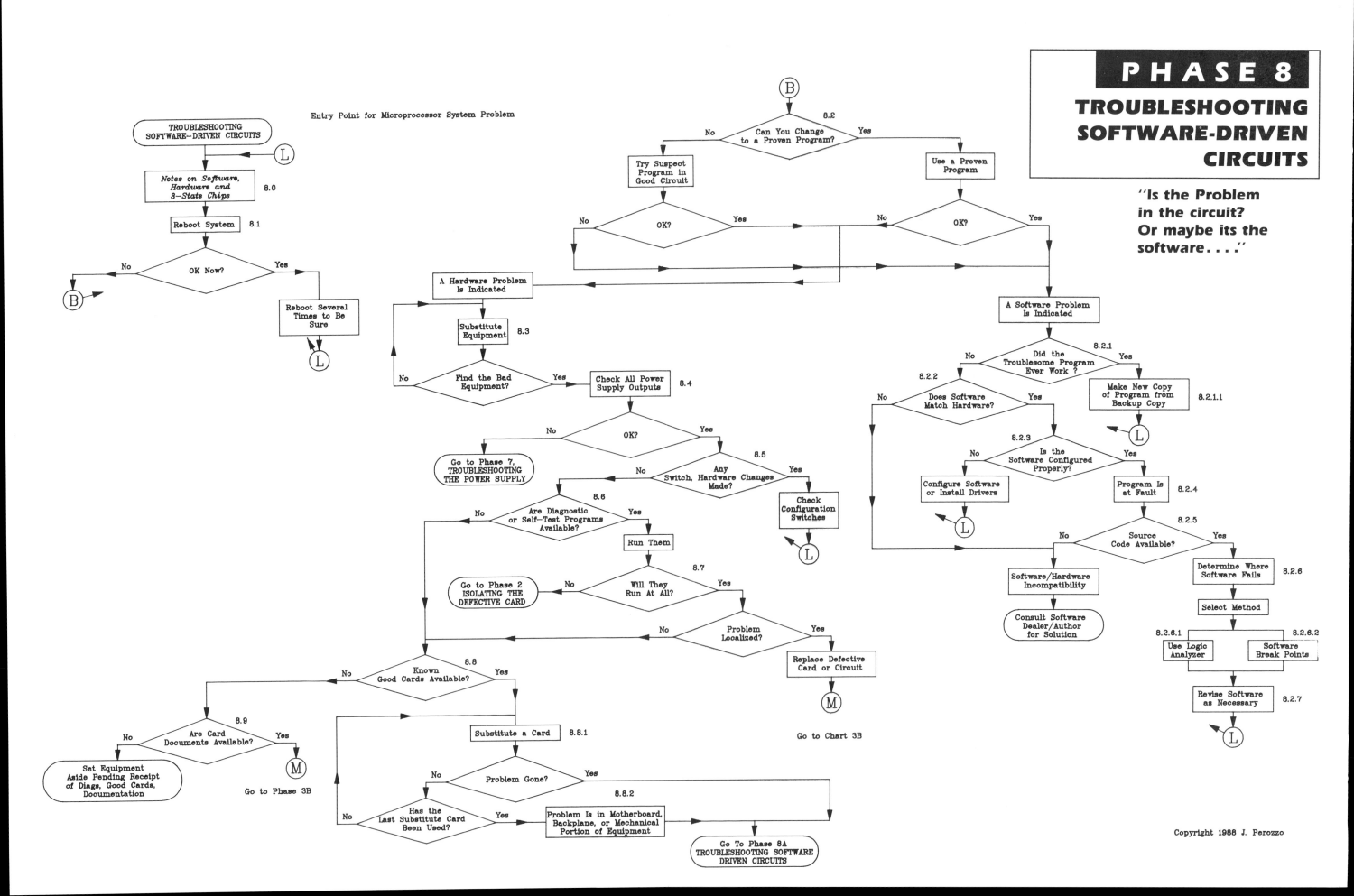

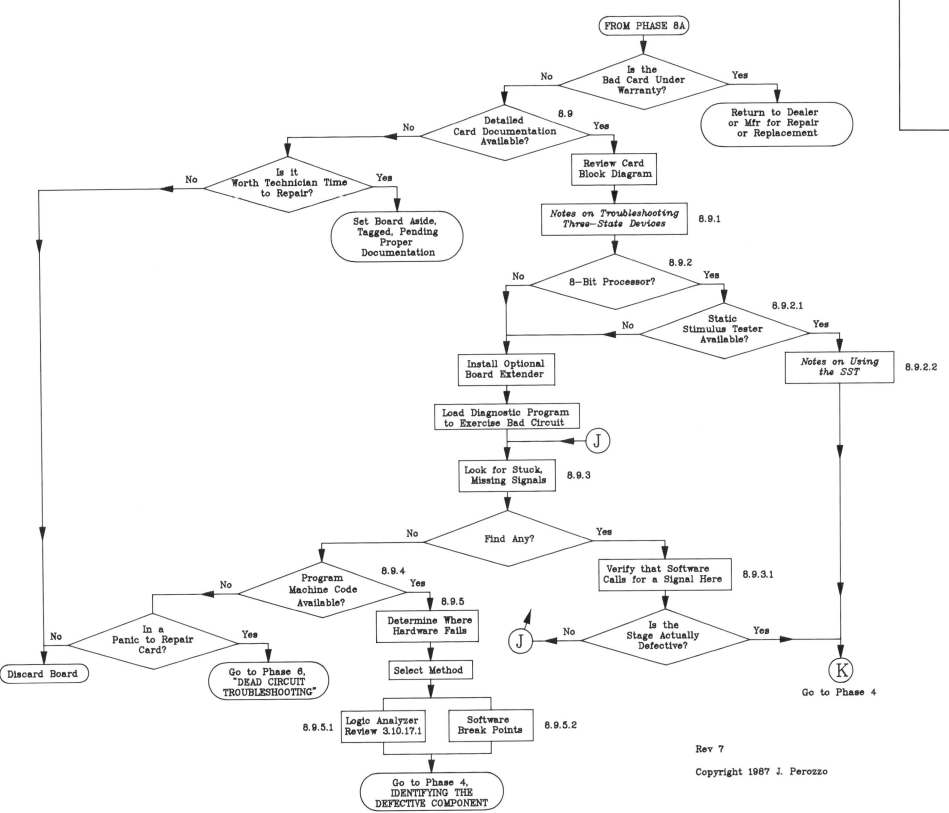

PHASE 8A
TROUBLESHOOTING SOFTWARE-DRIVEN CIRCUITS

Rev 7

Copyright 1987 J. Perozzo

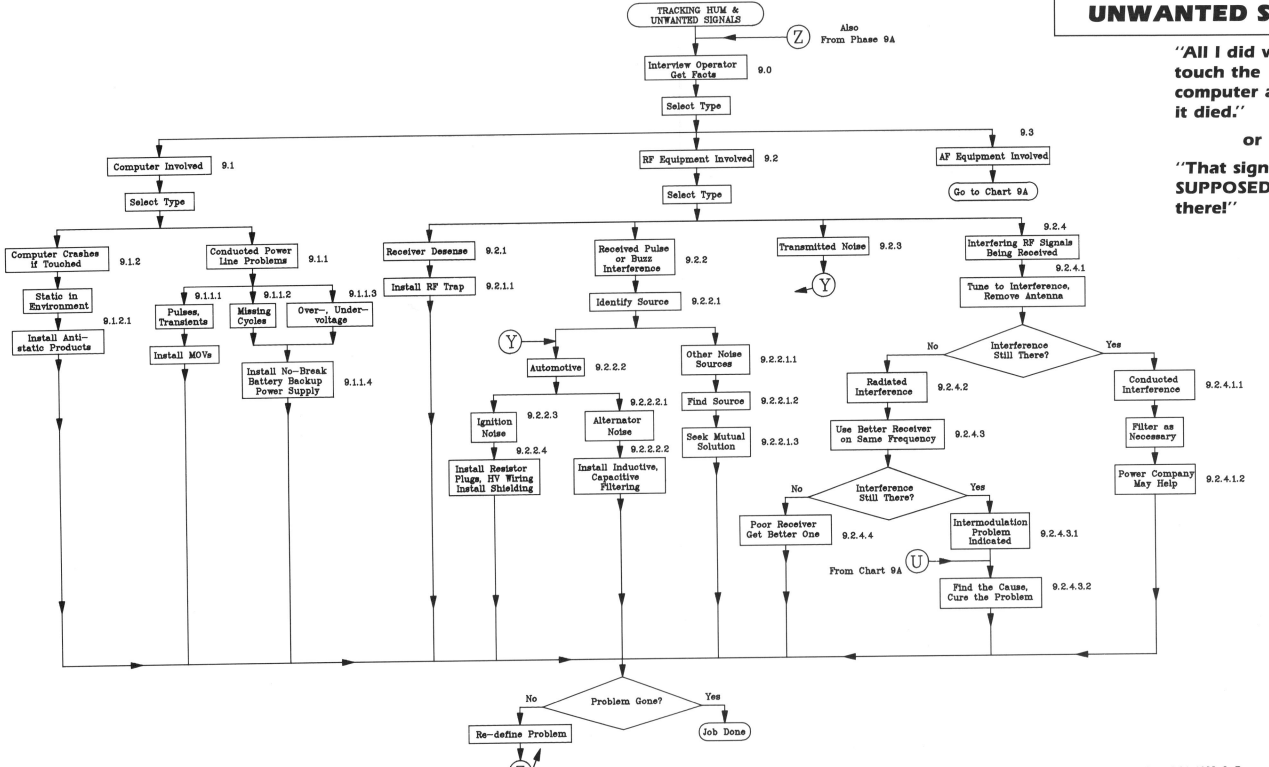

PHASE 9
TRACKING HUM AND UNWANTED SIGNALS

"All I did was touch the computer and it died."

or

"That signal isn't SUPPOSED to be there!"

Copyright 1988 J. Perozzo

PHASE 9A
TRACKING HUM & UNWANTED SIGNALS

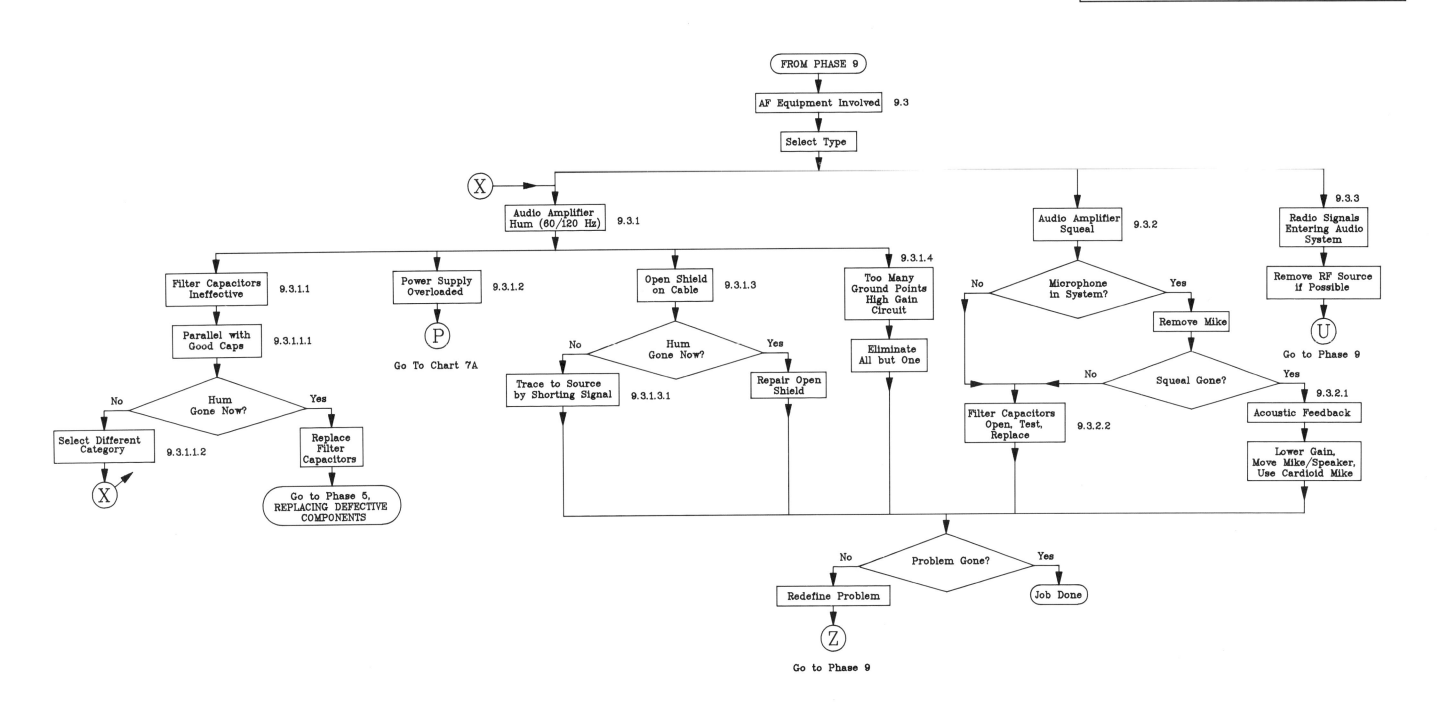

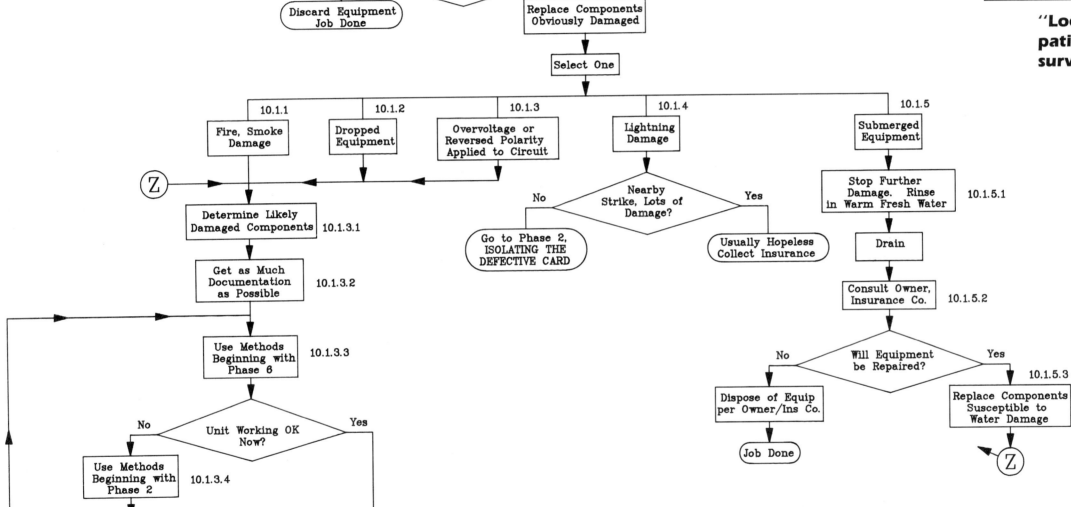

PHASE 10
REPAIRING PROFOUND FAILURES

"Looks like the patient might not survive..."

Copyright 1988 J. Perozzo